Combined Proceedings International Plant Propagators' Society 2022

IPPS Volume 72

Held 2022

Editors:

Robert L. Geneve
Clive Larkman
Charles W. Heuser, Jr.
Fokko Zur Mühlen
Masanori Tomita
Jill Reader
Elsa du Toit
Fred T. Davies, Jr.
Ryan Stewart

ISBN: 978-1-7138-7737-0

Printed from e-media with permission by:

Curran Associates, Inc.
57 Morehouse Lane
Red Hook, NY 12571

Some format issues inherent in the e-media version may also appear in this print version.

Printed with permission by Curran Associates, Inc. (2023)

For permission requests, please contact International Plant Propagators' Society (IPPS) at the address below.

International Plant Propagators' Society (IPPS)
174 Crestview Drive
Bellefonte, PA 16823-8516
USA

Phone: +01 814-355-2467

secretary@ipps.org

Additional copies of this publication are available from:

Curran Associates, Inc.
57 Morehouse Lane
Red Hook, NY 12571 USA
Phone: 845-758-0400
Fax: 845-758-2633
Email: curran@proceedings.com
Web: www.proceedings.com

THE INTERNATIONAL PLANT PROPAGATORS' SOCIETY

Combined Proceedings

Volume 72

2022

IPPS

Sharing Plant Production Knowledge Globally

Published 2022

Combined Proceedings International Plant Propagators' Society (ISSN 0538-9143) is a yearly publication of the International Plant Propagators' Society (IPPS) headquartered in Pennsylvania, U.S.A.
Email: Secretary@ipps.org
Website: www.ipps.org

Business Office. Memberships are processed at the regional level, with dues being allocated to both international and regional expenses. Contact the regional secretary / treasurer in your area of the world for membership information and applications.

Table of Contents

European Region

PROCEEDING'S PAPERS

SOUTHERN AFRICA REGION

Dr. Elsa du Toit, Regional Editor

Twenty-fourth Annual Meeting - 2022

Krugersdorp, South Africa

Combined Proceedings IPPS 72:1-7. 2022.

Welcome to the Walter Sisulu National Botanical Gardens

Andrew Hankey[1] and Alice Aubrey[2]

[1]Specialist Horticulturist – Walter Sisulu National Botanical Gardens
[2]Interpretation Officer – Walter Sisulu National Botanical Gardens

Keywords: black eagle, nature, conservation, biodiversity

Summary

Walter Sisulu National Botanical Garden is a large botanical reserve with an Environmental Education Centre, numerous feature gardens and plant collections. There are also areas for viewing native mammals, birds and reptiles.

INTRODUCTION

The Walter Sisulu National Botanical Garden and nature reserve is situated in Poortview, Roodepoort, Gauteng. Home to a breeding pair of Verreaux's eagles (*Aquila verreauxii*) on the cliffs adjacent the Witpoortjie waterfall, the centrepiece and backdrop of the Garden and source of the Crocodile River (**Fig. 1**). It was proclaimed as a National Botanical Garden in July 1982 by Prof Brian Rycroft, then director of the National Botanic Gardens (NBG).

The first curator of the gardens was Mr. Peter Chaplin, served from 1982-1999. The Botanical Garden was formerly a privately owned farm, acquired by the Roodepoort and Krugersdorp City Councils in the late 70's who made the establishment of the Garden possible by providing the land on a 99-year lease to the Institute. The Walter Sisulu National Botanical Garden is administered by South African National Biodiversity Institute (SANBI).

Figure 1. Aerial view of the Witpoortjie Falls in Walter Sisulu National Botanical Garden, Poortview, Roodepoort. (Photo SANBI)

The Garden was initially known as the Transvaal National Botanic Garden and could only be visited by special arrangement. The Garden was opened to the public in 1987 and was previously known as the Witwatersrand National Botanical Garden. In March 2004, the Garden was renamed the Walter Sisulu National Botanical Garden, in honour of the late ANC stalwart Mr Walter Sisulu (1912-2003) who, together with former president Nelson Mandela, led the struggle for a democratic South Africa. It was a well-known picnic spot since the 1800's and currently received ca. 320k visitors per annum.

Major infrastructural developments took place in the early 1990s. These included the construction of the Entrance Building, the Nestlé Environmental Education Centre, Sasol Dam and Bird Hide as well as the paving of the main walkway.

Over the years the following features have been developed in the Garden; a Succulent Rockery Garden (**Fig. 2**), Cycad Garden (**Fig. 3**), Water Garden, Fern Trail, Arboretum, Geological Garden (**Fig. 4**), People's Plant Garden, Birds and Butterfly Garden, Dell section, Waterwise Garden, Children's Garden, Wildflower Area, Visitors' Information Centre, restaurant and function venue (SANBI, 2022).

Figure 2. Succulent Garden (photo SANBI)

Figure 3. Cycad Garden (photo SANBI)

Figure 4. Geological Garden (photo SANBI)

Aspects of interest in the garden

Bird life: An abundant birdlife which ranges from the smallest, minute Fairy Fly-catcher (weighing in at a dainty 6 g) to the Verreaux's Eagles (*Aquila verreauxii*) (**Fig. 5**) which can weigh up to 4.5 kg and boast a wingspan of 2.8 m.

The eagles are a popular tourist attraction and have been since the 1970's. Every year they lay two eggs (usually around mid-April). The first chick kills the second chick (cainism) who is then raised by the adults. Once fledged the juvenile remains in the valley until it gets chased away by the parents usually around December -January.

Figure 5. Verreaux's Eagles (*Aquila verreauxii*) (photo Ernest Porter)

The super-speedy Peregrine Falcon hunts from the cliff faces, these birds are reputed to be one of the fastest of all, apparently capable of reaching up to 100 km per hour. By contrast, Helmeted Guineafowl are reluctant and slow fliers, being far more suited to their ground-based lives. They can be seen in flocks, sometimes shepherding tiny stripy chicks and always vocal, their unmistakeable calls part of the atmosphere of the Garden. In between these extremes is the wonderful assortment of birds making up the 241 listed species for the Garden (SANBI, 2022).

Mammals: The mammals tend to be far more secretive; they have adapted to human presence by lying low during the day and coming out at night. Mammals such as honey badger, serval and the aardvark (antbear) are still living in the urban wild. Mountain Reedbuck ewes with lambs have been 'camera-trapped' frequently, a sign that the ridge ecosystem is intact. Sengi (rock elephant shrew) have also been noted (SANBI, 2022).

Invertebrates: Invertebrates such as insects and spiders are vital to the functioning of the Garden ecosystem acting as pollinators, as well as cleaner, nutrient processors and serve as food sources for other wildlife (SANBI, 2022).

The nature reserve is also home to various naturally occurring wildlife such as Jackal, Porcupine, Rooikat, Serval, Common Diuker as well as a population of Southern Mountain Reedbuck (*Redunca fulvorufula* ssp. *fulvorufula*) which is listed as endangered on the red List of South African mammals (**Fig. 6**).

Figure 6. Southern Mountain Reedbuck (*Redunca fulvorufula* ssp. *fulvorufula*) in the Walter Sisulu National Botanical Garden (Photo Ernest Porter)

In late 2007, the WSNBG yielded one of its first surprises when a Red Data listed fish species, the Marico Barb (*Enteromius motebensis*) was discovered in the plunge pool of the waterfall. (SANBI, 2022; **Fig. 7**).

Figure 7. Marico Barb (*Enteromius motebensis)* (photo A. Hankey)

Vegetation: The cultivated garden is comprised of ca. 25ha of landscaped gardens and service areas, with an adjoining nature reserve, also managed by the garden, that comprises of 275ha. The natural vegetation is comprised of Afrotemperate Highveld forest, Gold Reef Mountain Bushveld and Egoli Granite Grassland (Mucina and Rutherford, 2006). The garden maintains a small herbarium of plants of the property and currently has ca. 600 species. The north-west ridge system known as Roodepoort Reef Mountain Bushveld (GP8) and is listed as Critically Endangered (Cr) in Threatened terrestrial ecosystems for South Africa (2011).

The Garden is at its best during late spring and summer when most of the flowering shrubs come into bloom. The Pride-of-De-Kaap (*Bauhinia galpinii*) with its unusual brick-red flowers makes a lovely show. The mauve-blue flowers of the wild phlox (*Jamesbrittenia grandiflora) and Dissotis* sp. are a delight for many weeks. One of the highlights of summer are the scarlet river lilies (*Hesperantha coccinea*) which line the stream near the restaurant and produce bright red flowers in mid-summer. The wild banana (*Ensete ventricosum*) (**Fig. 8**) with its giant leaves makes a magnificent show along the Water Garden. Arum lilies (*Zantedeschia aethiopica*) are also to be found in flower (SANBI, 2022).

Figure 8. Wild banana (*Ensete ventricosum*) (photo SANBI)

Common sugar bushes (*Protea caffra*) are abundant on the north-east facing slopes of the Garden. These naturally occurring trees produce attractive heads of flowers throughout the summer months and provide a constant source of nectar to a wide variety of birds. Cuckoos are an integral part of the Garden in summer and it seems as if there is no place to hide from their persistent calls. Listen out for the Red-chested Cuckoo, Diederik Cuckoo, Klaas's Cuckoo and the Black Cuckoo from the beginning of October. The striking Southern Red Bishops start to build their nests in the bulrushes and reeds around the Sasol Dam and wetland areas. The male birds can often be seen trying to impress a future mate (SANBI, 2022).

In the ponds in the wetland area, the large flowers of the indigenous water lily (*Nymphaea nouchali* var. *caerulea*) and the smaller star-shaped yellow flowers of the small yellow water lily (*Nymphoides indica*) is evident (SANBI, 2022).

The tremendous variety of spur-flowers (*Plectranthus*) burst into flower to herald the beginning of autumn. Many of these herbaceous shrubs and groundcovers make excellent garden subjects as they flower so profusely and have very attractive foliage.

Garden's Clivia (*Clivia gardenii*) start flowering in autumn near the Lion's Bridge with their unusual pendulous flowers; unlike those of the usual bush lily, *Clivia miniata*. These lesser known clivias come from the east coast province of KwaZulu-Natal. The bright orange flowers of the *Kalanchoe rotundifolia* and the red flowers of *Crassula alba* and *Crassula perfoliata var. minor* in the Succulent Rockery Garden also create a lovely show at this time of year.

By the end of autumn, many trees such as the white stinkwood (*Celtis africana*), river bushwillow (*Combretum erythrophyllum*) and the lavender tree (*Heteropyxis natalensis*) are all flowering in their respective yellows, reds and purples.

The Succulent Rockery is particularly beautiful during winter when most of the aloes are in full bloom (**Fig. 9**). A variety of birds, including the magnificent Malachite Sunbird, feed on the nectar from showy plants such as *Aloe cryptopoda* which has attractive bi-coloured flowers.

Figure 9. *Aloe arborescens* flowering in the Peoples plants garden (photo SANBI).

At the end of winter, the common wild pear (*Dombeya rotundifolia*) can be seen from afar by its axillary clusters of white to pale pink flowers. The sagewood (*Buddleja salviifolia*) promises the onset of spring and bursts into flower, filling the air with its fragrant scent.

Albertina Sisulu Orchid: In 2007 the Albertina Sisulu Orchid (**Fig. 10**) was re-discovered on the ridge behind the WSNBG. The staff at WSNBG initiated a search to document the extent of the population and the species was soon up listed on the Red List of South African Plants to Vulnerable (Vu).

The WSNBG garden staff together with several conservation partners initiated a search across Gauteng to visit all known historical locations of the species. After extensive searching it was concluded that the species was only remaining at four locations. Three of the four populations had less than ten individuals and the species was up listed to Critically Endangered (Cr). The WSNBG staff every year conduct population counts together with their conservation partners to monitor the extent of the population and search for new sub populations. In 2022 the count revealed 227 individuals. Negotiations with the Mogale City Council to incorporate the land on which the orchid is located under the management of WSNBG are under way.

Figure 10. Albertina Sisulu Orchid. (*Brachycorythis conica* subsp. *transvaalensis*). (photo A. Hankey)

Behind the scenes: Over and above what the public sees, the botanical garden is involved in various conservation and research activities, only indigenous plant species are being cultivated and planted in the National Botanical Gardens and living plant collections maintained. The Botanical Garden is also involved with provincial and national bodies in Threatened plant programs, conduct horticultural and botanical research, engage in field expeditions to collect, document and explore South African flora, host WIL students and interns in Horticulture & Nature Conservation students, participate in plant rescue operations and host shows and events.

Conclusion

The Walter Sisulu National Botanical Garden is a valuable resource for biodiversity conservation and biological research in South Africa. The garden forms part of a structured training ground for the green industry as well as provides valuable safe recreational open space where the community can experience nature, enjoy the hiking trails and have a peaceful picnic while meeting basic human relaxation needs and well-being. Botanical gardens play a vital role in society and should be valued as an asset to society.

LITERATURE CITED

Mucunia, L. and Rutherford, M.C. (2006). The vegetation of South Africa, Lesotho and Swaziland. *Strelitzia* 19. South African National Biodiversity Institute, Pretoria.

South African National Biodiversity Institute (SANBI). (2022). Walter Sisulu Botanical Garden.

Gauteng C-Plan Version 2. (2006). Gauteng Department of Agriculture, Conservation and Environment.

http://bgis.sanbi.org/ecosystems/showecosystem.asp?id=41 (Accessed 13/09/2022).

https://www.sanbi.org/gardens/walter-sisulu/garden-information/wildlife-and-biodiversity/ (Accessed 13/09/2022).

http://pza.sanbi.org/brachycorythis-conica-subsp-transvaalensis (Accessed 13/09/2022).

PROCEEDING'S PAPERS

JAPAN REGION

Dr. Masanori Tomita, Regional Editor

Twenty-seventh Annual Meeting - 2022

Gifu City, Japan

Combined Proceedings IPPS 72:8-12. 2022.

Effects of Light Quality during Cultivation and Cutting on Rooting of Cuttings of *Gynura bicolor* DC.

Chihiro Satoh, Arisa Noguchi, Hiroko Nakatsuka and Wakanori Amaki

Department of Agriculture, Tokyo University of Agriculture, 1737 Funako, Atsugi, Kanagawa 243-0034, Japan

amaki@nodai.ac.jp

Keywords: Asteraceae, supplemental light, LED, stock plant

Summary

Gynura bicolor DC. is a perennial plant belonging to the genus *Gynura* of the Asteraceae family, said to be native to Southeast Asia. Usually, this plant is propagated by cuttings. We investigated the light quality effects for mother plant cultivation before cutting and during cutting on the rooting of the cuttings in *Gynura bicolor*. Cuttings prepared from plants grown under white fluorescent light were placed under different light quality conditions, and the production of adventitious roots was compared. However, no statistically significant difference was observed. On the other hand, the light quality during cultivation greatly affected the rooting of cuttings. Rooting of cuttings taken from plants grown under white mixed light emitting diode (LED) and blue LED monochromatic lights were delayed. Cuttings of plants grown under red LED monochromatic light rooted faster, and the average root weight was more than three times that under mixed white light and blue light.

INTRODUCTION

Gynura bicolor DC. is a perennial plant in the Asteraceae native to Southeast Asia (Ikeda, 1988). In addition to being used as a traditional vegetable in southern China, in Japan it is cultivated as a local vegetable in Ishikawa Prefecture (Japanese local name: Kinji-sou), Aichi Prefecture (Shikibu-sou), Okinawa Prefecture (Handama), and Kumamoto Prefecture (Suizenjina). Rich in vitamin A, it also contains considerable amounts of vitamin B_2 and vitamin C. Therefore, *Gynura bicolor* is known as a functional health vegetable along with its antioxidant action (Do *et al*., 2020; Hsia *et al*., 2021). Abaxial side of the leaves is reddish purple with anthocyanin and has a characteristic scent associated with volatile components.

Usually, this plant is propagated by cuttings (Takeshita, 1998). The rooting of cuttings is relatively easy. However, the conditions necessary for rooting in this plant have been largely unexplored. In experiments using sweet basil (*Ocimum basilicum* L.), it is known that changing the light quality at the time of cultivation and the light quality at the time of subsequent cutting in water affects the rooting of cuttings (Yamada *et al*., 2015; Abe, 2019). Therefore, we investigated the light quality effects for mother plant cultivation before cutting and during cutting on the rooting of cuttings in *Gynura bicolor*.

MATERIALS AND METHODS

Cultivation of material plants

Cut-leaf vegetables of *Gynura bicolor* were purchased from a market, and the shoot apex was adjusted to a length of about 8 cm as for cuttings. The cuttings were inserted into a vermiculite (M size) filled a cell box (36 cell-type, Sakata Seed Co. Ltd.; 1 cell size is the top side 41 mm and depth 40 mm) and propagated. After rooting, they were transplanted into 6 cm pots and then 9 cm pots filled with a mixture of Metro Mix 360 JPN (Sungro Horticulture, Canada): akadama clay soil (S-size) = 1:1, according to their growth. For fertilization, 1.3 g of granular chemical fertilizer (Hardened IB-Compound (IB-S1); N-P_2O_5-K_2O-MgO = 10-10-10-1, Jcam Agri Co., Ltd.) was applied to the 6 cm pot and 2 g to the 9 cm pot once a month. The growth conditions were 23±1°C, 16 hours illumination/8 hours darkness with white fluorescent lamps (FLR40S-EX-N/M-H, Toshiba Lighting & Technology Co., Ltd.), photosynthetic photon flux density (PPFD) of 80 $\mu mol \cdot m^{-2} \cdot s^{-1}$.

Effect of light quality on rooting of plants cultivated under white fluorescent light

Pot plants where the main shoot had grown about 10 cm long were placed at 23 ± 1°C, under white fluorescent lamps. Illumination condition was at 80 $\mu mol \cdot m^{-2} \cdot s^{-1}$ PPFD and 16 hours light irradiation / 8 hours darkness. When the main shoot had grown to about 30 cm, cuttings of 8 cm in length were prepared on August 14, 2022. Five cuttings per treatment were inserted in a 200 ml tall glass beaker with tap water. There were four light conditions of white mixed light of blue + green + red LED, blue LED (peak wavelength: 470 nm) light, green LED (530 nm) light, and red LED (630 nm) light, 80 $\mu mol \cdot m^{-2} \cdot s^{-1}$ of PPFD and 16 hours light /8 hours dark at 24±2°C. These LEDs used in these experiments were manufactured by Stanley Electric Co., Ltd. for testing purposes and is not commercially available. On

August 29, 2022, the total weight of adventitious roots produced from each cutting was measured using a precision electronic balance (GH-200, A&D Co., Ltd.).

Effect of light quality on rooting of plants cultivated under different light quality

Plants were cultivated as previously described except the light source was LEDs with different light quality and that the white fluorescent lamp was used for light irradiation after the cuttings were inserted in water. However, the plants grown under green LED light grew poorly, and cuttings could not be taken, so this plot was omitted from this experiment.

Four cuttings per plot were inserted in a 200ml tall glass beaker with tap water. In this experiment, they were placed under dark at 23±1°C until August 21, 2022. After that, it was kept under white fluorescent light for 16 hours of illumination / 8 hours of darkness (PPFD: 80 $\mu mol \cdot m^{-2} \cdot s^{-1}$). On August 29, 2022, the total weight of adventitious roots produced from each cutting was measured.

RESULTS AND DISCUSSION

Cuttings taken from plants grown under white fluorescence light were placed under different quality of light. However, when the light quality after cutting was changed, there was no statistically significant difference in the fresh weight of roots (**Fig. 1**).

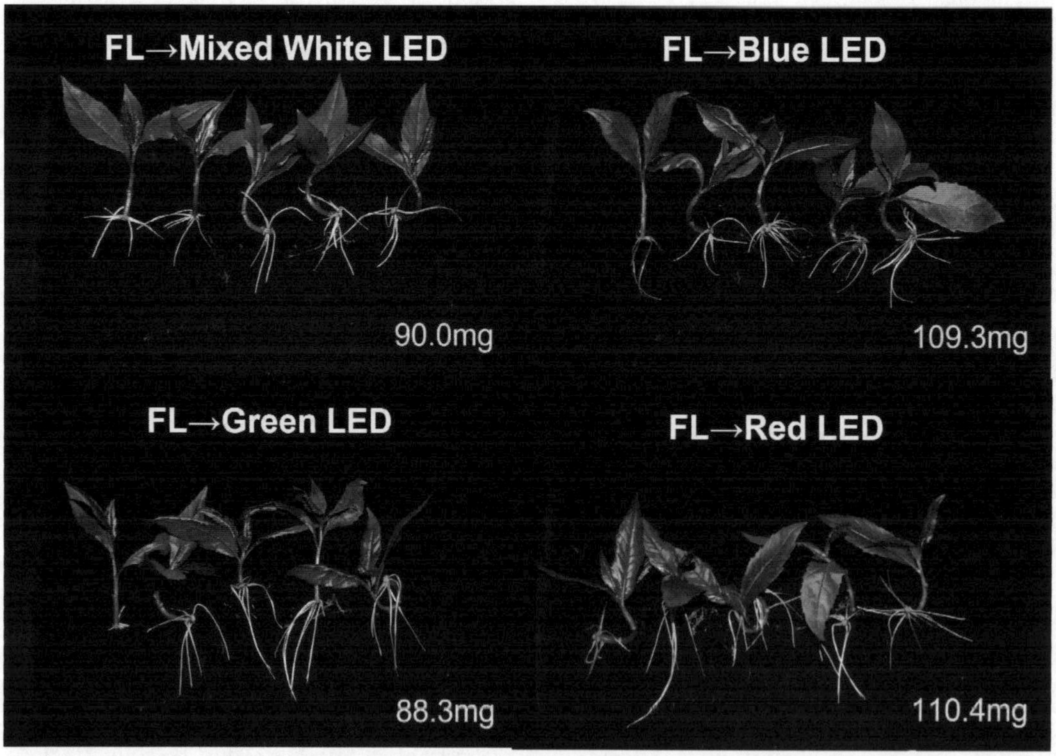

Figure 1. Rooting under different light qualities of cuttings prepared from plants grown under white fluorescent light. FL: White fluorescent light, LED: Light emitting diode. (The data numbers in the bottom right of each image indicate the average total root fresh weight of each plot).

The light quality during cultivation of mother plants greatly affected the rooting of cuttings (**Fig. 2**). Rooting of cuttings taken from plants cultivated under mixed white light and blue monochromatic light was delayed, and the rooting of individual cuttings was uneven. However, cuttings of plants grown under red light rooted faster and the average root weight was more than three times that under mixed white light and blue light.

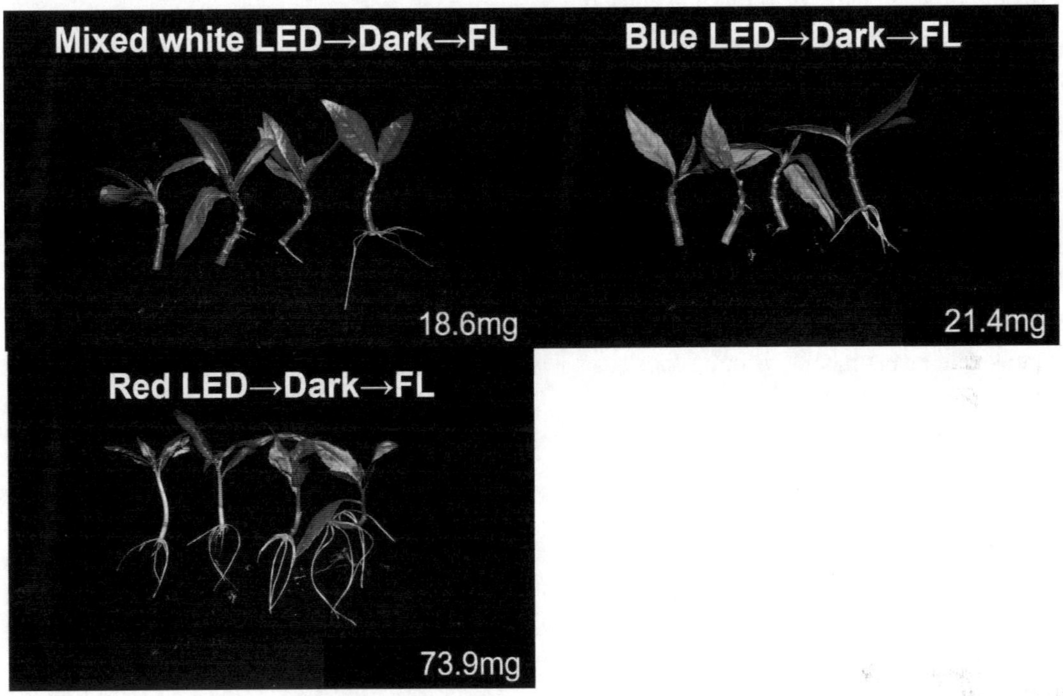

Figure 2. Rooting under white fluorescent light of cuttings prepared from plants grown under different light qualities. FL: White fluorescent light, LED: Light emitting diode, Dark: Meaning put under darkness for 1 week after cuttings. (The data numbers in the bottom right of each image indicate the average total root fresh weight of each plot).

Root fresh weight in cuttings from stock plants under different light qualities was lower compared to white light. In a preliminary experiment, cuttings of *Gynura bicolor* started rooting within a week when the cuttings were inserted on vermiculite, but when they were inserted in water, rooting started after 10 days. Assuming that this was caused by the exposure to light, we performed a one-week dark treatment. However, the dark treatment resulted in delayed rooting. It may be that the physical environment of the rooting part (cut end of stem) of the cuttings is inferior in water cuttings.

Rooting of cuttings was strongly affected by the light quality, and that the light quality during cultivation has a stronger effect. It is presumed that metabolic reactions such as plant hormones that promote rooting were activated in the plants cultivated under red monochromatic light. Similar results have been observed for rooting in sweet basil cuttings (Yamada, 2015; Abe, 2019). In sweet basil, blue light inhibited rooting and red light promoted it (Yamada, 2015). It has also been shown that the addition of appropriate concentrations of auxin to the tap water in which cuttings are caused to promote rooting, while the addition of gibberellin inhibits it. Furthermore, it has also been clarified that the addition of uniconazole, a gibberellin biosynthesis inhibitor, promotes rooting (Abe, 2019).

In conclusion, considering from the results mentioned above of sweet basil, it is suggested that the light quality of the irradiation light during the cultivation of *Gynura bicolor* plant changes in biosynthesis of auxin and gibberellin and /or sensitivity to such plant hormones.

LITERATURE CITED

Abe, H. 2019. Genetic variability of commercial basil seeds and heritability of light responses in sweet basil. Bachelor thesis, Faculty of Agriculture, Tokyo University of Agriculture. (In Japanese).

Do, T. V. T., Suhartini, W., Mutabazi, F. and Mutukumira, A. N. (2020). *Gynura bicolor* DC. (Okinawa spinach): A comprehensive review on nutritional constituents, phytochemical compounds, utilization, health benefits, and toxicological evaluation. Food Research International *134*:109222. https://doi.org/10.1016/j.foodres.2020.109222

Hsia, C.-H., Tung, Y.-T., Yeh, Y.-S. and Chien, Y.-W. (2021). Effects of *Gynura bicolor* on glycemic control and antioxidant ability in prediabetes. Appl. Sci. *11*:5066. https://doi.org/10.3390/app11115066

Ikeda, H. (1988). Suizenjina. pp.595-596. In: Tsukamoto, Y. (ed.). The Grand Dictionary of Horticulture Vol. 2. Shogakukan, Tokyo. (In Japanese).

Takeshita, S. (1998). Suizenjina. pp.177-178. In: Fujimaki, H. (ed.). Encyclopedia of Regional Biological Resources Utilization. Rural Village Cultural Association. Tokyo. (In Japanese).

Yamada, H., Watanabe, H. and Amaki, W. (2015). Effect of light quality on rooting of cuttings in sweet basil. Hort. Res. (Japan) 14 (Suppl.2):503. (In Japanese).

Combined Proceedings IPPS 72:13-18. 2022.

Effect of Day Length and Tuber Storage Temperature and Duration on Sprouting, Enlargement and Flowering of Tubers in *Pinellia ternata* (Thunb.) Makino

Masamichi Torii, Arisa Noguchi, Hiroko Nakatsuka and Wakanori Amaki

Faculty of Agriculture, Tokyo University of Agriculture, 1737 Funako, Atsugi, Kanagawa 243-0034, Japan

43321012@nodai.ac.jp

Keywords: Araceae, medicinal plant, aroid

Summary

Crowdipper (*Pinellia ternata* (Thunb.) Makino) is a perennial plant in the aroid family (Araceae), which grows naturally in various parts of East Asia. It can be propagated by dividing tubers. Large tubers that have been peeled and dried have long been used as a material in Kampo medicine called "Hange". We investigated the effects of temperature and light on sprouting, growth, and flowering of tubers. Tubers sprouted 100% in 3.2 weeks under long-day conditions (L). On the other hand, under the short-day condition (S), no sprouting occurred even after 7 weeks. From these results, it was expected that the photoperiod was the main factor for sprouting. The tubers cultivated for 20 weeks after planting swelled about 5 times under the L-L condition and S-L conditions. Flowering was only observed where the storage temperature was 4°C for 6 weeks. On the other hand, in the treatment plots stored at 16°C, the tuber enlargement rate tended to be higher in the 3-week storage than in the 6-week storage.

INTRODUCTION

Crowdipper (*Pinellia ternata* (Thunb.) Makino) is a perennial plant in the aroid family (Araceae), which grows naturally in various parts of East Asia. One leaf is usually developed from one tuber from spring to summer, and one spadix (inflorescence) peculiar to Araceae is formed at the top of a large tuber. It can be propagated by seeds (Nagao 1980), but the tuber forms a small daughter tuber on the surface, and in addition, it forms a propagule at the base of the petiole of the leaf and the base of the leaflet. Large size of tubers that have been peeled and dried have long been used as a material in Kampo medicine called "Hange". The main ingredient is homogentisic acid, which is also an ingredient in many traditional Kampo medicine (Harashima 2012). At present, materials imported from China are mainly used for the manufacture of Japanese Kampo medicines, but it is expected that these imports will become unstable due to the recent international situation (Tohoku Revitalization Research Center 2013).

Depending on the temperature of the cultivation area, sprouting and growth from tubers occurs once or twice a year in the Kanto region in Japan (Nagao 1979), and no method has yet been established for stably cultivating tubers suitable for Kampo medicine raw material. The authors have cultivated *Pinellia ternata* in a plant factory with a stable environment, and confirmed that under certain environmental conditions, the dormancy period is very short, and the growth and multiplication ratc can be increased up to about 4 growth cycles per year (Amaki *et al.*, 2015). However, although there are some studies that have examined the effects of temperature on growth and tuber enlargement (Eguchi 2019; 2020), there are not many studies on the effects of tuber storage temperature and light during cultivation (Nagao 1977; 1978). Therefore, we investigated the effects of temperature and light on sprouting, growth, flowering, and enlargement of tubers.

MATERIALS AND METHODS

Tubers growing wild in the sandy soil of Niigata City, Niigata Prefecture were collected and cultivated and multiplied in a greenhouse with a minimum temperature of 16°C at the Atsugi Campus of Tokyo University of Agriculture, Kanagawa Prefecture. Harvested tubers were cleaned, lightly dried, placed in plant boxes (Magenta Box G7, Magenta Co., Ltd), and stored at room temperature (20 - 24°C) until used for following experiments. In the daylength experiment, tubers were planted in 6 cm plastic pots with a mixture of Metro Mix 360 JPN (Sungro Horticulture, Canada): akadama clay soil (S-size) = 1:1. In the storage study, tubers were planted at a depth of 1 cm in 6 cm pots filled with Metro Mix 360. Cultivation was performed without fertilization, irrigation was performed with tap water, and temperature was 23 ± 1°C.

Effect of day length on sprouting and subsequent growth of tubers

Tubers stored at room temperature for about 6 months were planted, and 10 tubers for each treatment were cultivated under two conditions: long day (16h-light / 8h-dark: L) and short day (8h-light / 16h-dark: S). After 7 weeks, 5 unsprouted tubers from the short-day conditions were moved to long day conditions (S-L) or 12h-light/12h-dark conditions (S-M), and then cultivation was continued for 20 weeks. The light source used for cultivation was a white fluorescent lamp (FLR40S-EX-N/M-H,

Toshiba Lighting & Technology Co., Ltd.), and the irradiance (photosynthetic photon flux density: PPFD) was 100 μmol·m^{-2}·s^{-1}.

Effect of tuber storage temperature, duration, and cultivation light on sprouting and subsequent growth and flowering of tubers

Room temperature-stored tubers were moved to 4°C or 16°C for 3 months (3 plants per plot) or 6 months (2 plants per plot). After planting, tubers were grown under different LED light environments of mixed white (blue + green + red: W), blue (peak wavelength 470 nm: B), green (525 nm: G), or red (660 nm : R) with a PPFD set to 100 μmol·m^{-2}·s^{-1}.

RESULTS AND DISCUSSION

Effect of day length on sprouting and subsequent growth of tubers

Tubers sprouted 100% in 3.2 weeks under long-day conditions (**Table 1**). On the other hand, under the short-day condition (S), no sprouting occurred even after 7 weeks. From these results, it was expected that the photoperiod was the main factor for sprouting of *Pinellia ternata* tuber. The tubers cultivated for 20 weeks after planting swelled about 5 times under the L-L condition and S-L conditions. In the S-M condition with delayed sprouting, the average fresh weight of tuber was about 3 times larger than that of L-L and S-M (Table 1).

Table 1. Effects of day length on sprouting rate, tuber growth rate and sprouting time of *Pinellia ternata*

Daylength [z]			Sprouting (%)			
0 to 7 weeks	7 to 14 weeks		After 7 weeks	After 14 weeks	Sprouting time (weeks)	Tuber enlargement (%)[y]
16-L/8-D	16-L/8-D	L-L	100	-	3.2	467.0
8-L/16-D	16-L/8-D	S-L	0	100	9.8	488.4
8-L/16-D	12-L/12-D	S-M	0	100	14.2	296.6

[z]16-L/8-D means 16 hours of lighting and 8 hours of darkness (Same for others).
[y]Enlargement rate calculated by the fresh weight at the end of treatment divided by the weight at the start of treatment as 100.

Effect of tuber storage temperature, duration, and cultivation light on sprouting and subsequent growth and flowering of tubers

Flowering was observed only in tubers stored at 4°C for 6 weeks (**Table 2**). For tubers stored at 16°C, the tuber enlargement rate tended to be higher after 3-weeks of storage compared to 6-weeks storage. No significant difference was observed in the influence of light quality of cultivation light. Tuber enlargement after cold storage was generally smaller compared to tubers held at room temperature, but this is because the storage period at room temperature after digging is long and the cultivation period after planting is short (20 weeks vs. 12 weeks). In addition, almost no flowering was observed in the greenhouse at 16°C clarifing that 4°C for 6 months is a factor that promotes flowering.

Table 2. Effects of storage temperature, period of tubers and light quality during cultivation on tuber sprouting, flowering, and enlargement of *Pinellia ternata*.

Storage temperature (°C)	Storage period (month)	Light quality for cultivation	Sprouting time (weeks)	Tuber enlargement (%)[z]	Flowering (%)
4	3	Mixed White	1	77	0
		Blue	1	62	0
		Green	1	66	0
		Red	1	97	0
	6	Mixed White	1<	98	50
		Blue	1<	96	50
		Green	1<	107	50
		Red	1<	94	100
16	3	Mixed White	3	112	0
		Blue	3	103	0
		Green	3	76	0
		Red	3	82	0
	6	Mixed White	3	64	0
		Blue	1	68	0
		Green	2	75	0
		Red	3	82	0

[z] Enlargement rate calculated by the fresh weight at the end of treatment divided by the weight before storage treatment as 100.

Based on these results, if only tuber enlargement is to be promoted, white light LEDs or white fluorescent lamps should be used as the light source under long-day conditions. However, if this condition is continued, the shape of the tubers becomes unsuitable for materials for "Hange", as shown in **Figure 1**. Therefore, the tubers are divided into large and small size in a cycle of about half a year, and the large size tubers are continued to be cultivated. In order to increase the size of the tubers, it is thought that low-temperature treatment of the small tubers and the obtained propagules to promote sprouting and growth will lead to efficient tuber production.

Figure 1 Morphology of *Pinellia terunata* tubers cultivated for 18 months in a plant factory environment (24°C, white fluorescent lamps as light source, 150 μmolm^{-2}s^{-1}, 16 hours illumination per day) (Amaki *et al*., 2015). The planted parent tuber is in the center. The lateral buds were enlarged as secondary tubers one after another, making it an unsuitable form for processing as "Hange".

LITERATURE CITED

Amaki, W., Yamanishi, Y. and Watanabe, H. (2015). Effect of light quality on the growth of the medicinal plant, *Pinelia ternata*. Hort. Res. *14* (Suppl.1):387.

Eguchi, T., H. Tanaka, D. Moriuchi, S. Yoshida and K. Matsuoka. (2019). Temperature effects on the yield and quality of the Medicinal Plant *Pinellia ternata* Breit. Environ. Control Biol. *57*:83-85. https://doi.org/10.2525/ecb.57.83

Eguchi, T., Tanaka, H., Yoshid,a S. and Matsuoka, K. (2020). Temperature effects on the hotosynthesis by the medicinal Plant *Pinellia ternata* Breit. Environ. Control Biol. *58*:49-50. https://doi.org/10.2525/ecb.58.49

Harashima, H. (2012). Soyakutan -Pharmacognosy- (revised ed.) pp.212-213. NTS-BOOK, Tokyo.

Nagao, Y. (1977). On a few environmental conditions affecting the growth of corms and other species of *Pinellia ternata* Briet (1). J. Weed Sci. Tech. *22*:189-193. (In Japanese). https://doi.org/10.3719/weed.22.189

Nagao, Y. (1978). On a few environmental conditions affecting the growth of corms and other species of *Pinellia ternata* Briet. J. Weed Sci. Tech. *22*:69-73. (In Japanese) https://doi.org/10.3719/weed.23.75

Nagao, Y. (1979. Dormancy and sprout of *Pinellia ternata* Briet. J. Weed Sci. Tech. *24*:75-79. (In Japanese). https://doi.org/10.3719/weed.24.69

Nagao, Y. (1980). Growth of flower stalks and seed setting of *Pinellia ternate* Breit. J. Weed Sci. Tech. *25*:93-97. (In Japanese). https://doi.org/10.3719/weed.25.93

Tohoku Revitalization Research Center. (2013). Functional plants (bio and medicinal plants, etc.) marketing research report in plant factories (corrected version).

Combined Proceedings IPPS 72:19-25. 2022.

Investigation of Seed Germination Inhibitory Factors by Allelopathy of Purple Nutsedge Essential Oil

Arisa Noguchi, Misa Hatano and Hiroko Nakatsuka

Department of Agriculture, Tokyo University of Agriculture, 1773 Funako, Atsugi City, Kanagawa Prefecture, 243-0034 Japan.

a3noguch@nodai.ac.jp

Keywords: weeds, crop seed, *Cyperus rotundus*

Summary

The effect of allelopathic germination inhibitory effect of essential oil from tubers of purple nutsedge (*Cyperus rotundus* L.) on seeds of weeds and crops was investigated. Test plants were 5 weeds and 5 cultivated crops (Weeds: *Eleusine indica* (L.) Gaertn., *Digitaria ciliaris* (Retz.) Koeler, *Bidens pilosa* L. var. *pilosa*, *Galinsoga quadriradiata* (*ciliata*) (Raf.) Blake, *Trifolium repens* L., Crops: *Zea mays* L., *Daucus carota* L., *Lactuca sativa* var. *crispa*, *Raphanus sativus* var. *sativus*, *Brassica rapa* var. *perviridis*. Addition of 400 ppm concentration of purple nutsedge essential oil significantly reduced the germination percentage of weed seeds. In particular, the germination percentage was remarkably suppressed to 0% in *Galinsoga*. Among crop seeds, the germination rate decreased only in carrot. The germination rate tended to be lower for plants with smaller seed sizes. The seed germination rate of *Galinsoga*, which has the smallest seed size, decreased as the concentration of essential oil increased, even at concentrations of 40 ppm or less. On the other hand, the germination rate of komatsuna seeds did not decrease even at essential oil concentration of 600 ppm.

INTRODUCTION

Allelopathy is a phenomenon in which chemicals released by plants exert some inhibitory or stimulatory effect on other organisms. In agriculture, research is being conducted on its use for weeds suppression and crop growth promotion (Kobayashi and Ito, 1998; Scavo and Mauromicale, 2021). Among green manures, the effectiveness of white mustard and hairy vetch in controlling pests and weeds is thought to be caused by allelopathy. However, the effect is greatly influenced by the target plant species and environment, and for practical application, it is necessary not only to identify the causative agent of allelopathy but also to clarify the factors such as the sensitivity of the recipient plant (Islam and Hasan, 2021).

The purple nutsedge (*Cyperus rotundus* L.) belongs to Cyperaceae. It is a perennial herb that is native to southern Japan and sub-tropical to tropical areas of the world. This herb grows on maritime sands, roadside and shore of paddy field. Tubers of this plants used as a traditional herbal medicine for nervous disease or menstrual problem (Kilani et al., 2008). However, purple nutsedge has an underground tuber, and even if the above-ground part is removed, the remaining tubers continue to sprout. Therefore, in Japan, it is treated as a tough weed that is difficult to eradicate in upland fields. The purple nutsedge has been shown to exhibit growth-inhibiting allelopathic effects (Quayyum et al., 2000; Abo-Altemen et al., 2019). Therefore, in this study, we investigated the effect on the germination percentage of weed and crop seeds using the essential oil of purple nutsedge (Stoller and Sweet, 1987; Keeley, 1987; Komai et al., 1990; Dhillon et al., 1993), which has been reported to have an inhibitory effect on seed germination. Besides, we examined the relationship with the characteristics of the seeds and inhibitory effect of purple nutsedge essential oil.

MATERIALS AND METHODS

Tubers of purple nutsedge collected at the Atsugi Campus of Tokyo University of Agriculture, Kanagawa Prefecture. The essential oil of purple nutsedge was obtained by steam distillation of these tubers. In the germination test, 20 seeds and 2 mL of pure water-diluted purple nutsedge essential oil solution (contained 0.05 % Tween 20) were placed in a glass petri dish (φ60 mm×20 mm) lined with one layer of filter paper (No.2, Advantec Co., Ltd.). Water was used in a control. However, since sweet corn has a larger seed size than others, φ90 mm petri dishes were used and the amount of added water was increased (total amount was 10mL), resulting in an essential oil concentration of 80 ppm. The petri dishes were maintained in an incubator at 24°C and 18-hour light / 6-hour dark condition. The seed germination rate of each dish was measured.

Test plants were 5 weeds and 5 cultivated crops. Weeds: goosegrass (*Eleusine indica* (L.) Gaertn.), southern crabgrass (*Digitaria ciliaris* (Retz.) Koeler), hairy beggarticks (*Bidens pilosa* L. var. *pilosa*), hairy galinsoga (*Galinsoga quadriradiata* (*ciliata*) (Raf.) Blake), white clover 'Huia' (*Trifolium repens* L.), Crops: 'Golden Bantam' sweet corn (*Zea mays* L.), 'Tokinasi Sansun' carrot (*Daucus carota* L.), red-leaf lettuce (*Lactuca sativa* var. *crispa*), 'Kuroha mino wase' Chinese radish (*Raphanus sativus* var. *sativus*), 'Ajisai' komatsuna (*Brassica rapa* var. *perviridis*). The species and time to germination measurement are shown in Tables 1 and 2, respectively. The survey was repeated three times in each plot. Seed length, thickness and 1000-grain weight were also measured for each test plant.

Table 1. List of test weeds and incubation conditions for seed germination.

Common name	Scientific name	Family name	Incubating temperature (°C)	Light condition	Incubation time (hr)
Goosegrass	*Eleusine indica*	Poaceae	30	18h-L/6h-D	72
Southern crabgrass	*Digitaria ciliaris*	Poaceae	24	18h-L/6h-D	72
Hairy beggar-ticks	*Bidens pilosa* var. *pilosa*	Asteraceae	24	18h-L/6h-D	72
Hairy galinsoga	*Galinsoga quadriradiata*	Asteraceae	22	18h-L/6h-D	72
White clover 'Huia'	*Trifolium repens*	Fabaceae	24	18h-L/6h-D	72

Table 2. List of test vegetables and incubation conditions for seed germination.

Common name	Scientific name	Family name	Incubating temperature (°C)	Light condition	Incubation time (hr)
Sweet corn 'Golden Bantam'	*Zea mays*	Poaceae	24	18h-L/6h-D	72
Carrot 'Tokinasi sansun'	*Daucus carota*	Apiaceae	24	18h-L/6h-D	120
Red-leaf lettuce	*Lactuca sativa var. crispa*	Asteraceae	24	18h-L/6h-D	72
Chinese radish 'Kuroha mino wase'	*Raphanus sativus* var. *sativus*	Brassicaceae	24	Dark	72
Konatsuna 'Ajisai'	*Brassica rapa var. perviridis*	Brassicaceae	22	Dark	48

RESULTS AND DISCUSSION

Addition of 400 ppm concentration of purple nutsedge essential oil significantly reduced the germination rate of weed seeds. In particular, the germination rate was remarkably suppressed to 0% in hairy galinsoga (**Fig. 1**).

Among crop seeds, the germination rate decreased only in carrot (**Fig. 2**). The germination rate tended to be lower for plants with smaller seed sizes (**Figs. 3 and 4**). In particular, the correlation coefficient between germination rate and 1000-grain weight was as high as 0.73 in plants with 1000-grain weight of 3 g or less.

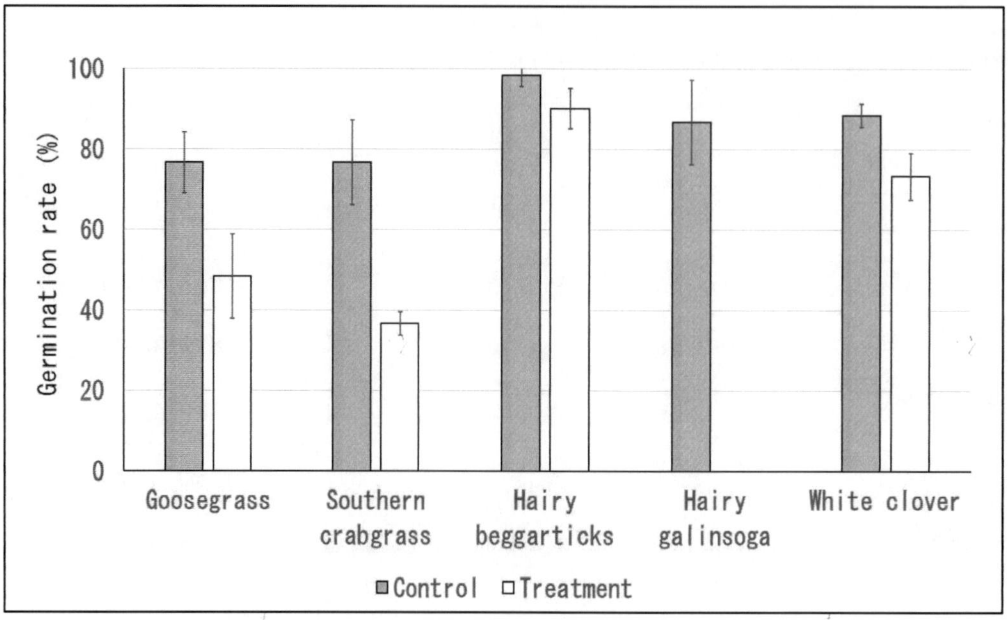

Figure 1. Effect of essential oil obtained from tubers of purple nutsedge at 400 ppm on germination rate of weed seeds (n = 3, error bars are standard deviation).

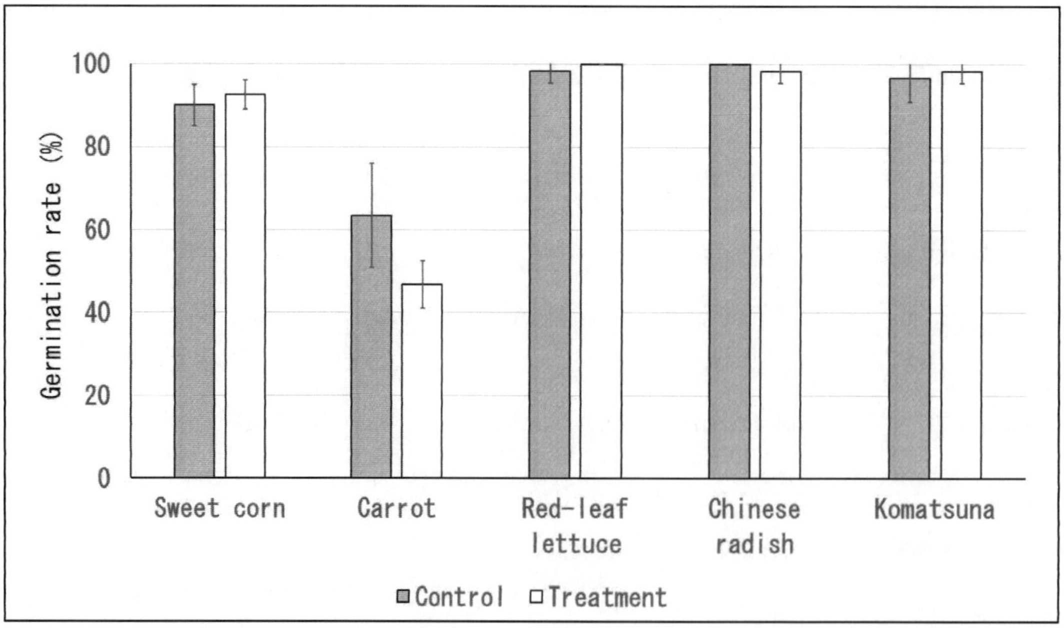

Figure 2. Effect of essential oil obtained from tubers of purple nutsedge at 400 ppm on germination rate of vegetable seeds (n = 3, error bars are standard deviation).

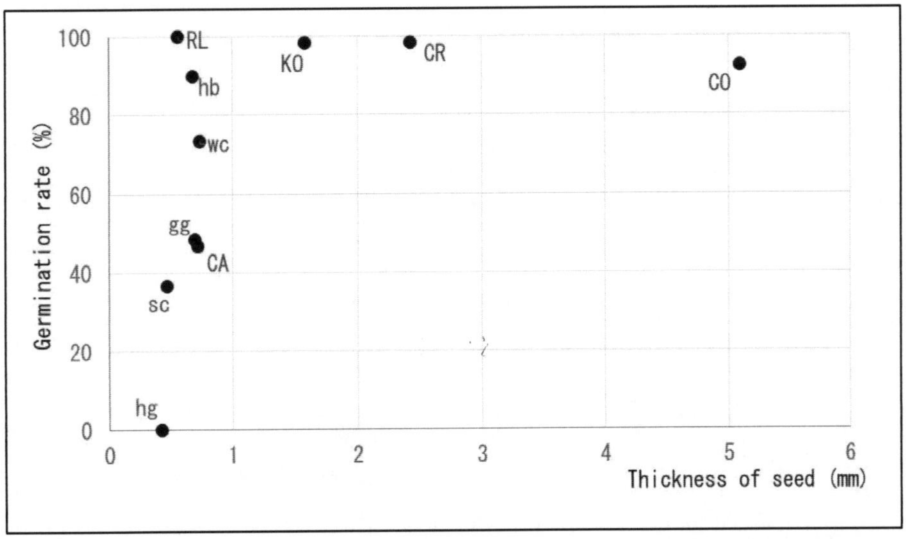

Figure 3. Relationship between thickness of seed and germination rate of seeds treated with essential oil obtained from tubers of purple nutsedge. Abbreviations: gg = goosegrass, sc = southern crabgrass, hb = hairy beggarticks, hg = hairy galinsoga, wc = white clover, CO = sweet corn, CA = carrot, RL = red-leaf lettuce, CR = chinese radish, KO = komatsuna.

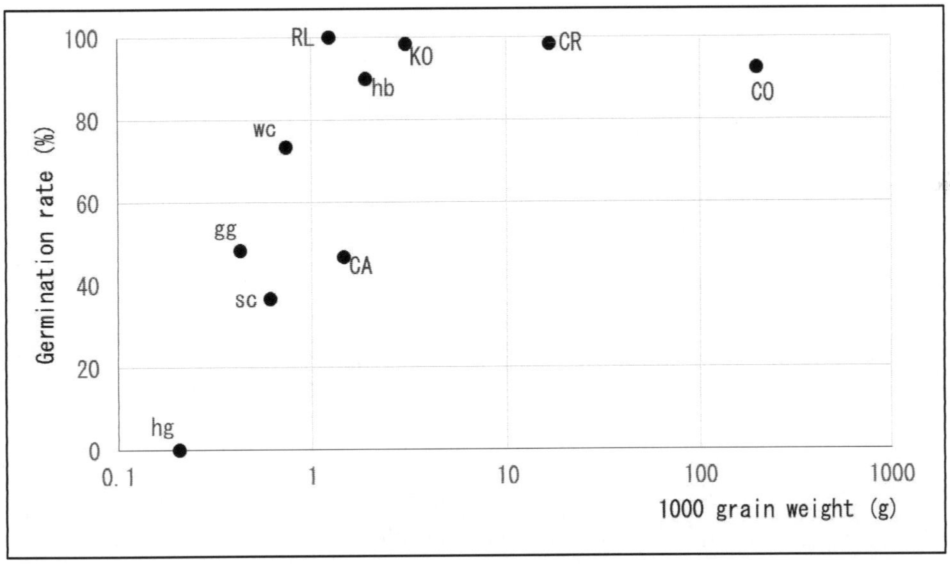

Figure 4. Relationship between weight of seed and germination rate of seeds treated with essential oil obtained from tubers of purple nutsedge. Abbreviations are the same as Figure 3.

The seed germination rate of hairy galinsoga, which has the smallest seed size, decreased as the concentration of essential oil increased, even at concentrations of 40 ppm or less (**Fig. 5**). On the other hand, the germination rate of komatsuna seeds did not decrease even at essential oil concentration of 600 ppm (**Fig. 6**), suggesting that germination of komatsuna seeds would not be affected until concentrations higher than this.

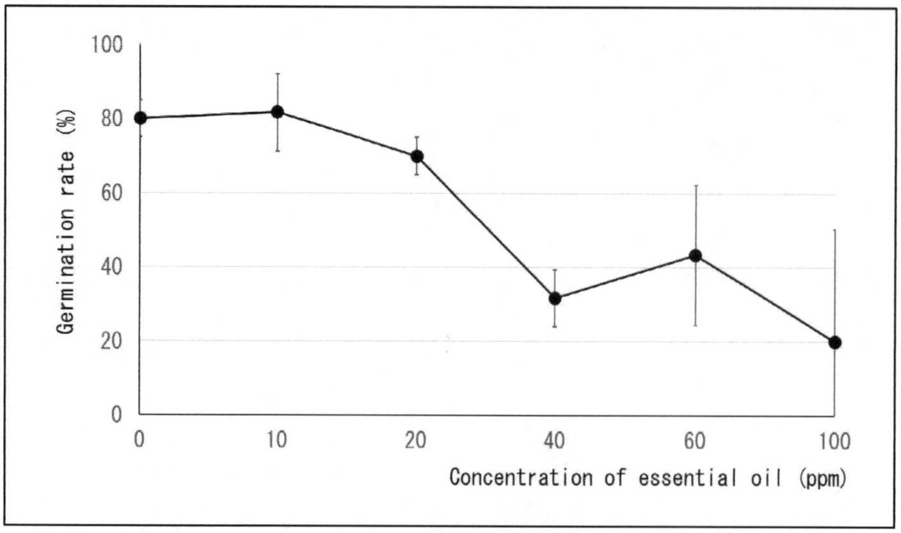

Figure 5. Effect of essential oil concentration of purple nutsedge on germination rate of hairy galinsoga (n = 3, error bars are standard deviation).

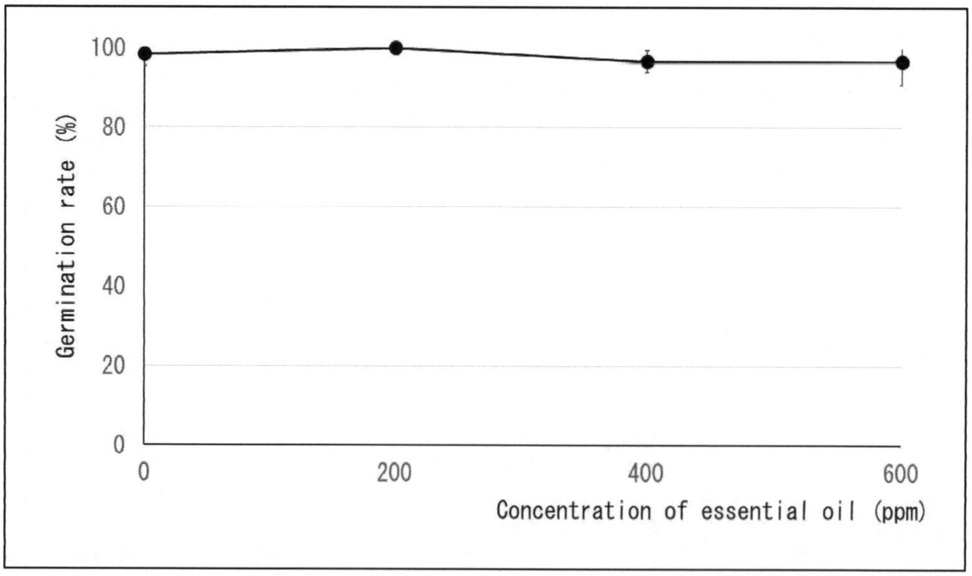

Figure 6. Effect of essential oil concentration of purple nutsedge on germination rate of komatsuna (n = 3, error bars are standard deviation).

From these results, it is considered that the weight of seed to be applied to the seed size greatly affects the suppression of germination by the essential oil of the nutgrass.

However, the difference in germination rate between carrot and lettuce was observed even though the seed size was similar, suggesting that not only the quantitative factor of the inhibitor but also the difference in susceptibility between plant species is involved.

In conclusion, from the above results, it is expected that the essential oil of purple nutsedge can be used as a natural pesticide for controlling weeds small seeds, although it inhibits the germination of weed seeds, but hardly inhibits the germination of crop seeds.

LITERATURE CITED

Abo-Altemen, R.A., Al-Shammari, A.M. and. Shawkat, M.S. (2019). GC-MS analysis and chemical composition identification of *Cyperus rotundus* L. from Iraq. Energy Procedia *157*:1462-1474. https://doi.org/10.1016/j.egypro.2018.11.311

Dhillon, R.S., Singh, S., Kundra, S. and Basra, A.S. (1993). Studies on the chemical composition and biological activity of essential oil from *Cyperus rotundus* Linn. Plant Growth Regulation *13*:89-93. https://doi.org/10.1007/BF00207597

Islam, M. and Hasan, M. (2021). Bioherbicidal properties of *Parthenium hysterophorus, Cleome rutidosperma* and *Borreria alata* extracts on selected crop and weed species. Agronomy *11*:643. https://doi.org/10.3390/agronomy11040643

Keely, P.E. (1987). Interferrence a d interaction of purple and yellow nutsedges (*Cyperus totundus* and *C. esculentus*) with crops. Weed Technology *1*: 74-81. https://doi.org/10.1017/S0890037X00029171

Kobayashi, Y. and Ito, M. (1998). Valuation of phototoxic activity of naturally occurring phenolic compounds. J. Weed Sci. Tech. *43*:341-348. https://doi.org/10.3719/weed.43.341.

Kilani, S., Ledauphin, J., Bouhle,l I., sghaier, M.B., Boubaker, J., Skandrani, I., Mosarti, R., Ghedira, K., Barillier, A.D., and Chekir-Ghedira, L. (2008). Comparative study of *Cyperus rotundus* essential oil by a modified GC/MS analysis method. Evaluation of its antioxidant, cytotoxic, and apoptonic effects. Chemistry & Biodiversity *5*:729-742. https://doi.org/10.1002/cbdv.200890069

Komai, K., Seto, N., Matsubayashi, K. and Hamada, M. (1990). Plant growth inhibition by a sesquiterpene ketolcohol, cyperolone, isolated from purple nutsedge tubers. J. Weed Sci. Tech. *35*:164-168. https://doi.org/10.3719/weed.35.164.

Quayyum, H.A., Mallik, A.U., Leach, D.M. and Gottardo, C. (2000). Growth inhibitory effects of nutgrass (*Cyperus rotundus*) on rice (Oryza sativa) seedlings. J. Chemical Ecology *26*:2221-2231. https://doi.org/10.1023/A:1005532802836

Scavo, A. and Mauromical, G. (2021). Crop allelopathy for sustainable weed management in agroecosystems: Knowing the present with a view to the future. Agronomy *11*:2104. https://doi.org/10.3390/agronomy11112104

Stoller, E.W. and Sweet R.D. (1987). Biology and life cycle of purple and yellow (*Cyperus rotundus* and *C. esculentus*). Weed Technology *1*:66-73. https://doi.org/10.1017/S0890037X0002916X

Combined Proceedings IPPS 72:26-35. 2022.

Tree Growth and Fruit Quality of Some Citrus Scion/Rootstock Combinations in Afghanistan

Ziaurrahman Hejazi[1, a], Rohullah Niazi[2], Abdul Latif Barakzai[3], Bahader Khan Karimi[3], Fida Mohammad Qurashi[3], Takuya Tetsumura[4]

[1]Department of Horticulture, Faculty of Agriculture, Nangarhar University, Afghanistan, [2]Department of Horticulture, Faculty of Agriculture, Paktika University, Afghanistan, [3]Directorate of Agriculture, Irrigation, and Livestock (DAIL), Jalalabad, Afghanistan, [4]Faculty of Agriculture, University of Miyazaki, Miyazaki 889-2192, Japan. [a]Present address: Interdisciplinary Graduate School of Agriculture and Engineering, University of Miyazaki, Miyazaki 889-2192, Japan

zia.hejazi@gmail.com

Keywords: Compatibility index, grafting, rootstock, seed number, sour orange

Summary

Improvement of citrus production was focused on during the rehabilitation process of Afghanistan. The *ex situ* germplasm center of citrus was established and a field trial in the subtropical climate of Nangarhar province was launched by grafting the eight citrus scions (Washington navel, Moro, Tarocco, Tardivo, Marsh seedless, Femminello Siracusano, Minneola tangelo, and Ortanique tangor) onto seven different rootstocks (Volkamer lemon, Carrizo citrange, Troyer citrange, X639, Rough lemon, Trifoliate orange, and Sour orange). The objective of the present study was to measure the performance of these combinations in the local climate. The rootstocks didn't significantly affect the canopy growth, while they affected the compatibility index (CI) of some citrus scions. Tarocco grafted on Volkamer and X639 rootstocks showed significantly lower CI values in comparison to the value (1.0) obtained by the Troyer, Trifoliate, and Sour orange. The CI of Sour orange (1.0) on Tardivo was significantly higher than of the Trifoliate, X639 and Carrizo. The X639 and Trifoliate produced the lowest CI in the Ortanique tangor at 0.77. The fruit quality was influenced by the scion/rootstock combinations. The Washington navel, Moro, and Femminello produced the heaviest fruit on Volkamer, whereas the Tardivo, Marsh seedless, and Minneola tangelo yielded the biggest fruits on Carrizo. The fruit of Tarocco was the largest on Rough lemon and the smallest on Troyer citrange. In a similar manner, the fruit rind of the Washington navel, Moro, Tarocco, and Femminello was the thickest on Volkamer. The thinnest fruit rind was observed on Sour orange, Rough lemon or Trifoliate. A noticeable difference in seed number was indicated by the fact that Tardivo produced the largest number of seeds (10.33) on Volkamer and the smallest (2.50) on X639 citrange. The juice percentage, TSS, and TA were also affected by the rootstocks. Washington navel and Tardivo produced the highest percent juice on Carrizo and Rough lemon, respectively. Most scions grafted on Trifoliate or Volkamer had the highest TSS contents and the lowest TA. Overall, the results provide the first empirical-based insights for the local researchers to have future exploration adding the yield parameter for the specific scion/rootstock combinations.

INTRODUCTION

Afghanistan has long been noted for many kinds of high value fruits. Such crops are spread over the country, generate revenue and provide sustenance to a significant portion of the population. Despite conflicts, strife and drought, the fruits remained the main source of the country foreign exchange earnings.

Eastern Afghanistan secures the Mediterranean climate and brings a conducive environment to grow subtropical fruits. Among them, commercial production of citrus particularly Sweet orange (*Citrus sinensis* (L.) Osbeck), Sour orange (*Citrus aurantium* L.) and lemon (*Citrus limon* (L.) Burm. f.), is predominantly practiced in this region; nevertheless, some other cold parts of the country is recently started lemon cultivation in the protected structures (Glozer and Ferguson, 2007). Because of the unstable political conditions, citrus fruit production experienced fluctuations in the last four decades. However, as per ministry of agriculture report, the acreage and production tended to increase again in the recent years ending at 13,243 tons in 2020. Based on the reported figures, Nangarhar ranked the top citrus producing province followed by Kunar, Laghman and Khost provinces.

Demand for citrus fruits is very high in the domestic market but has always been fulfilled by the imports which mostly constituted Kinnow mandarin (*Citrus reticulata* Blanco) from Pakistan. The market size of the fresh citrus is estimated at 200,000 tons per annum, while the local produce occupies 6.6% of its share (Afghanistan National Horticulture Development Organization, 2013). During the last twenty years, efforts have been made again to boost the local citriculture. The national germplasm center of fruits was established in Jalalabad, where with other fruit species, 75 accessions of citrus including new varieties of sweet oranges and mandarins (*Citrus* spp. (Sect. *Acrumen*)) are preserved.

In Afghanistan, the sour orange is used to be the most common rootstock for the citrus nursery production (Glozer and Ferguson, 2007); however, due to the slow growth of the seedlings and insufficient seed supply, it can't fulfill the current need of saplings in the market. Hence, the Afghan nurseries are now widely relied on imported seeds of Rough lemon to produce the rootstocks. In order to locally build a strong foundation of sustainable citrus production, it has always been suggested to test the long-term performance of the available citrus scions against various rootstocks. Although a trial has been run evaluating some citrus scion/rootstock combinations, no single report has been published yet on the relevant topic. Therefore, this study aimed to present the performance of the tree growth and fruit quality of eight citrus scions budded to seven rootstocks.

MATERIALS AND METHODS

Plant materials and growing conditions

In Feb 2012, the citrus trial was run by planting the grafted saplings in Jalalabad perennial horticulture development and research center (PHDC). Jalalabad is the capital of Nangarhar province which also serves as a regional center for eastern Afghanistan. The location of the experimental site is 566 m above sea level where the annual low and high average temperature respectively falls 3 ^{0}C and 40 ^{0}C. The weather is mostly sunny and dry with <500 mm annual rainfall. Three trees assigned to each scion-stock combination were distanced 5 m apart and headed north to south in a straight row. The distance between the adjacent rows was also 5 m and the soil type was sand clay loam having 8.9 pH. The eight scions are Washington navel, Moro, Tarocco from *C. sinensis*, Tardivo di Ciaculli mandarin (*C. reticulata* Blanco), Marsh seedless grapefruit, (*C. paradisi* Macfad.) Femminello Siracusano lemon, (*Citrus* × *limon* L.) Minneola tangelo (*Citrus* × *tangelo*), and Ortanique tangor (*Citrus reticulata* × *sinensis*). The seven rootstocks are Volkamer lemon (*C. limonia* Osbeck), Carrizo citrange (*C. sinensis* × *P. trifoliata*), Troyer citrange (*P. trifoliata* × *C. sinensis*), X639 (*P. trifoliata* × *C. reticulata*), Rough lemon (*Citrus* × *jambhiri* Lush.), Trifoliate orange (*P. trifoliata* (L.) Raf.), and Sour orange.

Determination of tree vegetative growth and quality attributes of fruit

Tree growth and fruit quality characteristics were measured on the existing trial during late December 2020. According to Zhu et al. (2020), circumferences of the trunk were measured at 5 cm above and below the graft union, and termed as Cs and Cr, respectively. Compatibility index (CI) was calculated with the equation CI = Cs/Cr. The height of the tree and diameters in both parallel and perpendicular directions of the tree to the row were measured as H, Dl and Dr, respectively. The canopy volume was estimated with the equation: V= (π/6) x H x Dl x Dr.

Totally twelve fruits from four directions of three plants were randomly sampled on each scion-stock combination and subjected to the fresh weight, diameter, height, rind thickness, seed number, juice content, total soluble solids (TSS), and titratable acid (TA) measurements. The weight of the fruit was determined by using a digital balance. The fruit diameter, height and rind thickness were measured with a Vernier caliper. The fruit rind was gently peeled from the flesh and referred to a high precision micro-digital caliper for thickness determination. Only fully developed mature seeds were counted as seed numbers. Juice was squeezed with the help of a manual squeezer and then weighed for each fruit. The juice percentage was determined by following formula (juice % = (juice weight x 100) / whole fruit weight). The TSS content of juice was measured by a digital refractometer (PAL-1, Atago, Japan) and expressed as a percent. The TA was determined by titrating 10 mL of juice dilution (10x dilution) with 0.1 N NaOH (pH = 8.1).

Data analysis

Data on tree growth and fruit quality parameters were prepared in Microsoft Excel spreadsheet and then subjected to one-way analysis of variance using SPSS 16.0 statistical software. Means were compared with Tukey's HSD test at p-value 0.05.

RESULTS AND DISCUSSION

Tree canopy and compatibility index

In the present study, the rootstocks did not significantly affect the growth of the tree canopy evaluated for the individual citrus scions, but some slight changes were evident (**Fig. 1**).

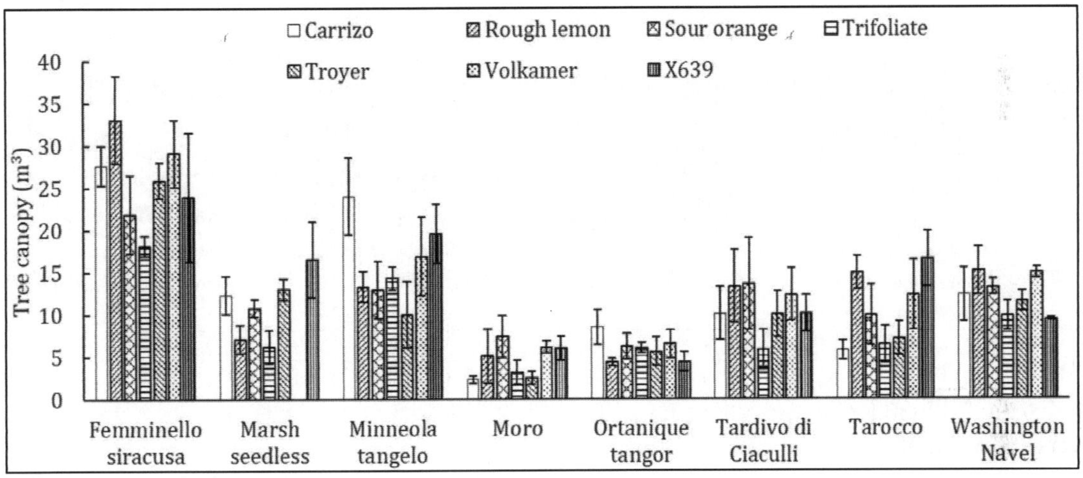

Figure 1. Tree canopy of the eight scions grafted on seven different rootstocks. Error bars show SE of the means of replicates

Although the differences were nonsignificant, Washington navel grafted onto Rough lemon demonstrated vigorous growth at 15.1 m^3 canopy volume, while the same scion onto X639 tended to be smaller of the canopy at 9.4 m^3. Femminello also had greater canopy growth (33.1 m^3) on Rough lemon but lowered (18.2 m^3) when Trifoliate was used as rootstock. Moro sweet orange and Tardivo mandarin showed vigorous growth on Sour orange, whereas the insignificant lower of their growth was noticed on Carrizo and Trifoliate, respectively. The trees of Tarocco and Marsh seedless were larger on X639, while they were smaller when grafted respectively on Carrizo and Trifoliate. Minneola tangelo and Ortanique tangor showed vigorous trees on Carrizo rootstock, whereas the canopy was reduced to the minimum on Troyer, X639 or Rough lemon, respectively. The present findings agree with the studies of Gora et al. (2022) and Zhu et al. (2020). Zhu et al. (2020) evaluated the performance of three late-ripening navel oranges on seven rootstocks. Among them, Carrizo citrange proved the most vigorous rootstock and Trifoliate the smallest for the canopy volume.

Gora et al. (2022) referred the superior canopy growth of the scion at relevant rootstock to the well-adapted characteristics to soil conditions such as an effective root system. Because of the dwarf growth, Trifoliate might be suggested for high dense planting, while other vigorous rootstocks would be considered for sparse planting (Zhu et al., 2020). The citrus rootstocks are probably regulating hormonal induction, anatomy and physiology of the tissue, and through affecting the canopy growth (Gaona-Ponce et al., 2018; Liso et al., 2004; Noda et al., 2000).

The rootstocks significantly affected the compatibility of Tarocco grapefruit, Tardivo mandarin, and Ortanique tangor (**Fig. 2**).

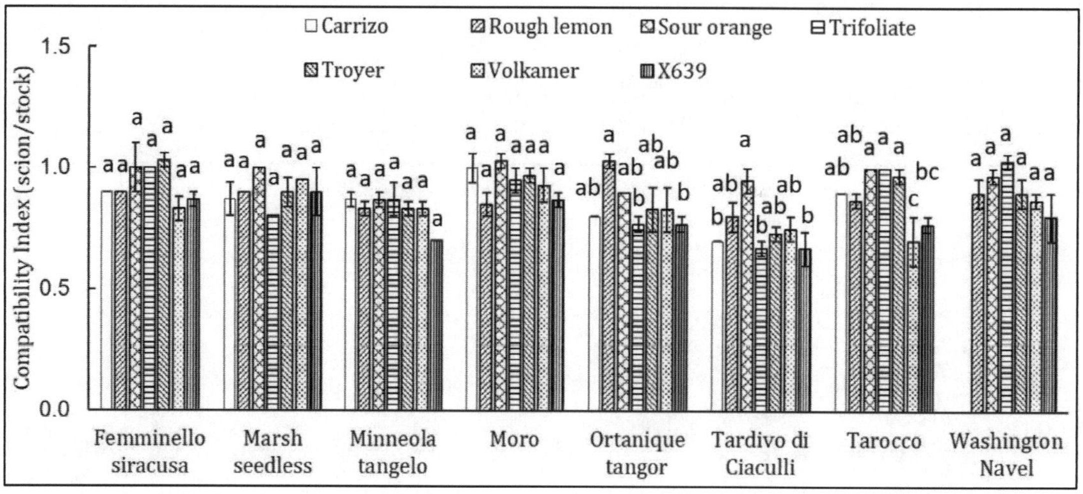

Figure 2. Compatibility index of the eight scions grafted on seven different rootstocks. Values with different lowercase letters are significantly different in an individual scion. Error bars show SE of the means of replicates.

The means of the compatibility index (CI) of all tested combinations fell in the range of 0.67 to 1.0. In Tarocco scion, the CI values of the Troyer, Trifoliate, and Sour orange were at 1.0, significantly higher than of Volkamer and X639. In Tardivo mandarin, the highest CI (1.0) obtained by Sour orange, significantly greater than the values recorded for Trifoliate, X639, and Carrizo. Compared to the highest value of Rough lemon and Sour orange, X639 and Trifoliate produced a significantly lower compatibility index in Ortanique tangor at 0.77. Zhu et al. (2020) also conferred lower CI of three navel orange varieties to Swingle and Trifoliate rootstocks. In general, the closer the index to 1, the high would be the compatibility or affinity. The better compatibility might be associated with the closer genetic relationship between the scion variety and rootstock (Nito et al., 2005).

Fruit quality

The effects of seven rootstocks on the fruit quality of eight citrus scions were shown in Table 1. All scions were significantly affected in fruit weight by the rootstocks except of the Ortanique tangor which was insignificant not only in fruit weight but also in fruit diameter, height, rind thickness, seed number, and juice percentage. In Washington navel, the Volkamer rootstock produced the highest fruit weight (285 g) significantly different from the lowest (189 g) of the Trifoliate. The Volkamer rootstock also produced the heaviest fruit in Moro and Femminello lemon. The fruit of Tarocco was the biggest (253 g) on Rough lemon, significantly different than the lowest (161 g) weighed on Troyer citrange. Tardivo, Marsh seedless, and Minneola Tangelo produced the biggest fruits on Carrizo rootstock, while the significantly smallest fruits of them respectively resulted on Sour orange, Trifoliate, and X639 citrange.

Rootstocks significantly affected diameter of the fruits. More in a similar pattern to the result of the fruit weight, fruit diameters of Washington navel, Moro, and Femminello were remarkably the greatest on Volkamer. The fruit diameter of Tarocco was the biggest on Rough lemon followed by Volkamer in a significant difference with the lowest on Troyer citrange. Tardivo, Marsh seedless, and Minneola Tangelo produced the biggest fruit diameters on Carrizo rootstock, whereas the significantly lowest of them were obtained on Sour orange and Trifoliate, respectively.

The height of fruits of the individual scion was significantly different on various rootstocks but did not change for the Minneola tangelo and Ortanique tangor fruits. Washington navel, Moro, and Femminello followed a similar trend for the fruit height as they formed the tallest on Volkamer rootstock and the shortest on Trifoliate or Rough lemon. Conversely, Tarocco shaped the tallest on Rough lemon and the shortest on Trifoliate. The Trifoliate rootstock also yielded the shortest for the fruit of Tardivo and Marsh, while the tallest of these scions were occurred on Carrizo. Yildiz et al. (2013) also reported the heaviest fruit of 'Valencia Late' sweet orange onto Carrizo rootstock.

Except of Tarocco, Femminello, and Ortanique, the rootstocks significantly affected the rind thickness of the rest of other scions. In a similar manner, the fruit of Washington navel, Moro, Tarocco, and Femminello developed the thickest rind on Volkamer, while the significantly thinner of the first two of them were recorded for the Sour orange and Rough lemon, respectively. In Moro, Rough lemon reduced the rind thickness (3.55 mm) to almost half of that of Volkamer (6.21 mm). Tardivo produced a thicker rind on Troyer compared to significantly thinner on Rough lemon. Marsh seedless and Minneola Tangelo developed the thickest fruit rind on Carrizo, while the thinnest of them was observed on Trifoliate and Sour orange, respectively. On the other hand, a positive correlation was found between the fruit weight and rind thickness at $R^2 = 0.5799$ (**Fig. 3**). The thicker rind might preserve better post-harvest life of fruits but would be considered detrimental if it tends to lower the juice content and pulp (Gora et al., 2022).

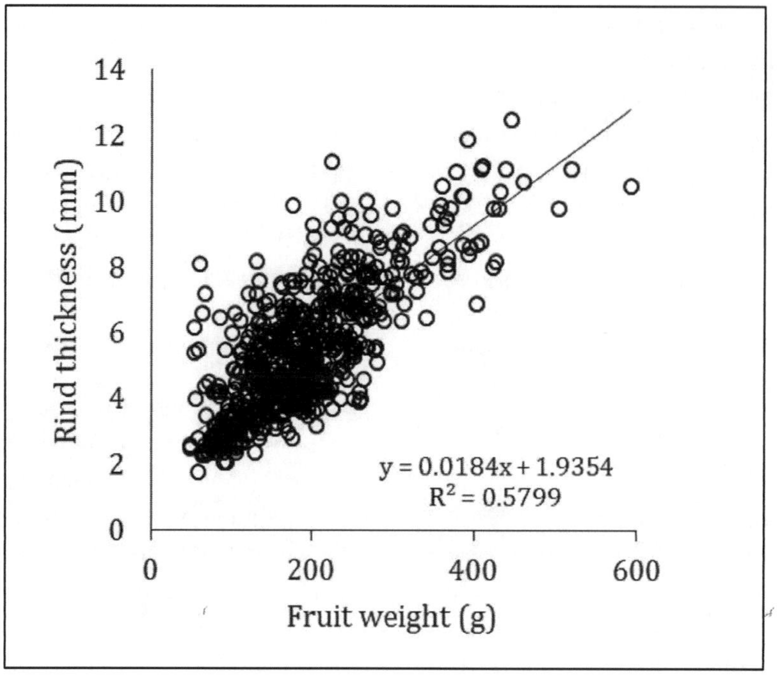

Figure 3. Relationship between fruit weight and rind thickness.

The seed number of the scions was not affected by the rootstocks, with the exception of Tardivo mandarin that significantly produced the highest numbers of seeds (10.33) on Volkamer and the lowest (2.50) in combination with X639 citrange.

The quality of fruit including juice content, TSS, and TA, was affected by the rootstocks (**Table 1**). The significant effects in juice content were observed only for the Washington navel and Tardivo fruits. The fruits of Washington navel on Carrizo possessed the highest percent of juice at 26.62 %, while that on Rough lemon was the least at 7.67 %. In contrast, Tardivo produced the greatest juice percentage (42.03%) on Rough lemon compared with the lowest (29.99%) on Troyer citrange.

TSS was greatly affected by the rootstocks. Except for the Tardivo and Femminello, the highest content of TSS of the scions was found on Trifoliate, whereas the significantly lowest was on Volkamer, except for Moro and Minneola tangelo. The highest TSS of Tardivo and Femminello was observed on Sour orange and rough lemon, respectively.

Rootstocks demonstrated significantly different impact over TA of the fruits of individual scions. TA of Moro, Tarocco, Tardivo, Femminello, and Minneola tangelo were the highest on Sour orange, while the lowest of Moro, Tarocco, Marsh seedless, Minneola tangelo, and Ortanique tangor recorded on Volkamer rootstock. The lowest TA values of Tardivo and Femminello were observed on Troyer citrange and the highest of Marsh seedless and Ortanique were on Trifoliate and X639 citrange, respectively. In Washington navel, the TA value of the fruits was the highest ($0.84\ g \cdot 100\ mL^{-1}$) on both Carrizo and Troyer, but not significantly differed from the TA ($0.82\ g \cdot 100\ mL^{-1}$) of Rough lemon and Trifoliate. However, the significantly lowest TA of Washington navel had noted when budded on X639 citrange, followed by Volkamer and Sour orange.

Table 1. Fruit quality of eight citrus scions grafted on seven rootstocks.

	Rootstock	Scion: Femminello Siracusa	Marsh seedless	Minneola tangelo	Moro	Ortanique tangor	Tardivo di Ciaculli	Tarocco	Washington navel
Fruit weight (g)	Carrizo	197 ab	402 a	223 a	104 bc	163 a	118 a	221 abc	248 ab
	Rough lemon	161 b	280 bc	204 ab	87 c	182 a	87 cd	253 a	222 ab
	Sour orange	192 ab	339 ab	182 ab	110 bc	190 a	79 d	230 ab	224 ab
	Trifoliate	169 b	204 c	177 b	129 ab	178 a	90 bcd	177 bc	189 b
	Troyer citrange	172 ab	267 bc	198 ab	87 c	176 a	107 ab	161 c	224 ab
	Volkamer	217 a	299 b	188 ab	163 a	189 a	106 abc	235 ab	285 a
	X639 citrange	197 ab	282 bc	174 b	112 bc	179 a	97 bcd	173 bc	224 ab
Fruit diameter (mm)	Carrizo	68.3 ab	100.1 a	74.8 a	55.3 b	71.0 a	62.7 a	73.4 abc	83.9 ab
	Rough lemon	64.0 b	86.0 bc	71.2 ab	51.5 b	72.2 a	55.7 bc	77.4 a	79.6 abc
	Sour orange	67.5 ab	92.8 ab	69.3 ab	57.3 b	74.8 a	53.4 c	75.4 ab	76.5 bc
	Trifoliate	64.3 b	77.5 c	68.3 b	59.5ab	72.8 a	56.3 bc	68.3 bc	72.7 c
	Troyer citrange	65.8 ab	86.2 bc	71.3 ab	52.4 b	71.8 a	60.1 ab	66.6 c	77.7 abc
	Volkamer	70.2 a	90.3 ab	70.0 ab	66.0 a	74.7 a	59.0 ab	77.3 a	86.7 a
	X639 citrange	68.6 ab	86.3 bc	68.4 ab	58.0ab	73.4 a	59.8 ab	67.8 bc	79.0 abc
Fruit height (mm)	Carrizo	87.7 ab	84.5 a	76.6 a	57.7bc	60.6 a	56.3 a	73.8 abc	83.1 ab
	Rough lemon	79.0 b	76.4 ab	76.8 a	54.0 c	62.5 a	49.4 b	79.3 a	80.7 ab
	Sour orange	84.1 ab	79.7 ab	72.1 a	60.2 bc	63.2 a	49.4 b	77.5 ab	76.7 ab
	Trifoliate	85.3 ab	65.8 c	72.3 a	63.8 ab	62.1 a	51.0 b	68.0 c	74.3 b
	Troyer citrange	83.6 ab	71.4 bc	73.2 a	54.2 c	61.3 a	53.6 ab	68.5 bc	78.8 ab
	Volkamer	91.0 a	78.4 ab	71.5 a	71.1 a	62.8 a	52.4 ab	75.1 abc	87.5 a
	X639 citrange	86.6 ab	74.9 bc	70.0 a	58.4 bc	60.9 a	52.8 ab	66.5 c	79.3 ab
Rind thickness (mm)	Carrizo	5.86 a	9.71 a	4.96 a	4.38 bc	4.32 a	3.17 ab	6.37 a	7.21 ab
	Rough lemon	4.96 a	7.95 bc	4.21 ab	3.55 c	4.57 a	2.64 c	7.14 a	6.89 ab
	Sour orange	5.58 a	8.73 ab	3.76 b	5.02	4.65 a	2.58 c	5.98 a	5.72 b
	Trifoliate	5.57 a	6.78 c	4.52 ab	5.17 ab	4.88 a	2.83 bc	5.32 a	5.75 b
	Troyer citrange	5.38 a	7.44 bc	4.62 ab	5.22 ab	4.37 a	3.39 a	5.53 a	6.52 ab
	Volkamer	5.88 a	8.68 ab	4.61 ab	6.21 a	4.55 a	3.00 abc	6.76 a	7.67 a
	X639 citrange	5.49 a	7.64 bc	3.98 b	5.17 b	4.10 a	3.23 ab	5.30 a	7.04 ab
Seed number	Carrizo	4.67 a	3.08 a	2.50 a	1.00 a	14.00 a	5.00 bc	0.08 a	0.17 a
	Rough lemon	4.83 a	2.00 a	4.67 a	2.10 a	12.25 a	5.75 bc	0.17 a	0.75 a
	Sour orange	3.25 a	2.75 a	8.08 a	1.25 a	13.08 a	7.50 ab	0.17 a	0.50 a
	Trifoliate	4.92 a	3.33 a	6.50 a	0.92 a	11.83 a	6.92 ab	0.17 a	0.08 a
	Troyer citrange	3.67 a	2.17 a	2.75 a	0.50 a	14.25 a	3.75 bc	0.08 a	0.17 a
	Volkamer	4.75 a	2.58 a	6.00 a	1.33 a	13.33 a	10.33 a	0.42 a	0.25 a
	X639 citrange	4.33 a	1.83 a	7.42 a	1.17 a	14.25 a	2.50 c	0.08 a	0.67 a
Juice (%)	Carrizo	33.65 a	35.36 a	44.15 a	32.20 a	43.93 a	32.05 cd	32.18 a	26.62 a
	Rough lemon	36.85 a	35.61 a	48.80 a	39.15 a	42.30 a	42.03 a	25.06 a	7.67 c
	Sour orange	34.71 a	38.14 a	48.23 a	33.83 a	39.99 a	39.24 ab	30.57 a	21.62 ab
	Trifoliate	34.90 a	38.03 a	45.85 a	36.50 a	42.64 a	38.13	28.53 a	12.71 bc
	Troyer citrange	35.08 a	37.62 a	48.10 a	28.37 a	41.22 a	29.99 d	30.85 a	10.88 bc
	Volkamer	34.07 a	36.10 a	47.39 a	31.07 a	40.49 a	33.27	29.29 a	9.31 bc
	X639 citrange	36.01 a	36.46 a	46.57 a	35.45 a	42.5 a	31.62 d	27.87 a	16.65 abc
TSS (%)	Carrizo	7.10 de	8.65 bcd	8.45 b	7.60 b	8.4 bc	8.90 cd	9.15 a	7.80 a
	Rough lemon	7.85 a	8.75 bc	7.60 c	7.30 b	7.9 d	9.65 bc	8.80 ab	8.05 a
	Sour orange	7.35 bc	9.05 b	9.25 a	7.85 b	8.2 bcd	10.60 a	9.15 a	7.95 a
	Trifoliate	7.55 b	10.45 a	9.35 a	8.95 a	9.35 a	10.35 ab	9.50 a	8.05 a
	Troyer citrange	6.90 ef	8.75 bc	8.60 b	7.45 b	8.50 b	10.05 ab	8.00 c	8.00 a
	Volkamer	6.75 f	7.95 d	7.70 c	7.55 b	7.35 e	8.55 d	8.00 c	6.75 b
	X639 citrange	7.15 cd	8.15 cd	8.60 b	8.00 b	8.05 cd	10.30 ab	8.40 bc	7.80 a

	Carrizo	5.70 b	1.72 e	1.24 f	1.33 c	2.00 b	1.96 e	0.77 bc	0.84 a
TA	Rough lemon	5.58 c	1.87 c	1.32 e	1.31 c	1.77 f	2.24 c	0.76 cd	0.82 a
(g·100	Sour orange	5.76 a	1.92 b	1.82 a	1.56 a	1.80 e	2.56 a	0.88 a	0.77 b
mL^{-1})	Trifoliate	5.50 d	2.22 a	1.65 b	1.31 c	1.98 c	2.50 b	0.88 a	0.82 a
	Troyer citrange	5.24 f	1.76 d	1.56 c	1.42 b	1.88 d	1.56 g	0.74 d	0.84 a
	Volkamer	5.52 d	1.65 f	1.21 g	1.15 e	1.54 g	2.12 d	0.65 e	0.77 b
	X639 citrange	5.40 e	1.67 f	1.40 d	1.21 d	2.04 a	1.64 f	0.79 b	0.75 b

[1)] TSS, Total Soluble Solid; TA, Titratable Acidity.
Means followed by the same letter in a column do not differ significantly for a trait according to Tukey's HSD.

The results of the fruit weight of Washington navel, Moro, and Femminello were similar to Ahmad et al. (2007), who reported significantly highest fruit weight per tree of 'Kinnow' on Volkamer and Sour orange rootstocks. In the present study, TSS of the six out of eight scions was the highest on Trifoliate rootstock, similar to the result of the TSS content of three navel orange varieties which was the highest on *Poncirus* (Zhu et al., 2020). Khalifa and Hamdy (2015) compared Volkamer and Sour orange rootstocks for the yield and fruit quality of two mandarin varieties. Volkamer showed higher fruit yield, fruit weight, height, diameter, fruit pulp weight, fruit rind weight, rind thickness, juice, number and weight of seeds, and fruit firmness than those budded on Sour orange. However, trees budded on Sour orange had higher TSS, total acidity, and vitamin C compared with those on Volkamer. Similar results were reported when these rootstocks (Volkamer and Sour orange) were evaluated with Ruby Red, Marsh grapefruit (Ramin and Alirezanezhad, 2005) and 'Hamlin' orange (Al-Hosni et al., 2011). Gora et al. (2022) stated that large fruit with thick and rough peel, low juice percent, and lower concentrations of TSS and ascorbic acid in the juice would be often related to varieties budded on fast-growing and vigorous rootstocks.

CONCLUSION

In this study, the scions belonged to different citrus groups that produced varied results, so the results were difficult to generalize except to consider individual findings on scion/rootstock combinations. The present report is the first published insight on the citrus trial that has been launched a decade ago in Jalalabad Afghanistan condition. It would further help the Afghan researchers to have more exploration about the citrus scion/rootstock combinations in the local environment, but data on yield and productivity over several years is highly desired.

ACKNOWLEDGMENT

The authors would like to thank students of the agriculture faculty, Aminullah, Jahanzib and Amjad khan, for the help in data collection. We also extend our gratitude to Mohammad Aziz Saeedi for generously sharing his personal experience about local citrus production.

LITERATURE CITED

Afghanistan National Horticulture Development Organization, 2013. Afghanistan citrus market survey report. Retrieved June 6, 2022, from http://anhdo.org.af/ .

Ahmed, W., M.A. Nawaz, M.A. Iqbal and M.M. Khan. 2007. Effect of different rootstocks on plant nutrient status and yield in

Kinnow mandarin (*Citrus reticulata* Blanco). Pak. J. Bot. *39*:1779-1786.

Al- Hosni, A.S., S. Mustafa, K. Al-Busaidi, M. Al-Jabri and H. Al-Azri. 2011. Effects of different citrus rootstocks on growth, yield, quality and granulation of 'Hamlin' orange in Oman. Acta Hortic. *903*:563-568.

Food and Agriculture Organization, 2019. FAOSTAT Statistical Database of the United Nation Food and Agriculture Organization (FAO) statistical division. Rome. http://www.fao.org/faostat/en/#data/QC. Accessed 3 May 2021.

Gaona-Ponce, M., G. Almaguer-Vargas, A.F. Barrientos-Priego. 2018. Relationship of rootstock xylem anatomy with the initial growth of 'Tahiti' lime (Citrus. latifolia Tanaka ex Q. Jiménez). Revista Chapingo Serie Ciencias Forestales y del Ambiente. *24*:359-370.

Glozer, K. and L. Ferguson 2007. Citrus growing in Afghanistan. Mark Bell international programs. University of California, Davis, pp. 1-51. http://www.aes.ucdavis.edu/IntProg/Default.htm

Gora, J.S., R. Kumar, B.D. Sharma C. Ram, M.K. Berwal, D. Singh, R.S. Bana, and P. Kumar. 2022. Performance evaluation of Fremont mandarin on different rootstocks under the hot arid environment of India. S. Afr. J. Bot. *144*:124-133.

Khalifa, S.M. and A.E. Hamdy. 2015. Effect of some citrus rootstocks on yield and fruit quality of two mandarin varieties. 6th Inter. Sci. Agric. Sympo., Agrosym. Pp. 182-190.

Lliso I, J.B. Forner and M. Talon. 2004. The dwarfing mechanism of citrus rootstocks FA 418 and #23 is related to competition between vegetative and reproductive growth. Tree Physiol. *24*:225–232

Nito, N., S.H. Han and Y. Katayama. 2005. Evaluation of graft compatibility for taxonomically relationships among species of the orange subfamily. Acta Hortic. *692*:85–89.

Noda, K., H., Okuda and I. Iwagaki. 2000. Indole acetic acid and abscisic acid levels in new shoots and fibrous roots of citrus scion-rootstock combinations. Sci. Hortic. *84*:245–254.

Ramin, A.K. and A. Alirezanezhad. 2005. Effects of citrus rootstocks on fruit yield and quality of Ruby Red and Marsh grapefruit. Fruits. *60*:311-317.

Yıldız, E., T.H. Demirkeser and M. Kaplankıran. 2013. Growth, yield, and fruit quality of 'Rhode Red Valencia' and 'Valencia Late' sweet oranges grown on three rootstocks in eastern Mediterranean. Chil. J. Agric. Res. *73*: 142-146.

Zhu, S., T. Huang, X. Yu, Q. Hong, J. Xiang, A. Zeng, G. Gong and X. Zhao. 2020. The effects of rootstocks on performances of three late-ripening navel orange varieties. J. Integr. Agr. *19*:1802–1812.

Combined Proceedings IPPS 72:36-38. 2022.

Morphological Characteristics of Tetraploid *Rosa multiflora* Obtained by Diploid Breeding

Masaki Ochiai[1], Daiki Maruyama[2], Hirokazu Fukui[1] and Kunio Yamada[1]

[1]Graduate School of Natural Science and Technology, Gifu University, 1-1 Yanagido, Gifu, 501-1193, Japan. [2]Faculty of Applied Biological Sciences, Gifu University, 1-1 Yanagido, Gifu, 501-1193, Japan.

ochiai.masaki.h3@f.gifu-u.ac.jp

Keywords: morphology, multiflora rose, hybrid, flower size

Summary

This paper reports the results of a survey of morphological traits from derived tetraploid compared to diploid *R. multiflora* plants. Leaf length did not change between diploid and tetraploid plants, but leaf width and petal size did show significant increases.

INTRODUCTION

Rosa multiflora Thunb. is a wild rose species that grows wild from China to Japan. Although it is considered as one of the wild species that contributed to the establishment of horticultural varieties of rose, it still has value to be utilized again as a genetic resource to provide disease resistance and environmental stress tolerance. However, since most rose cultivars are tetraploid while *R. multiflora* is diploid, it is necessary to produce tetraploids of *R. multiflora* by polyploidization for obtaining fertile hybrid progeny. In this presentation, we report on the results of a survey of traits of tetraploid *R. multiflora*, which we have already obtained through our previous research, compared them with diploid.

RESULTS AND DISCUSSION

No clear differences were observed in budding and flowering time. In terms of leaf shape, diploids had thin, pointed leaflets with many spaces between leaflets, whereas tetraploids had rounded leaflets with overlapping other leaflets (**Fig. 1**).

Figure 1. Leaves of diploid and tetraploid of *R. multiflora*

Focusing on the floral organs, the petals showed an increase in width only, as did the leaflets. As a result, the inflorescence of the diploid appeared to have many gaps between flowers, whereas that of the tetraploid appeared to be densely packed with flowers, giving a luxurious impression (**Fig. 2**).

Figure 2. Flowers of diploid and tetraploid of *R. multiflora*. A. Inflorescence, B. Flower.

In plants in general, tetraploids do not change significantly in length of leaves and petals, while the width of leaves and petals increases compared to diploids. These changes were also conserved in *R. multiflora*.

This tetraploid was originally bred for the purpose of introducing disease resistance to rose cultivars. Our previous research suggested that the diploid and tetraploid tested in this study may differ in the degree of disease resistance, too. Future studies should compare not only morphology but also disease resistance and tolerance to environmental stresses.

PROCEEDING'S PAPERS

SOUTHERN REGION OF NORTH AMERICA

Dr. Fred Davies, Jr., Regional Editor

Forty-sixth Annual Meeting - 2022

Athens, Georgia U.S.A.

Combined Proceedings IPPS 72:39-41. 2022.

Technical Sessions of International Plant Propagators' Society-Southern Region of North America Annual Meeting

Bobby Green

Green Nurseries & Landscape Design
415 North Greeno Road, Fairhope, Alabama 36532, USA

bobby.green@greennurseries.com

Keywords: Annual Meeting, Southern Region of North America (SRNA)

Summary

The 46th Annual Meeting of the International Plant Propagators' Society-Southern Region of North America (SRNA) convened at 8:00 am on 18 October 2022 at the University of Georgia Center for Continuing Education & Hotel, Athens, Georgia, with President Bobby Green presiding.

PRESIDENT BOBBY GREEN

President Green welcomed everyone to Athens, Georgia for the 46th Annual Meeting of the SRNA. It is so awesome to be here in Athens to "seek and share" with one another! The relationships and experiences forged at our meetings are what distinguishes the IPPS.

He thanked Local Site Committee Chair, Dr. Tim Smalley and his committee and volunteers for their outstanding work in arranging the excellent tours, hotel, other planning activities and all their attention to detail. He also thanked Mike and Bonnie Dirr for hosting the SRNA at their home and garden. The SRNA remains financially strong, with the due diligence of Sec-Tres, Donna Foster, and is the largest international region with

242 active, paying members, and 19 student/military veteran members for a total of 261 members.

Green thanked the Executive Committee, and the Sponsorship Committee of Dr. Cheryl Boyer, Tom Saunders, Tim Smalley, Leanne Kenealy Atkins and Michael Roe who raised $56,657 – which is outstanding! Green encouraged the membership to thank, visit and show their support of our sponsors during the meeting. The SRNA is deeply indebted to our loyal sponsors who make our annual meeting financially possible. He encouraged all members to make new members and first-time attendees feel welcome — share with them and seek from them.

Green announced that the SRNA is in its fifth year of the Southern Region Educational Endowment, with a base donation of $20,000 from an anonymous donor. The Education Endowment balance is now at $103,000 – and growing. It will greatly enhance our region's ability to support students and early career professionals – and ensure continued quality of the outstanding educational, out-reach programs our region is known for. All of this year's contributions to the silent and live auction are to go to the Endowment Fund – so please contribute! He thanked Kevin Gantt for leading the endowment effort.

Two years ago - the SR-IPPS initiated the *Margie Jenkens Industry Scholarship* to support industry professionals attending our conference for the 1st time.

For the *Early-Career IPPS Exchange* program between the SRNA and the European Region, Erika Ramos of J. Berry Nursery, represented the Southern Region at the European Conference and toured some of the European nursery industry. Louise Heissel of the European Region is being hosted by Brie Arthur who will accompany her visiting the Southern Region green industry after the conference.

This is the eleventh year the SRNA is doing the *Vivian Munday Young Horticultural Professional Scholarship Work Program (formerly Vivian Munday Scholarship).* We currently have four interns: Kaitlin Swiantek of the University of Georgia, Runshi Xie of Texas A&M University, Teagan Young of the University of Florida and Kristopher Criscione of Louisiana State University. These young professionals are making a strong contribution to this year's program.

This year the SRNA will vote on its *Constitution and Bylaws*, which are being revised after some 20-plus years. There is also a revised, updated *SRNA Operations Manual* which is posted on the SRNA website. The SRNA has grown to 20 standing committees to run the organization and its Green Industry educational, outreach activities!

Green thanked Program Chair and 1st Vice-President, Dr. Judson LeCompte for the excellent program and incredible group of speakers he assembled!

PROGRAM CHAIR DR. JUDSON LECOMPTE

Program Chair Judson LeCompte welcomed all members, guests and students. He acknowledged President Green for his leadership and very capably serving as President. He thanked the membership for the opportunity to serve them, and then reviewed the scheduled program. There were thirteen paper submissions for the Charlie Parkerson Student Research Competition which is outstanding. There will be four students competing in the oral competition, and 12 poster presentations for the membership to visit with student presenters during the meeting. The Question Box, scheduled for Tuesday evening, was to be moderated by Michael Roe. He acknowledged Scott Langlois who seamlessly runs the audiovisuals for our speakers and meeting. He then introduced the first moderator, Laura Ney from the University of Georgia.

Figure 1. President Bobby Green (right) with Dr. Judson LeCompte (left), Program Chair of the 2022 Athens, Georgia, 46th annual conference.

Combined Proceedings IPPS 72:42-49. 2022.

Pesticide Application Method and Timing Influences Contamination of Nectar in *Salvia*

Vanesa Rostán[1a], P. Christopher Wilson[1] and Sandra B. Wilson[2]

[1]Soil, Water, and Ecosystem Sciences Dept., University of Florida, PO Box 110290, Gainesville, FL 32611, USA
[2]Dept. of Environ. Horticulture, PO Box 110675, Gainesville, FL 32611, USA

vrostan@ufl.edu

[a]First Place – Charlie Parkerson Graduate Student Research Paper Competition

Keywords: thiamethoxam, clothianidin, pollinators, *Salvia* × 'Indigo Spires', best management practices (BMPs)

Summary

Exposure to pesticides is one potential factor contributing to the loss of pollinators and pollinator diversity observed over the recent past. This project evaluated the influence of pesticide application method (drench vs. spray) and timing (relative to flowering) on contamination of nectar of *Salvia* × 'Indigo Spires' (*Salvia longispicata* × *S. farinacea*). The systemic insecticide thiamethoxam (Flagship) was used as a model pesticide and applied at the lowest label-rate. The concentrations of thiamethoxam and its metabolite clothianidin in nectar were highest in drench applications, regardless of application timing, and exceeded published toxicity thresholds for native bees and/or honeybees in the case of thiamethoxam. In contrast, concentrations in nectar were below toxicity thresholds for both spray applications before and after flowering. Concentrations were lower for spray and drench applications made before flowering relative to applications made after flowers began opening.

INTRODUCTION

Pollinators are important not only for maintaining biodiversity and ecosystems, but also for the provision of food for humans, which would be greatly limited without pollination (Kearns et al., 1998). World-wide pollinator populations and diversity have declined over the past hundred years due to several potential factors, including exposure to pesticides (Biesmeijer et al., 2006). Pollinators may be exposed to pesticides by ingestion when collecting nectar and pollen from contaminated flowers, as well as from contact on contaminated flower and leaf surfaces.

Ornamental plant production and allied industries are an important economic engine. Hall et al. (2020) reported that the output for the U.S. green industry in 2018 was $159.6 billion. Ornamentals marketed as pollinator friendly are important sources of food and refuge in urbanized landscapes, as well as urban areas bordering agricultural areas (Marquardt et al., 2021). Given this critical role, minimization of unnecessary contact with pesticides is important. Lentola et al. (2017) surveyed ornamental plants at retail centers, detecting neonicotinoid insecticides in 70% of the plants. While "pollinator- friendly" plants promote and support pollinators with pollen and nectar, they may be harmful if pesticide levels are above toxicity thresholds.

Given that pesticides are necessary tools for economically producing ornamental plants, pest management practices should be optimized to reduce exposures of pollinators as much as possible, while simultaneously providing adequate control of pests. Before practices can be optimized, an understanding of the influence of pesticide application and cultural practices on contamination of floral resources is needed. Few studies have specifically focused on the relationship between pesticide management during ornamental plant production and contamination of nectar.

This research focused on the systemic neonicotinoid insecticide thiamethoxam. It is an antagonist of insect nicotinic acetylcholine receptors, and is metabolized to clothianidin, which is also highly toxic to pollinators with the same mode of action (Nauen et al., 2003). The objectives of this project were to evaluate and quantify the influence of thiamethoxam application methods (spray vs. drench) and timing (relative to flowering) on contamination of nectar using *Salvia* × 'Indigo Spires' as a model ornamental species.

MATERIALS AND METHODS

Salvia × 'Indigo Spires' is a vigorous and non-stop bloomer, producing dark violet flowers from early summer through fall (Clebsch and Barner, 2003) (**Fig. 1A, 1B**). It is valued for its long spikes of flowers that produce large volumes of nectar and attract a wide variety of pollinators including hummingbirds, honeybees, and native bees. Plants were purchased as rooted cuttings from Hatchett Creek Farms (Gainesville, FL) and transplanted into 9.9 cm (4.5-in) pots filled with a general-purpose soilless media (Promix BX, Premier Tech Horticulture, Quebec). Plants were grown under controlled conditions in a greenhouse on the University of Florida campus for six weeks, with periodic pruning (3x) to promote branching and maintain plant size. Afterwards, plants were repotted into 11.4 L (3-gal) containers using the same media, fertilized with 15-ml (1-tbls) of Osmocote

14N-4P-14K per plant, and moved to a shade house (**Fig. 1A**). Plants were manually irrigated at least once a day.

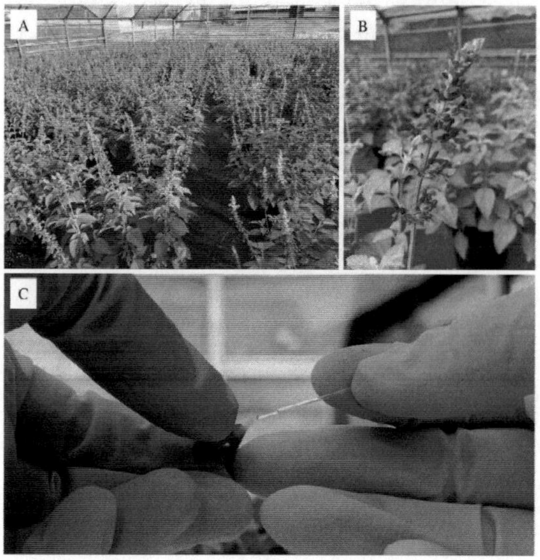

Figure 1. A) *Salvia* × 'Indigo Spires' grown in shade house during the study. B) Representative flower spike of *Salvia* × 'Indigo Spires'. C) Extraction of nectar from floret using a 50 µL glass microcapillary (note nectar in tip of microcapillary).

This study used eight replicate plants per treatment (including controls) and a 2 x 2 factorial design to evaluate the influence of application method (2 levels: drench vs. spray) and timing (2 levels: pre-bloom vs. post-bloom opening) on contamination of flower nectar. Commercially available Flagship 25WG (a water dispersible granule containing 25% thiamethoxam) was applied as a foliar spray using a hand-operated spray bottle (0.15 g/L, 155 mL per plant) or soil drench (0.30 g/L, 1 L per pot) at the lowest labeled rate for ornamentals. Applications for the "pre-bloom" treatment were made 21 days after transplanting. Two weeks later, applications for the "post-bloom" treatment were made, when half of the flowers were open. Stratified random sampling was used to select plants for each treatment.

Nectar samples were collected seven days after the last pesticide application using 50 µL glass microcapillaries (**Fig. 1C**). Nectar volume was estimated by measuring the length of liquid in the tube. Each sample was stored separately in an Eppendorf tube on ice in the field followed by storage at -80°C until analysis. Samples were diluted with H_2O:ACN (9:1), thoroughly mixed using a vortex, and centrifuged (13,000 RCF, 10 min) prior to analysis on the same day. Thiamethoxam and its' metabolite clothianidin were analyzed using a 1290 Infinity II ultra high-pressure liquid chromatograph (uHPLC, Agilent) equipped with a C18 reversed-phase column (Zorbax Eclipse C18, Rapid resolution HD, 150 × 2.1 mm, 1.8 µm) and coupled to an Agilent 6495 tandem mass spectrometer (MS/MS). The MS/MS was operated in electrospray ionization (ESI) positive mode, with nitrogen as the source and collision gas. Mobile phase solvent composition and gradient (uHPLC), and multiple reaction monitoring (MRM) transitions used for identification and quantification are shown in **Table 1**. Concentrations in the samples were determined using external calibration curves.

Statistical analyses were conducted using the software R (R Core Team, 2021). To test if method and/or timing of pesticide application influenced the concentration of thiamethoxam and/or clothianidin detected in nectar, two-way analysis of variance (ANOVA, P=0.05) was conducted after log transforming the thiamethoxam and log transforming +1 the clothianidin data to meet the assumptions of ANOVA.

Table 1. Mobile phase gradient conditions and multiple reaction monitoring (MRM) transitions (m/z) for identification and quantification of thiamethoxam and clothianidin in nectar.

Mobile Phase Gradient			
Time (min)	*A (%)	**B (%)	Flow (mL/min)
0.00	90	10	0.400
1.00	90	10	0.400
7.00	10	90	0.400
7.50	90	10	0.400

MRM Transitions for MS/MS Analysis			
Analyte	Precursor m/z	Quantifier m/z	Qualifier m/z
Thiamethoxam	292.03	211.11	181.1
Clothianidin	250	169	131.9

*Solvent A: 95% Optima LC-MS water, 5% Optima LC-MS ACN, with 0.1% Optima formic acid, 5 mM ammonium formate.

**Solvent B: 95% Optima LC-MS ACN, 5% Optima LC-MS water, with 0.1% Optima formic acid, 5mM ammonium formate.

RESULTS

Application method significantly influenced concentrations of thiamethoxam (*P*-value:<2E-16) and clothianidin (*P*-value:<2E-16) in nectar. When thiamethoxam was applied before flowering, concentrations in nectar were 421.6±71.6 ng/mL with drench applications (**Fig. 2A**), and 3.5±1.3 ng/mL when applied as a spray (**Fig. 2B**). Concentrations in nectar were higher in both cases when thiamethoxam was applied after flowering (**Fig. 2 A, B**). In this case, thiamethoxam concentrations were 820.4±192.9 ng/mL with the drench applications and 13.7±2.6 ng/mL with the spray applications. Clothianidin concentrations followed a similar pattern where concentrations in nectar were significantly higher when thiamethoxam was applied as a drench (**Fig. 3A, B**). However, concentrations were lower than its' precursor thiamethoxam. In this case, clothianidin concentrations in nectar were 27.2±2.9 ng/mL (drench applied before flowering), <0.5 ng/mL (spray applied before flowering), 44.8±12.1 ng/mL (drench applied after flowering), and 2.2±0.6 ng/mL (spray applied after flowering).

Timing of applications also significantly influenced concentrations of thiamethoxam (*P*-value:<2E-16) and clothianidin (*P*-value:<2E-16) in nectar. Regardless of application method (spray or drench), concentrations were higher when applications were made after flowering. For drench applications, thiamethoxam concentrations when applied after blooming (820.4±192.9 ng/mL) were almost twice the concentrations detected in nectar when the pesticide was applied before flowering (421.6±71.6 ng/mL) (**Fig. 2A**).

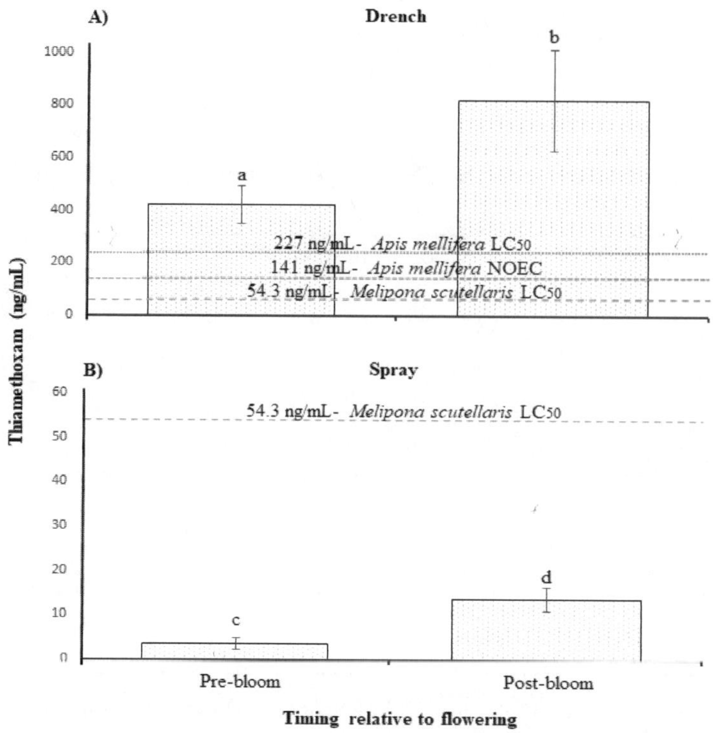

Figure 2. Thiamethoxam concentrations (± standard deviation) in nectar associated with application by A) drench applied pre- and post-bloom and B) spray applied pre- and post-bloom. Different letters indicate significant differences between timings and application methods based on 2-way ANOVA (P=0.05). Horizontal reference lines indicate median lethal concentration (LC_{50}) values for honeybees (*A. mellifera*) and native bees (*Melipona scutellaris*) (Miotelo et al., 2021) and 'no observable effects concentration' (NOEC) for honeybees (Overmeyer et al., 2018).

For spray applications, thiamethoxam concentrations from applications after blooming (13.7±2.6 ng/mL) were three times the concentrations when the pesticide was applied before flowering (3.5±1.3 ng/mL) (**Fig. 2B**). For clothianidin drench applications, the concentrations in nectar were 1.5x higher when thiamethoxam was applied after flowers opened (44.8±12.1 ng/mL) relative to applications made before flowering (27.2±2.9 ng/mL) (**Fig. 3A**). For spray applications, clothianidin concentrations were 4.4x higher in plants treated after flowering (2.2±0.6 ng/mL) relative to before (<0.5 ng/mL) (**Fig. 3B**).

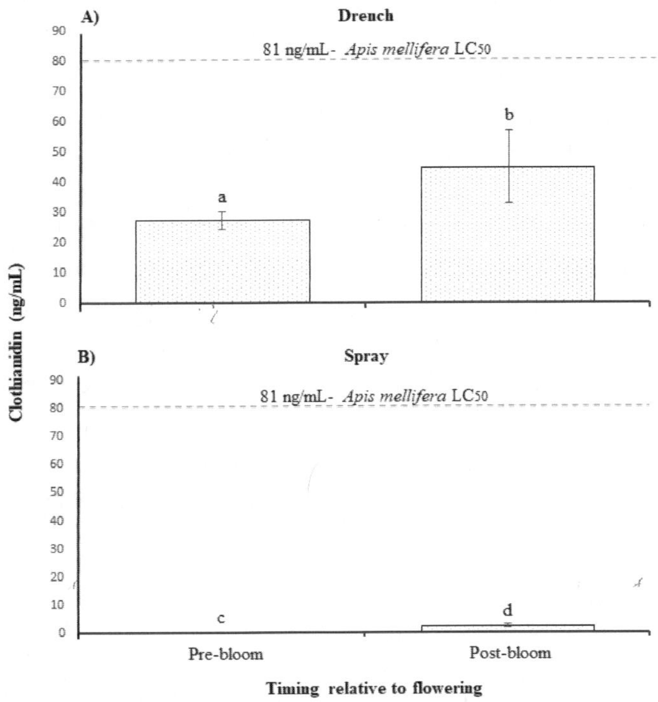

Figure 3. Clothianidin concentrations (± standard deviation) in nectar associated with application of thiamethoxam by A) drench applied pre- and post-bloom and B) spray applied pre- and post-bloom. Different letters indicate significant differences between timings and application methods based on 2-way ANOVA (P=0.05). Horizontal reference line indicates median lethal concentration (LC_{50}) values for honeybees (*A. mellifera*) (Laurino et al., 2011).

DISCUSSION

Results presented herein clearly reveal that pesticide application method and timing influence contamination of nectar with thiamethoxam and clothianidin. Concentrations of thiamethoxam were 120x higher when it was applied as a drench (relative to spray application) before flower initiation, and 58x higher when applied as a drench after flowering (relative to spray applications after flowering). Relative to the timing of each application method the differences were not as great but were significant. In this case concentrations in nectar from the drench applications were 1.9x greater when applied after flowering. Concentrations from the spray applications were 4x greater in treatments applied after flowering. These results are consistent with findings of Cowles and Eitzer (2017) who evaluated contamination of pollen and nectar in sunflower and swamp milkweed associated with spray and drench applications with the neonicotinoid insecticides dinotefuran, imidacloprid, and thiamethoxam. They reported that concentrations were generally higher with drench applications, and that concentrations of dinotefuran and thiamethoxam tended to increase as applications approached flowering. The higher concentrations detected in the drench treatments are likely related to actual amounts of active ingredient applied to each plant. Based on label recommended rates, 0.075 g of thiamethoxam was applied in the drench treatments, as compared to 0.0058 g for the spray applications. Nevertheless, while the plants in the drench treatments received 12.9x more thiamethoxam/plant compared with spray treatments, results showed that residues of thiamethoxam in nectar were 120x higher for drench relative to spray when applied before blooming, and 58x higher for drench relative to spray when applied after blooming. Clothianidin behaved similarly, though concentrations were much lower.

To assess ecological risks of both compounds, concentrations in nectar were compared with published median lethal concentrations (LC_{50}) in nectar/feeding solutions and no observable effects concentrations (NOEC) for pollinators. Thiamethoxam concentrations exceeded LC_{50} values of 227 ng/mL (European honeybee, *Apis mellifera*) and 54.3 ng/mL (native bee, *Melipona scutellaris*) with drench application treatments before and after blooming, indicating significant risks for acute toxicity (**Fig. 2A**) (Miotelo et al., 2021). The NOEC reported for thiamethoxam was 141 ng/mL for *Apis mellifera* (Overmyer et al., 2018), indicating significant toxicity would likely occur with the drench treatments. In contrast, these toxicity thresholds were not exceeded with either spray application (**Fig. 2B**). Concentrations of clothianidin in nectar for all treatments (drench or spray, before or after flowering) were below the LC_{50} for *Apis mellifera* (81 ng/mL) (Laurino et al., 2011) (**Fig. 3**), indicating low risks of acute toxicity.

CONCLUSIONS

Application method and timing significantly impacted contamination of nectar with thiamethoxam and clothianidin, with thiamethoxam concentrations from drench treatments likely being toxic to pollinator species. Spray treatments resulted in the least contamination of nectar. Future research should evaluate other species, pesticides, and application parameters for develop best management practices (BMPs) for pollinator protection.

LITERATURE CITED

Biesmeijer, J.C., Roberts, S.P.M., Reemer, M., Ohlemüller, R., Edwards, M., Peeters, T., Schaffers, A.P., Potts, S.G., Kleukers, R., Thomas, C.D., Settele, J., and Kunin, W.E. (2006). Parallel declines in pollinators and insect-pollinated plants in Britain and the Netherlands. Science *313*: 351-354.

Cowles, R.S. and Eitzer, B.D. (2017). Residues of neonicotinoid insecticides in pollen and nectar from model plants. J. Environ. Hort. *35*:24–34.

Hall, C.R., Hodges, A.W., Khachatryan, H., and Palma, M.A. (2020). Economic contributions of the green industry in the United States in 2018. J. Environ. Hort. *1:38:*73–79.

Clebsch, B. and Barner, C.D. (2003). The New Book of Salvias: Sages for Every Garden. Timber Press, Portland, Oregon.

Kearns, C.A., Inouye D.W., and Waser. N.M. (1998). Endangered mutualisms: the conservation of plant-pollinator interactions. Ann. Rev. Ecol. Syst. *29*: 83–112.

Laurino, D., Porporato M., Patetta, A., Manino, A. (2011). Toxicity of neonicotinoid insecticides to honeybees: laboratory tests. Bull. Insectol. *64*:107-113.

Lentola, A., David, A., Abdul-Sada, A., Tapparo, A., Goulson A., and Hill, E.M. (2017). Ornamental plants on sale to the public are a significant source of pesticide residues with implications for the health of pollinating insects. Environ. Pollution. *228*: 297-304.

Marquardt, M., Kienbaum L., and Kretschmer L.A. (2021). Evaluation of the importance of ornamental plants for pollinators in urban and suburban areas in Stuttgart, Germany. Urban Ecosystems *24:*811–825.

Miotelo, L., Mendes dos Reis A.L., Malaquias J.B., Malaspina, O., and Roat, T.C. (2021). *Apis mellifera* and *Melipona scutellaris* exhibit differential sensitivity to thiamethoxam. Environ. Pollution. *268*: doi: 10.1016/j.envpol.2020.115770.

Nauen, R., Ebbinghaus-Kintscher, U., Salgado, V.L., and Kaussmannb, M. (2003). Thiamethoxam is a neonicotinoid precursor converted to clothianidin in insects and plants. Pesticide Biochem. Physiol. *76*:55–69.

Overmyer, J., Feken, M., Ruddle, N., Bocksch, S., Hill, M., and Thompson, H. (2018). Thiamethoxam honeybee colony feeding study: linking effects at the level of the individual to those at the colony level. Environ. Toxicol. Chem. *37*:816–828.

R Core Team (2021). R: A language and environment for statistical computing. R Foundation for Statistical Computing, Vienna, Austria. <https://www.R-project.org/. 1.1.1717>. (Accessed 25 Aug. 2022)

Combined Proceedings IPPS 72:50-57. 2022.

Effects of Auxin and Taxa on Rooting Performance of Vegetatively Propagated Wild Coffee (*Psychotria* sp.)

Teagan Young[1a], Sandra B. Wilson[1] and Mack Thetford[2]

[1]University of Florida, Department of Environmental Horticulture, P.O. Box 110670, Gainesville, FL 32611, USA. [2]University of Florida, Department of Environmental Horticulture, West Florida Research and Education Center, 5988 Hwy. 90, Bldg. 4900 Milton, FL 32583, USA

Teagan.young@ufl.edu

[a]Second Place – Charlie Parkerson Graduate Student Research Paper Competition

Keywords: Bahama coffee, dwarf wild coffee, soft leaf wild coffee, Rubiaceae, propagation

Summary

Wild coffee (*Psychotria nervosa*), softleaf wild coffee (*Psychotria tenuifolia*), and Bahama wild coffee (*Psychotria ligustrifolia*) are evergreen shrubs with attractive foliage, fragrant white flowers, and colorful fruit. A cutting propagation study was conducted to evaluate the effects of auxin concentration on rooting of these three species, and a dwarf form, *P. nervosa* 'Little Psycho'. Percentage of rooting, root system quality, root length, and root number were determined for semi-hardwood cuttings treated with one of three indole-3-butyric acid (IBA) talc treatments: 0, 8000 or 16000 mg/kg (ppm) - placed under mist for 8 wk. Among all four taxa, at least one of these measures of rooting performance was improved with a talc formulation containing 8000 mg/kg IBA. The percentage of cuttings with roots was greatest for cuttings treated with auxin (88±4% and 88±2% for 8000 and 16000 mg/kg auxin, respectively) compared to cuttings not treated with auxin (control) (71±10%). Results of this study reveal that all four *Psychotria* taxa evaluated can be easily rooted with or without auxin in a relatively short production cycle and show promise for widened commercial production and use in Florida. For improved root quality, application of auxin is recommended for more efficient liner propagation systems.

INTRODUCTION

Native plants have been historically overlooked in their value to urban landscapes and gardens. However, the appeal of using native plants in landscapes has increased (Kalaman et al., 2022) and homeowners are willing to pay more for a landscape that includes native plants (Gillis and Swim, 2020). The underutilization of native plants can be attributed to inefficient or unknown propagation systems, insufficient marketing and promotion, and limited availability in consumer markets (Wilson, 2020). Moreover, a majority of native species reported in natural areas are not available commercially (Tangren et al., 2022). Propagation knowledge is key to increasing the diversity within a native plant palette (Campbell-Martínez et al., 2022).

Psychotria, a member of the family Rubiaceae, is one of the largest genera of flowering plants with economic, medicinal, and ornamental importance. There are three *Psychotria* species native to Florida: wild coffee (*Psychotria nervosa)*, softleaf wild coffee (*Psychotria tenuifolia*), and Bahama wild coffee (*Psychotria ligustrifolia*) (Gann et al., 2022) with cold hardiness zones of 9a-11 (USDA, 2012) (Fig. 1A-D).

Characteristics common among these taxa include: leaves that are simple, oppositely arranged, with entire margins and prominent veins (**Fig. 1A-D**); white inflorescences that are cymose and fragrant (**Fig. 1E**); fruits occurring as drupes with a fleshy pericarp that turns red in the fall (**Fig. 1F**), each with 2-seeded, longitudinally ribbed seeds (**Fig. 1G**).

Native to Florida's peninsula, *P. nervosa* and *P. tenuifolia* occur in mesic and rockland hammocks. *P. nervosa* 'Little Psycho' is a dwarf cultivar first identified by Brightman Logan in a west-central Florida hammock and is now in commercial micropropagation (AgriStarts, 2022). *Psychotria ligustrifolia* is a rarer species native to mainland and Florida Keys Rockland hammocks. These *Psychotria* taxa can be readily distinguished from each other by their leaves and fruit size.

All four *Psychotria* taxa can be used interchangeably in the landscape under part shade to sun conditions, with minimal care once established. Yet, their availability is limited to a handful of native nurseries who typically propagate by seed when available. Even then, seed germination is inconsistent and erratic (S. Wilson, unpublished data). The overall goal of this study is to widen the year-round availability and use of *Psychotria* taxa in landscapes by developing practical methods for commercial cutting propagation. Specific objectives were to determine the effects of taxa and auxin concentration on optimal rooting responses of *Psychotria*.

Figure 1. Four taxa included in the study were: *Psychotria ligustrifolia* (A), *Psychotria nervosa* 'Little Psycho' (B), *Psychotria nervosa* (C), and *Psychotria tenuifolia* (D) displaying prominently veined, elliptical leaves. White, cymose, terminal inflorescences (E) are followed by ovoid drupes eventually turning red in color (F), and each containing 2 ribbed seed.

MATERIALS AND METHODS

Cuttings of all taxa were taken on the morning of 23 October 2020 from plantings located on the University of Florida Campus. A total of 54 semi-hardwood cuttings from each taxon were selected that had a terminal bud, a minimum of five nodes, was pest and disease-free, and lacked fruits. Cutting lengths were 8 to 9.5 cm for *P. ligustrifolia and P. tenuifolia,* 5 to 8.5 cm for *P. nervosa,* and 4.5 to 9.5 cm for *P. nervosa* 'Little Psycho'. The basal leaves were removed, and the 1.3 cm basal portion of each stem dipped in tap water prior to commercial talc rooting hormone containing either 0, 8000, or 16000 mg/kg (ppm) talc Indole-3-butyric acid (IBA; Hormex, Mainland, PA). For the control treatment (0 mg/kg IBA), cuttings were dipped into water only. After treatment application, cuttings were stuck into 6-cell trays (width 3.8 cm x length 3.8 cm x depth 5.8 cm) (T.O. Plastics, Clearwater, MN) filled with Metro-Mix 852 (6:3:1 bark:Canadian peat:perlite) (Sun Gro Horticulture, Agawam, MA). Each auxin treatment was applied to 6 cuttings replicated 3 times in a randomized complete block design. Overhead mist was provided every 5 min at 5 sec time intervals for 8 wks. The average, maximum, and minimum temperatures in the mist house were 25.1 °C, 36.4 ° C, and 13.7 ° C, respectively. Cuttings were checked weekly for root emergence, and foliage loss.

At the end of the experiment (after 8 weeks), rooting was evaluated by gently pulling cuttings out of individual cells to determine a root system quality value using a scale from 1 to 4 with 1 = alive cuttings with no roots; 2 = roots forming but do not hold medium; 3 = root ball partially holds plug medium; and 4 = fully formed root ball entirely holding the medium. Cuttings that did not survive the study were scored as a zero and removed from the means. Once a root quality value was determined, the substrate was gently removed from roots to measure root length (mean of the two longest roots) and record the number of roots (up to 25).

RESULTS

The data was analyzed in a two-step process, the first part was to test for interactions between taxa and auxin. Based on those results nonparametric tests were used to determine statistical differences in auxin, either across or for each taxon. This two-step method was used due to non-normality, right censoring, and boundary issues that made a traditional method inaccurate. In this analysis sub samples were assumed independent. Root system quality value, rooting percentage, root length, and root number data were analyzed using a linear mixed model with JMP v. 16 (SAS Institute Inc., Cary, NC).

The root system quality value reflects the combined effects of rooting percentage, root length and root number of all living cuttings. This overall assessment value (scale of 1 to 4) was influenced by auxin application and these effects differed among taxa. Root quality values for *P. ligustrifolia* and *P. nervosa* (2.2 to 2.9 and 1.9 to 2.4, respectively) were similar among auxin treatments, whereas root quality values for *P. nervosa* 'Little Psycho' and *P. tenuifolia* were significantly improved with an auxin application (**Table 1**). As such, *P. nervosa* 'Little Psycho' had nearly twice the root quality score when treated with auxin compared to the control (1.6 vs 2.9 to 3.6; **Table 1**). *Psychotria tenuifolia* treated with 8000 mg/kg IBA also had higher root quality (3.6) than the control cuttings (2.7) and cuttings treated with 16000 mg/kg IBA had a similar root quality (3.1) compared to cuttings treated with 0 or 16000 mg/kg IBA (**Table 1**).

There was a significant auxin effect for percent rooting ($P<0.0001$). However, rooting percentage did not statistically differ by taxa ($P=0.0860$) (rooting percentages were 87.0%, 87.0%, 85.2%, and 98.1% for *P. ligustrifolia*, *P. nervosa*, *P. nervosa* 'Little Psycho' and *P. tenuifolia*, respectively) nor was there a taxa × auxin interaction. The percentage of cuttings with roots was greatest for cuttings treated with auxin [88±4% and 88±2% for 8000 and 16000 mg/kg IBA, respectively] compared to cuttings not treated with auxin (control) (71±10%) (**Table 2**).

Table 1. Mean root system quality rating and root length ± standard error of four *Psychotria* taxa subjected to three auxin treatments [0, 8000, and 16000 mg/kg (ppm) indole-3-butyric acid (IBA)].

Taxa[z]	IBA conc.	Root system quality[y] (1 to 4 scale)	*n*[x]	Root length[w] (mm)	*n*[x]	Root no.[v]	*n*[x]
P. ligustrifolia							
	0 mg/kg	2.22 ± 0.24 a	18	23.21 ± 3.58 b	14	10.50 ± 2.16 b	14
	8000 mg/kg	2.83 ± 0.25 a	18	43.06 ± 4.83 a	16	19.81 ± 2.21 a	16
	16000 mg/kg	2.88 ± 0.18 a	18	33.82 ± 5.75 ab	17	18.82 ± 1.85 a	17
P. nervosa							
	0 mg/kg	1.89 ± 0.18 a	18	18.08 ± 4.28 a	13	7.31 ± 1.55 b	13
	8000 mg/kg	2.44 ± 0.23 a	18	27.72 ± 3.73 a	16	24.25 ± 0.49 a	16
	16000 mg/kg	2.44 ± 0.21 a	18	29.06 ± 3.52 a	17	17.00 ± 1.97 a	17
P. nervosa 'Little Psycho'							
	0 mg/kg	1.61 ± 0.12 b	18	10.05 ± 1.36 c	11	7.73 ± 1.30 c	11
	8000 mg/kg	2.89 ± 0.18 a	18	32.06 ± 4.53 b	18	23.17 ± 0.91 b	18
	16000 mg/kg	3.59 ± 0.15 a	18	44.21 ± 2.03 a	17	25.00 ± 0.00 a	17
P. tenuifolia							
	0 mg/kg	2.67 ± 0.24 b	18	36.74 ± 4.95 b	17	14.76 ± 2.33 a	17
	8000 mg/kg	3.64 ± 0.18 a	14	55.25 ± 2.81 a	14	23.93 ± 0.94 a	14
	16000 mg/kg	3.06 ± 0.22 ab	18	43.78 ± 5.22 ab	18	22.00 ± 1.43 a	18

[z] For each taxa, means within a column followed by the same letter are not significantly different according to a Steel-Dwass test at $P \geq 0.05$.
[y] Cuttings were evaluated using a visual root system quality scale from 1 to 4 with 1 = alive cuttings with no roots; 2 = roots forming but do not hold medium; 3 = root ball partially holds plug medium; and 4 = fully formed root ball entirely holding the plug medium when removed from the tray.
[x] Total number (*n*) of cuttings included in analysis are designated in columns after each measured trait.
[w] Rooting lengths of the two longest roots were averaged.
[v] Root number was counting up to 25 roots.

Table 2. Mean percent rooting ± standard error based on the Agresti–Coull Percent estimates method of the combined four *Psychotria* taxa subjected to three auxin treatments [0, 8000, and 16000 mg/kg (ppm) indole-3-butyric acid (IBA)]. Means within a column followed by the same letter are not significantly different according to a Steel-Dwass test at P≥0.05. Total number (*n*) of cuttings included in analysis are designated in column after the rooting percent.

IBA concentration [z]	Rooting percent	*n*
0 mg/kg	70.60 ± 9.53 b	72
8000 mg/kg	87.77 ± 4.38 a	72
16000 mg/kg	87.77 ± 2.29 a	72

[z] Rooting percentage did not differ by taxa (*P*=0.0861) and there was no taxa × auxin interaction (*P*=0.2849). Therefore, main treatment effects of auxin are reported for the combined taxa.

Root length and root number were influenced by auxin application and the effects of auxin differed among taxa (taxa × auxin interaction). Cuttings of *P. ligustrifolia* and *P. tenuifolia* treated with 8000 mg/kg IBA produced roots that were ~2.0 and 1.5 times longer (43 and 55 mm), respectively, than control cuttings (23 and 37 mm) while cuttings treated with 16000 mg/kg IBA had a similar root length (34 and 44 mm) compared to cuttings treated with 0 and 8000 mg/kg IBA (Table 1). For *P. nervosa*, auxin treatment did not improve root length regardless of auxin level (18 to 29 mm). However, for *P. nervosa* 'Little Psycho' auxin application did result in longer roots compared to the control and a higher auxin concentration further increased root length. *Psychotria nervosa* 'Little Psycho' cuttings treated with 16000 mg/kg IBA had the longest roots (44 mm), followed by 8000 mg/kg IBA (32 mm), while control cuttings had the shortest roots (10 mm) (Table 1). Root number was influenced by auxin application for three of the four taxa (Table 1). Auxin application increased root number compared to the non-treated control for *P. ligustrifolia*, *P. nervosa*, and *P. nervosa* 'Little Psycho' while root number for *P. tenuifolia* did not differ regardless of auxin concentration. For *P. ligustrifolia* and *P. nervosa* root number was similar among cuttings treated with 8000 or 16000 mg/kg IBA and these cuttings had more roots than control cuttings. *Psychotria nervosa* 'Little Psycho' had the greatest number of roots when treated with 16000 mg/kg followed by 8000 mg/kg, compared to the control.

DISCUSSION

Results herein show that all four *Psychotria* taxa can be propagated by stem cuttings within ranges considered acceptable for native nursery production (Cartabiano and Lubell, 2013; Green Isle Gardens Nursery, personal communication). While marginally acceptable rooting percentages (70%) were obtained without the application of auxin, rooting percentage was increased

~1.2 times and there was a positive effect on root quality, root number and root length for cuttings treated with 8000 mg/kg IBA with only minimal improvement with the use of 16000 mg/kg IBA. This study utilized fall cuttings and remarkably high IBA concentrations at 8000 or 16000 mg/kg, typically reserved for hard to root species (Davies et al., 2018). Spring and summer cuttings of a closely related Brazilian wild coffee (*P. nuda*) rooted to a comparable percentage but with a much lower IBA concentration (3000 mg/kg) (Nery et al., 2014). Likewise, our own subsequent anecdotal investigations revealed that *P. nervosa* can also be rooted in late spring using 3000 mg/kg talc IBA. Thus, it is probable that all four *Psychotria* taxa could have high rooting performance when treated with a similarly low IBA concentration and that propagation is not seasonally dependent.

Results also emphasize the value of assessing not only the rooting percentage of treated cuttings, but the overall root system performance, reflecting the combined effects of visual root quality (ability to hold medium), root length and root number. For example, *P. nervosa* 'Little Psycho' cuttings had 85.2% rooting, but very few cuttings held the medium in the absence of auxin. The application of auxin increased both root length and root number, resulting in higher quality root systems. Likewise, *P. tenuifolia* had 98.1% rooting but the addition of auxin improved visual quality (ability to hold medium) and length resulting in a finished liner sooner.

CONCLUSIONS

This serves as a first report of cutting propagation of four *Psychotria* taxa presented herein. Propagation results demonstrate these taxa are relatively easy to produce in a greenhouse under intermittent mist from fall stem cuttings leading to a finished liner within 8 weeks. Although clonal propagation of native plants such as *Psychotria* may not be ideal for restoration purposes, it serves as a reliable alternative to seed propagation for ornamental landscape use.

LITERATURE CITED

AgriStarts. (2022). *Psychotria nervosa* 'Little Psycho™.' URL: https://www.agristarts.com/index.cfm/fuseaction/plants.plantDetail/plant_ID/374/index.htm (Accessed 5 Aug. 2022). Apopka (FL): AgriStarts.

Campbell-Martínez, G.E., Steppe, C., Wilson, S.B., and Thetford, M. (2022). Effect of temperature, light and seed provenance on germination of *Paronychia erecta* (squareflower): a native plant with ornamental potential. Native Plants J. *23*:56-64.

Cartabiano, J. and Lubell, J. (2013). Propagation of four underused native species from softwood cuttings. HortScience 48:1018-1020, doi:10.21.273/HORTSCI.48.8.1018

Davies, F.T., Geneve, R.L., and Wilson, S.B. (2018). Hartmann and Kester's Plant Propagation: Principles and Practices. 9th ed. New York City (NY), Pearson Education.

Gann, G.D., Stocking, C.G., Brennan, K.M., Hines, K.N., and Collaborators. (2022). Natives for your neighborhood. URL: https://regionalconservation.org/beta/nfyn/PlantList.asp (Accessed 5 Aug. 2022). Delray Beach (FL): Floristic Inventory of South Florida Team.

Gillis, A. and Swim, J.K. (2020). Adding native plants to home landscapes: The roles of attitudes, social norms, and situational strength. J. of Environ. Psych. *72*:1-11.

Kalaman, H., Wilson, S.B., Mallinger, R.E., Knox, G.W., and van Santen, E. (2022). Evaluating native and non-native ornamentals as pollinator plants in Florida: I. Floral abundance and insect visitation. HortScience. *57*:126-136.

doi:10.21273/HORTSCI16123-21.

Nery, F.S.G., Zuffellato-Ribas, K.C., and Koehler, H.S. (2014). Rooting of *Psychotria nuda* (Cham. & Schltdl.) Wawra (Rubiaceae) in the four seasons of the year. Ciencia Florestal *24*:243-250.

Tangren, S., Toth E, Siegel S. 2022. A survey of native plant materials use and commercial availability in the Eastern United States. Native Plants J. *23*:17-32, doi.org/10.3368/npj.23.1.17.

USDA. 2012. Plant hardiness zone map. Agricultural Research Service, United States Department of Agriculture. URL: https://planthardiness.ars.usda.gov/(Accessed 5 Aug. 2022). Stillwater (OR): Prism Climate Group.

Wilson, S.B. (2020). Expanding our plant palette: The role of native and non-invasive cultivars. Proc. Fla. State Hort. Soc. *133*:1-4.

Wunderlin, R.P. and Hansen, B.F. (2011). Guide to the vascular plants of Florida. 3rd ed. Gainesville (FL): University Press of Florida.

Combined Proceedings IPPS 72:58-66. 2022.

Effect of Cutting Time and Auxin Application Method on Propagation of *Magnolia grandiflora* 'Southern Charm'

Anthony T. Bowden[a], Patricia R. Knight, Jenny B. Ryals, Christine E.H. Coker, Scott A. Langlois, and Eugene K. Blythe

Coastal Research and Extension Center, South Mississippi Branch Experiment Station, P.O. Box 193, Poplarville, MS 39470, USA

anthonybowden@gmail.com

[a]Third Place – Charlie Parkerson Graduate Student Research Paper Competition

Keywords: Adventitious rooting, indole-3-butyric acid (IBA), surfactant, season

Summary

Use of foliar auxin applications are increasing in the nursery and greenhouse industry. However, previous research has shown that insufficient auxin is being absorbed or translocated to the site of action. Research was conducted to determine whether cutting time, addition of surfactant to the auxin solution, or auxin application method had an impact on the propagation of *Magnolia grandiflora* 'Southern Charm'. Terminal cuttings of 'Southern Charm' magnolia (*M. grandiflora* 'Southern Charm') were taken at two times of the year: spring and fall and sprayed to the drip point using a solution of Hortus IBA Water Soluble Salts™ at concentrations of 0, 500, 1,000 or 1,500 ppm or dipped for 3-sec in a solution of 2,500 ppm IBA. For many of our tested parameters, fall cuttings were better than spring cuttings. Fall cuttings treated with 1,500 ppm foliar IBA solution plus 0.85 ppm Regulaid® had greater root numbers than spring cuttings treated with less than 1,500 ppm IBA with or without Regulaid®, or fall cuttings treated with a 2,500 ppm basal quick dip, a foliar application of 1,000 ppm or less without Regulaid®, or fall cuttings receiving a foliar application of 0.85 ppm Regulaid®.

INTRODUCTION

Research into foliar auxin applications methods over the past decade indicated that one-time applications are the industry standard (Blythe et al., 2007; Kroin, 2014). When applied post-sticking, much lower concentrations (50 to 100 ppm) of rooting hormones are required compared to other conventional application methods (Dole and Gibson, 2006). Overhead applications of water soluble IBA are increasing in the nursery industry. Bailey Nurseries Inc. in Minnesota and Oregon has been conducting repetitive on-farm trialing for the fifteen years. Their results indicated that many of the taxa commonly propagated respond similarly to foliar-applied auxin compared to a traditional basal quick-dip. At Bailey Nurseries, propagation trays and beds are treated with a single application of a water-soluble IBA solution ranging from 250 to 2,000 ppm (Drahn, 2007).

Decker's Nursery in Ohio uses a battery-powered backpack sprayer to treat their cuttings since it atomized the auxin solution similarly to the mist from the mist system and applied a very small droplet with excellent coverage over both the top and bottom of the cutting (Decker, 2016). Since propagation areas vary in size, overhead applications are applied via a backpack sprayer for small houses and reel-and-hose sprayers for larger production areas.

When being applied overhead, Kroin (2014) of Hortus USA recommends to "spray the solution evenly over the cuttings until drops fall onto the media". To do this, Bailey Nurseries aims to deliver 1 L of solution per 60 ft^2. Currently, both Decker's Nursery and Bailey Nurseries generally treat their cuttings within 24-h of being stuck, either at the end of each day or the following morning, but application occurring during the day in conjunction with frequent mist intervals has not reduced efficacy (Drahn, 2007; Decker, 2016). Cuttings are treated in the early morning or late afternoon due to both lower light levels and reduced misting requirements. For both nurseries, the switch to overhead auxin application led to a decrease in plant material handling and the time cuttings spend in cold storage and the preparation room, where problems associated with lengthened exposure to low temperatures, high humidity, and/or handling can occur (Drahn, 2007). In 2003, 99.6% of cuttings at Bailey Nurseries were quick dipped and 0.4% were treated with foliar applications. By 2007, the percentages had reversed, with 95% of all propagated material being treated with overhead applications and 5.2% of material being quick-dipped (Drahn, 2007). Currently, overhead applications of water-soluble IBA are used to treat the following genera at Bailey Nurseries Minnesota operation: *Acer, Berberis, Cornus, Diervilla, Euonymus, Forsythia, Hydrangea, Juniperus, Lonicera, Philadelphus, Physocarpus, Rhus, Rosa, Spiraea, Symphoricarpos, Syringa, Thuja, Viburnum, and Weigela*. (Drahn, 2007).

Surfactants are common in agricultural production as penetration of the leaf cuticle is required for efficacy of foliar-applied compounds (Robertson and Kirkwood, 1969). Effectiveness of foliar-applied compounds depends on its ability to penetrate through the cuticle and translocate to the site of action (White et al., 2002). Surfactants enhance penetration of these chemicals by increasing the wetting capacity up to the critical micelle concentration (CMC), defined as the concentration above which any added surfactant molecules appear with high probability as micellar aggregates (Ruckenstein and Nagarajan, 1975; Lownds et al., 1987). Research was conducted by Lownds et al. (1987) to determine the effects surfactants would have on foliar penetration of NAA and NAA-induced ethylene production by cowpea [*Vigna unguiculata (*L.) Walp. subsp. *unguiculata* cv. Dixielee]. This research indicated that foliar penetration of NAA was increased when co-applied with a surfactant (Pace®, Regulaid®, or Tween 20®) and all three induced similar qualitative changes in surface tension, contact angle, and droplet: leaf interaction. All three surfactants increased the droplet: leaf ratio. However, Regulaid® was the only surfactant tested that showed a correlation between NAA penetration and interface area (Lownds et al., 1987).

When choosing a surfactant for plant production, several factors should be considered: (1) it should be non-toxic to both the plant and the environment; (2) it should be a small molecule that is water soluble; (3) it should be non-ionic; (4) and relatively effective at decreasing surface concentration at a relatively low concentration (Colwell and Rixon, 1961). While anionic and cationic surfactants are labeled and frequently used in plant production (Dobozy and Bartha, 1976), using non-ionic surfactants are preferable since they do not affect water hardness, nutrient balance, or enzymatic activity and are compatible with most herbicides due to lack of activity with the foliar-applied chemical (Bayer and Foy, 1982).

The objective of this research was to evaluate whether addition of surfactants to foliar auxin solutions increased root growth and uniformity compared to the industry-standard basal quick dip for 'Southern Charm' magnolia (*Magnolia grandiflora* 'Southern Charm').

MATERIALS AND METHODS

An experiment was performed to evaluate the effect of four foliar auxin concentrations [0, 500, 1,000, and 1,500 ppm IBA indole-3-butyric acid (IBA) (Hortus IBA Water Soluble Salts™; Phytotronics Inc., Earth City, MO)] each at two concentrations (0 and 85 ppm Regulaid®) on rooting of 'Southern Charm' magnolia. Additionally, a basal quick dip of 2,500 ppm IBA was used as an industry-based control. Five-node terminal cuttings of *Magnolia grandiflora* 'Southern Charm' [12.7cm (5-in)] were harvested from established landscape plants and stuck to a depth of 1.3 cm (0.5-in) on two dates: 14 April 2021 and 15 November 2021. During cutting preparation, a 2.5 cm (1-in) wound was applied to opposite sides of the basal end of the cutting. Propagation medium was 100% pine bark placed into 8.9 cm (3.5-in) square production pots (T.O. Plastics, Inc., Clearwater, MN). Cuttings receiving foliar applications of auxin were sprayed once to runoff with a 3.75-L battery operated sprayer (One World Technologies, Inc., Anderson, SC). Pine bark for this experiment was sourced from Eakes' Nursery Supply (Seminary, MS) and delivered as a mix of 50% aged and 50% fresh bark passed through a 3/8" (0.95 cm) screen. After treatment, cuttings were placed under intermittent mist applied for 4 sec/4 min during daylight hours and adjusted as needed for the duration. Experimental design was a completely randomized design. Treatment structure was a complete factorial (5 auxin rates × 2 surfactant concentrations × season). Data collected after 120 days included rooting percentage, shoot height, total root number, average root length (three longest roots), and root quality (1-5, with 1=callused cuttings without roots and 5= ≥ 10 roots). Additionally, net photosynthetic rate (A) and stomatal conductance (g_{sw}) values were sampled between the hours of 7:30 A.M. and 11:30 A.M. using the LiCOR 6000 Portable Photosynthesis System (LI-COR Biosciences; Lincoln, NE).

Data were analyzed using linear models and generalized linear models with the GLIMMIX procedure of SAS (ver. 9.4; SAS Institute Inc., Cary, N.C.) by first testing for an interaction between treatment factors (auxin rate, surfactant concentration, and season). When the three-way interaction was significant, auxin rates were compared within surfactant concentration by season. When the three-way interaction term was not significant ($p > 0.10$), main effects means for levels within each treatment factor were compared. Mean separation was performed using the Holm-Simulated Method for multiple comparisons to maintain an overall significance level of $\alpha = 0.05$.

RESULTS

Rooting percentage of Teddy Bear® Southern magnolia (*Magnolia grandiflora* 'Southern Charm') ranged from 26-87% but neither use of surfactant, auxin rate, nor season impacted rooting percentage (**Table 1**).

Use of surfactant, auxin rate, or season had no effect on stomatal conductance. The three-way interaction between surfactant concentration, auxin rate, and season was significant for root number. Fall cuttings treated with a 1,500 ppm foliar IBA plus 0.85 ppm Regulaid® resulted in greater root numbers than spring cuttings treated with less than 1,500 ppm IBA with or without Regulaid®, or fall cuttings treated with a 2,500 ppm basal quick dip, a foliar application of 1,000 ppm or less without Regulaid® or fall cuttings receiving a foliar application of 0.85 ppm Regulaid®. Auxin rate impacted the average length of the three longest roots (**Table 1**). Cuttings treated with a 1,000 ppm foliar application of IBA had greater root lengths compared to cuttings treated with the 2,500 ppm IBA quick dip or cuttings receiving a 0 ppm foliar application of IBA. The interaction between auxin rate and season was significant for shoot height (**Table 2**). For spring cuttings, a 2,500 basal quick dip led to greater shoot length than cuttings receiving a foliar spray of either a 0, 500 ppm or 1,000 ppm IBA (**Fig. 1**). However, for fall cuttings, a 1,500 ppm IBA applied foliarlly led to greater shoot length than cuttings treated with 0, 500, or 2,500 ppm solution applied as a basal quick dip. Auxin rate and season impacted net photosynthesis rate (A). Cuttings treated with foliar applications of 1,500 ppm IBA had greater photosynthetic rates compared to cuttings treated with 0 ppm IBA. Cuttings taken during the spring season had greater photosynthetic rates than cuttings taken in the fall.

Table 1. Influence of surfactant, auxin and season on rooting in Teddy Bear® Southern magnolia (*Magnolia grandiflora* 'Southern Charm').

Regulaid	Auxin (ppm)	Season	Rooting (%)	Root (no.)	Avg. Length of three longest roots (cm)	Root Quality Rating[z]
			Least squares means for main effects			
0 ppm			28%	1.1 b[z]	7.6 a	2.3 b
0.85 ppm			28%	1.4 a	7.6 a	2.7 a
	Foliar IBA 0		50%	0.9 C	6.4 C	2.0 C
	Foliar IBA 500		43%	1.1 BC	7.8 ABC	2.3 BC
	Foliar IBA 1,000		62%	1.2 B	8.9 A	2.4 BC
	Foliar IBA 1,500		70%	1.7 A	8.3 AB	3.2 A
	Basal dip 2,500		56%	1.5 A	6.5 BC	2.9 AB
		Spring	54%	1.1 B	7.6 a	2.3 b
		Fall	59%	1.5 A	7.6 a	2.8 a
Regulaid	Auxin (ppm)	Season	Least squares means grouped by surfactant concentration within auxin rate by season			
	Foliar IBA 0			0 G	8.4	1.6
	Foliar IBA 500		33%	0.4 FG	8.5	1.6
0 ppm	Foliar IBA 1000		60%	0.8 EFG	8.2	2.1
	Foliar IBA 1500		53%	1.6 ABCDE	8.9	2.6
	Basal dip 2,500	Spring	60%	1.7 ABCD	6.2	3.1
	Foliar IBA 0		73%	0.9 DEFG	5.3	2
	Foliar IBA 500		26%	0.3 FG	7	1.7
0.85 ppm	Foliar IBA 1000		80%	1.0 CDEFG	9.7	2.2
	Foliar IBA 1500		73%	1.7 ABC	8	3.1
	Foliar IBA 0		40%	1.3 BCDEF	6.4	2.3
	Foliar IBA 500		60%	1.2 BDCEFG	7.3	2.4
0 ppm	Foliar IBA 1000		46%	1.2 BCDEFG	8.6	2.3
	Foliar IBA 1500		66%	1.6 ABCD	7.4	3
	Basal dip 2,500	Fall	53%	1.3 BCDEF	6.7	2.4
	Foliar IBA 0		53%	0.9 EFG	6.7	2.1
	Foliar IBA 500		60%	1.8 AB	8.2	3.2
0.85 ppm	Foliar IBA 1000		60%	1.6 ABCD	8.7	3
	Foliar IBA 1500		87%	2.0 A	8.9	3.9
			Significance of treatment factors			
Surfactant			NS	0.006	NS	0.0145
Auxin Rate			NS	<0.0001	0.0018	<0.0001
Season			NS	<0.0001	NS	0.0045
Regulaid * Auxin Rate			NS	NS	NS	NS
Auxin Rate * Season			NS	<0.0001	NS	NS
Regulaid * Season			NS	NS	NS	NS
Auxin Rate * Regulaid *Season			NS	0.0026	NS	NS

[z]Root Quality (1-5, with 1 = callused without roots and 5 = ≥ 10 roots); [y]means followed by the same lower-case or upper-case letter for either main or interaction effects were not significantly different using the Holm-Simulated method for multiple comparisons (α = 0.05), otherwise, the treatment means are presented without letter groupings for informational purposes. NS = not significant.

Table 2. Influence of surfactant, auxin and season on shoot growth, photosynthesis and stomatal conductance in Teddy Bear® Southern magnolia (*M. grandiflora* 'Southern Charm').

Regulaid	Auxin (ppm)	Season	Shoot Height (cm)	Net Photosynthesis (A) ($\mu mol \cdot m^{-2} \cdot s^{-1}$)	Stomatal Conductance (gs_w) ($mol \cdot m^{-2} \cdot s^{-1}$)
			Least squares means for main effects		
0 ppm			1.1 b[z]	3.6 a	0.1 a
0.85 ppm			2.5 a	4.3 a	0.1 a
	Foliar IBA 0		0.4 C	3.2 B	0.1 A
	Foliar IBA 500		1.4 BC	4.0 AB	0.1 A
	Foliar IBA 1,000		2.6 AB	4.2 AB	0.1 A
	Foliar IBA 1,500		3.4 A	4.7 A	0.13 A
	Basal dip 2,500		1.1 BC	3.6 AB	0.1 A
		Spring	0.66 B	6.0 a	0.1 a
		Fall	2.9 A	1.9 b	0.1 a
Regulaid	Auxin (ppm)	Season	Least squares means grouped by surfactant concentration within auxin rate by season		
0 ppm	Foliar IBA 0	Spring	0	5.1	0.10
	Foliar IBA 500		0	7.4	0.12
	Foliar IBA 1000		0	6.2	0.08
	Foliar IBA 1500		0.6	6.2	0.11
	Basal dip 2,500		2.0	7.4	0.12
0.85 ppm	Foliar IBA 0		0.7	4.9	0.07
	Foliar IBA 500		0	5.6	0.06
	Foliar IBA 1000		0.9	5.6	0.06
	Foliar IBA 1500		1.8	7.7	0.10
0 ppm	Foliar IBA 0	Fall	0	1.3	0.08
	Foliar IBA 500		0.9	1.1	0.07
	Foliar IBA 1000		4.5	2.1	0.14
	Foliar IBA 1500		3.9	2.2	0.14
	Basal dip 2,500		0.3	1.0	0.06
0.85 ppm	Foliar IBA 0		1.5	1.3	0.08
	Foliar IBA 500		4.0	3.0	0.17
	Foliar IBA 1000		4.6	2.6	0.14
	Foliar IBA 1500		6.6	2.9	0.15
Significance of treatment factors					
Surfactant			0.0013	NS	NS
Auxin Rate			<0.0001	0.048	NS
Season			<0.0001	<0.0001	NS
Regulaid * Auxin Rate			NS	NS	NS
Auxin Rate * Season			<0.0001	NS	NS
Regulaid * Season			NS	NS	NS
Auxin Rate * Regulaid *Season			NS	NS	NS

[z]means followed by the same lower-case or upper-case letter for either main or interaction effects were not significantly different using the Holm-Simulated method for multiple comparisons ($\alpha = 0.05$), otherwise, the treatment means are presented without letter groupings for informational purposes. NS = not significant.

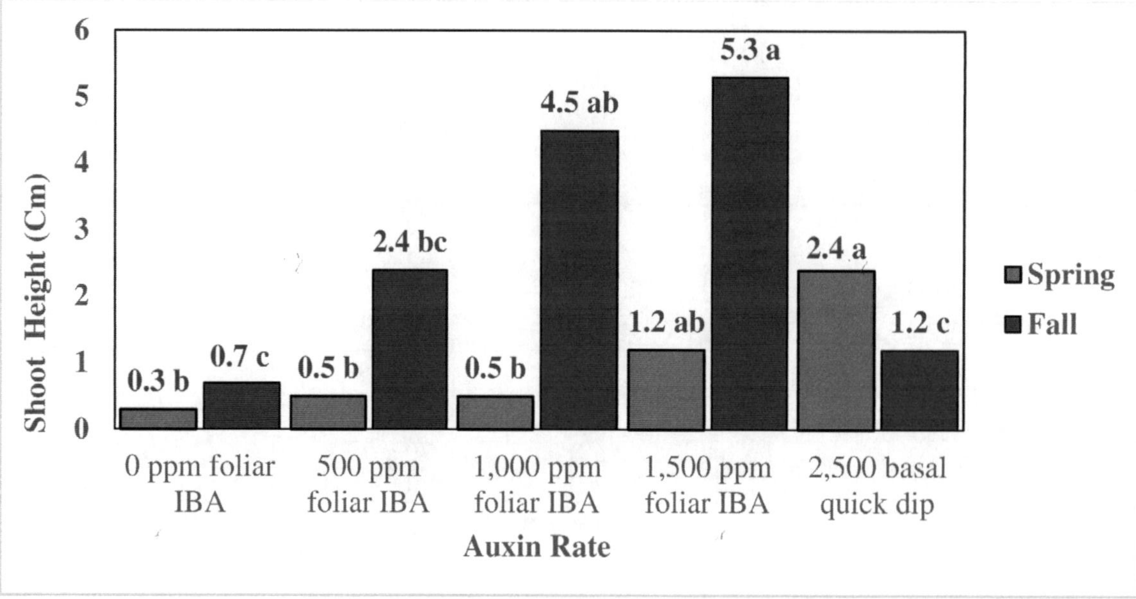

Figure 1. Shoot length of *Magnolia grandiflora* 'Southern Charm' as influenced by auxin rate and season.

DISCUSSION

The best rooting response for *M. grandiflora* 'Southern Charm' was obtained using a foliar spray of 1,500 ppm IBA or a basal quick dip of 2,500 ppm compared to foliar applications of lower concentrations. When taking cuttings in the fall, as recommended in the literature, a foliar application of 1,500 ppm IBA resulted in greater root numbers than cuttings treated with a 2,500 ppm basal quick dip. The season that cuttings were taken had a significant impact on root number, root quality ratings, and shoot growth; however, fall cuttings had lower net photosynthesis compared to spring cuttings. Previous research into photosynthetic rate during adventitious root formation suggested that the mass flow of hormones from newly formed roots were the principal contributing factor to photosynthetic rate (Svenson et al., 1995). It has further been theorized there is a correlation between a cutting's photosynthetic rate and the ability of the root initial to produce endogenous hormones (i.e., cytokinins) prior to the root elongating and penetrating the cutting surface (periderm) (Svenson et al., 1995). Physiologically, seasonal variability in endogenous hormone concentration (i.e., higher hormone concentrations in spring compared to fall) could potentially explain differences in photosynthetic rates seen in this study.

For shoot growth, fall cuttings treated with a 1,500 ppm foliar application of IBA resulted in longer shoot lengths than the 2,500 ppm IBA applied as a basal quick dip or a foliar application of 0 ppm IBA or 500 ppm IBA; however, spring cuttings treated with a 2,500 ppm basal quick dip had greater shoot lengths than cuttings treated with a 1,000 ppm foliar application of IBA or less. Nursery owners utilizing foliar auxin applications have reported slowed vegetative growth in several woody species compared to using a basal quick dip. Our results suggest that no slowed vegetative growth occurred when utilizing a foliar application of IBA compared to a basal quick dip as shoot growth was similar between a 1,500 ppm foliar application of IBA and a 2,500 ppm basal quick dip.

Our results from this trial and similar trials on foliar applications of auxin suggests that benefits of foliar applications are species dependent (Blythe et al., 2004; Bowden et al., 2021). For propagation of *Magnolia grandiflora* species and cultivars, the literature recommends higher hormone rates (5,000 to 10,000 ppm) (Dirr and Heuser, 2006). Our results suggest that sufficient auxin was absorbed from foliar applications and translocated to the site of root initiation so that root response is comparable to a basal quick dip for 'Southern Charm' magnolia. By using a foliar application of 1,500 ppm IBA on a crop of 'Southern Charm' magnolia, growers can eliminate the use of a basal quick dip for propagation of this species.

Using current methods, propagators handle a cutting multiple times before the flat even enters the propagation house, which results in higher labor costs for the producer. In addition, having several people treating and sticking cuttings can result in a wide variability in rooting, which can lead to production issues further down the process. In situations where foliar auxin applications yield equal or better results compared to traditional quick-dips, propagators only need to handle the cutting once while a licensed applicator can treat cuttings using a backpack sprayer. This helps reduce the potential for wide rooting variability. Additionally, by eliminating the need to basally treat cuttings with a stock IBA solution, the propagator can also remove the potential cross-contamination that can occur from potential pathogens that may be present on cutting stems and leaves - or from prolonged exposure to storage prior to use – when using traditional quick-dips.

LITERATURE CITED

Bayer, D.E. and Foy, C.L. (1982). Action and fate of adjuvants in soils, p. 84–92. In Weed Sci. Soc. Amer. Adjuvants for herbicides. Weed Sci. Soc. Amer. Champaign, Ill.

Blythe, E.K., Sibley, J.L., Ruter, J.M., and Tilt, K.M. (2004). Cutting propagation of foliage crops using a foliar application of auxin. Scientia Hort. *103*:31-37.

Blythe, E.K., Sibley, J.L., Tilt, K.M., and Ruter, J.M. (2007). Methods of auxin application in cutting propagation: A review of 70 years of scientific discovery and commercial practice. J. Environ. Hort. *25*:166-185.

Bowden, A.T., Knight, P. R., Coker, C.E.H., Langlois, S.A., Broderick, S.R., Blythe, E. K., Ryals, J.B., Sakhanokho, H.F., and Babiker, E.M. (2021). Evaluation of surfactants for use in one-time foliar auxin applications in the propagation of woody ornamentals. Comb. Proc. Intl. Plant Prop Soc. *71*:100-106.

Colwell, C.E. and Rixon, W.E. (1961). Consideration in the use of nonionic surface-active agents. Am. Dyest. Rep. *50*:679–682.

Decker, B.M., and Graff, D. (2016). Foliar applications of rooting hormone at Decker Nursery. Comb. Proc. Intl. Plant Prop. Soc. *66*:191-196.

Dirr, M.A. (2009). Manual of woody landscape plants: Their identification, ornamental characteristics, culture, propagation, and uses. Stipes Publishing. Champaign, IL.

Dirr, M.A. and Heuser, C.W. Jr. (2006). The reference manual of woody plant propagation: From seed to tissue culture-Second Edition. Varsity Press, Inc. Athens, GA.

Dole, J.M. and Gibson, J.L. (2006). Cutting Propagation: A guide to propagating and producing floriculture crops. Ball Publishing. Batavia, IL.

Dobozy, O.K. and Bartha, B. (1976). Non-pollutant surfactants stimulating the growth of plants. Tenside Detergents *13*:139–144.

Drahn, S.R. (2007). Auxin application via foliar sprays. Comb. Proc. Intl. Plant Prop. Soc. *57*:274-277.

Kroin, J. (2014). Propagate plants from cuttings using foliar-applied aqueous (water-based) IBA rooting solutions: Tips-do's and don'ts. Comb. Proc. Intl. Plant Prop. Soc. *64*:149-159.

Lownds, N.K., Leon., J.M., and Bukovac, M.J. (1987). Effect of surfactants on foliar penetration of NAA and NAA-induced ethylene evolution in cowpea. J. Amer. Soc. Hort. Sci. *112*:554-560.

Robertson, M.M. and Kirkwood, R.C. (1969). The mode of action of foliage-applied translocated herbicides with particular reference to the phenoxyacid compounds: 1. The mechanism and factors influencing herbicide absorption. Weed Res. *9*:224-240.

Ruckenstein, E. and Nagarajan, R. (1975). Critical micelle concentration: A transition point for micellar size distribution. J. Phys. Chem. *79*:2622-2626.

Svenson, S. E., Davies, F.T. and Duray, S.A. (1995). Gas exchange, water relations, and dry weight partitioning during root initiation and development of Poinsettia cuttings. J. Amer. Soc. Hort. Sci. *120*: 454-459.

White, A. D.,. Heaverlo, C. A, and Owen, M.D.K. (2002). Evaluation of methods to quantify herbicide penetration in leaves. Weed Tech. *16*:37-42.

Combined Proceedings IPPS 72:67-78. 2022.

High-efficiency plant regeneration via callus-induced organogenesis from leaf explants of Queen's crapemyrtle (*Lagerstroemia speciosa*)

Bin Wu[1,2], Benjamin Dixon[3], Ivan Sierra[3], Natalia Mesa[3], Qiansheng Li[4], Nicholas Zhang[2], Mengmeng Gu[5], Hongmin Qin[2]

[1] Department of Horticultural Sciences, Texas A&M University, College Station, TX 77843, USA. [2] Department of Biology, Texas A&M University, College Station, TX 77843, USA. [3] Department of Plant Pathology and Microbiology, Texas A&M University, College Station, TX 77843, USA. [4] Horticultural Sciences Department, University of Florida, FL 32611, USA. [5] Department of Horticulture and Landscape Architecture, Colorado State University, Fort Collins, CO 80523 USA.

bin.wu@tamu.edu

Fourth Place – Charlie Parkerson Graduate Student Research Paper Competition

Keywords: Queen's crapemyrtle, *Lagerstroemia speciosa*, callus-induced regeneration, leaf-derived organogenesis, micropropagation, plantlet rooting

Summary

Crapemyrtle (*Lagerstroemia* sp.) is the best-selling flowering tree and provides excellent pollen sources for pollinators in the U.S. However, the market's most commercially available crapemyrtle cultivars are easily infested by a recently invasive insect, crapemyrtle bark scale (CMBS; *Acanthococcus lagerstroemiae*), which jeopardizes the production and esthetic value of crapemyrtles anticipated by the Green Industry. Therefore, breeding CMBS-resistant cultivars is in great need on the market. Our previous study revealed that Queen's crapemyrtle (*L. speciosa*) was resistant to CMBS among all available crapemyrtle species. Establishing a highly efficient regeneration system for Queen's crapemyrtle is essential to molecular plant breeding for resistance improvement. Here, our study found that 97.9 ± 1.0% of leaf explants were induced callus when cultured on Lloyd & McCown woody plant medium (WPM) supplemented with 0.20 mg/L 2,4-D and 1.00 mg/L 6-BA. After transferring to the WMP medium supplemented with 10.00 mg/L 6-BA with 0.50 mg/L NAA, 32.4 ± 3.2% of callus successfully differentiated as the largest number of adventitious buds (23.4 ± 3.4) at the highest differentiation ratio (3.9 ± 0.1). The WPM medium supplemented with 1.00 mg/L 6-BA and 0.02 mg/L NAA induced 94.6 ± 4.0% of nodal segments of the regenerated shoots to produce 80.40 ± 15.16 new shoots (4.1 ± 0.9 cm in length) at the proliferation ratio of 4.5 ± 0.3. Half-strength WPM supplemented with 0.20 mg/L IBA induced 100.0 ± 0.0% of the regenerated shoots to produce 10.4 ± 1.1 roots (3.6 ± 0.7 cm in length) per shoot, and 98.3 ± 1.7% of the rooted plantlets survived after transplanting into the pots containing Jolly Gardener ® Pro-Line C/GP soil and 30% perlite for acclimatization. The successful establishment of the highly efficient callus-induced regeneration system lays a critical foundation for the genetic engineering of crapemyrtle to improve plant resistance or other desired traits, which meet priority needs of the nursery production or Green Industry.

INTRODUCTION

Crapemyrtle bark scale (*Acanthococcus lagerstroemiae*) is a recently invasive sap-sucking hemipteran initially reported on crapemyrtle (*Lagerstroemia* sp.) in the U.S. Heavy infestations of *A. lagerstroemiae* significantly reduce crapemyrtle growth and flowering, negatively impacting the production and landscape aesthetic value of crapemyrtles (Marwah et al., 2021; Merchant et al., 2018). A new crapemyrtle species or cultivar resistant to *A. lagerstroemiae* is in great need for plant resistance improvement and commercial on the U.S. market (Boutigny et al., 2020; Datta, 2021; Smith, 2021).

Greenhouse assays for host confirmation and plant susceptibility/resistance evaluation found that *Lagerstroemia speciosa* was relatively resistant to *A. lagerstroemiae* among the tested crapemyrtle species (Wu et al., 2021; Wu et al., 2022). *Lagerstroemia speciosa*, commonly known as 'Queen's crapemyrtle', is a tropical deciduous tree native to southeast Asia (Gilman and Watson, 2014; Klein et al., 2007; Rojas-Sandoval, 2017). Currently, most of the research on *in vitro* propagation of Queen's crapemyrtle focused on improving the micropropagation system by investigating different plant growth regulator (PGR) combinations for shoot proliferation and rooting, such as thidiazuron (TDZ), 6-benzyladenine (6-BA), α-naphthalene acetic acid (NAA), and N6-(3-hydroxybenzylamino purine) (meta-Topolin) (Ahmad et al., 2022a; Ahmad et al., 2022b; Lim-Ho and Lee, 1985; Vijayan et al., 2015). Although one publication mentioned shoot organogenesis-based regeneration from leaf callus, it is beneficial to have a comprehensive investigation to optimize the PGR combination for callus initiation, differentiation, micropropagation, and rooting, which allows to establish a highly efficient and stable callus-induced regeneration system of Queen's crapemyrtle (Rahman et al., 2010).

In this study, aseptic leaf explants of Queen's crapemyrtle were utilized to induce callus, and different types and concentrations of auxins, including 2,4-dichlorophenoxyacetic acid (2,4-D), indole-3-butyric acid (IBA), and NAA, and cytokinin, including 6-BA, TDZ, zeatin (ZT) were inspected to determine the most efficient PGR combination for each step of callus-induced plant regeneration.

MATERIALS AND METHODS

Collection and disinfection of plant materials. Queen's crapemyrtle plants were provided by Dr. Gary Knox at North Florida Research and Education Center (Quincy, FL 32351). The plants were transplanted in 11.4 L pots containing potting soil (Jolly Gardener® Pro-Line C/GP growing mixture, Oldcastle Lawn & Garden Inc., Poland Spring, ME) and maintained in the Greenhouse (at 25 ± 5 °C, 50% ± 10% relative humidity (RH), and a photoperiod of 10.5:13.5 (light: dark) h) at the Department of Horticultural Sciences of Texas A&M University (lat. 30°36'31.9"N, long. 96°21'1.9"W).

Healthy and juvenile stems were collected as explants from the potted Queen's crapemyrtle in early June 2021 (**Fig. 4A**). The explants were disinfected using 70% (v/v) ethanol for 10 s and 20% commercial bleach solution (Clorox®, 6% sodium hypochlorite, Oakland, CA) for 15 min followed by being thoroughly rinsed with sterile distilled water for 4-6 times. The disinfected explants were cut into nodal segments and cultured onto Lloyd and McCown woody plant medium (WPM; (Lloyd and McCown, 1980)) supplemented with 3.0% (w/v) sucrose, 0.65% (w/v) agar, and 0.5 mg/L 6-BA to acquire and maintain aseptic plantlets from axillary shoot formation and micropropagation/proliferation. Each culture jar contained 2 to 5 nodal segments.

Callus initiation. Aseptic leaf explants collected from the aseptic plantlets were cut into smaller pieces and cultured on WPM medium supplemented with different PGR treatments (**Table 1**). The callus initiation experiment was repeated thrice and 10 replicates (culture jars/vessels) for each PGR treatment per time. And each culture jar/vessel contained 4-9 explants. The percentage of leaf pieces successfully formed callus (total number of leaf pieces that formed callus divided by the total leaf pieces) 2 weeks after the 1st initiation were recorded to optimize the PGR combination for callus initiation/formation from leaf pieces of Queen's crapemyrtle.

Callus differentiation. Leaf-derived callus was subcultured on WPM medium supplemented with different PGR combinations (**Table 2**). The callus differentiation experiment was repeated at least thrice and 4 replicates (culture jars/vessels) for each PGR treatment per time. Each culture jar/vessel contained 3-6 callus. The total number of newly developed adventitious buds per differentiated callus within each PGR combination was recorded to optimize the PGR combination for callus differentiation from the leaf-derived callus.

Micropropagation/shoot proliferation. When the differentiated buds developed as 2-3 cm long shoot clumps, the clumps were split, cut into 1-1.5 cm long nodal segments, and proliferated on WPM medium supplemented with four different PGR combinations (**Table 3**). The micropropagation experiment was repeated 5 times and 5-33 replicates (culture jars/vessels) for each PGR treatment per time.

Each culture jar/vessel contained 3-6 nodal segments. The number and length of newly developed shoots per nodal segment within each PGR combination were recorded 4 weeks after transfer to optimize the PGR combination for micropropagation/shoot proliferation.

Rooting and acclimatization. After elongating and hardening the shoots on WPM medium added with 0.5 mg/L 6-BA for 4 weeks, 3-5 cm long shoots were excised and cultured on half-strength WPM medium supplemented with 3.0% (w/v) sucrose, 0.65% (w/v) agar, 0.05% (w/v) ascorbic acid, 0.05% (w/v) PVP-40, and different concentrations of IBA. Each well-rooted plantlet was gently washed off the medium residuals and separately transplanted into 6-cell plug trays capped with a humidify dome for hardening and acclimatization. The plug trays were potted with soil (Jolly Gardener® Pro-Line C/GP growing mixture, Oldcastle Lawn & Garden Inc., Poland Spring, ME) and 30% perlite (Dicaperl®, Dicalite Management Group, Inc., West Conshohocken, PA).

The rooting experiment was repeated 5 times and 2-14 replicates (regenerated shoots) for each PGR treatment per time. The number and length of newly developed roots (longer than 1 cm) per regenerated plantlet, and the total number of rooted plantlets survive 4 weeks after transplanting in covered chambers were recorded to optimize the IBA concentration for rooting regenerated plantlets of Queen's crapemyrtle *in vitro*.

Culture condition and statistical analysis. Unless further clarification, after inoculation, all plant materials were incubated in a walk-in growth chamber (CONVIRON®, Controlled Environments Ltd., Winnipeg, Manitoba, Canada) at 25 ± 1 °C under 60 ± 5 % RH and a 16:8 h (L:D) light intensity of 1500 – 2000 lux. All culture media, PGRs, agar, and other reagents used for crapemyrtle regeneration experiment were purchased from PhytoTech Labs, Inc., Lenexa, KS, and all the culture media were WPM media supplemented with 3.0% (w/v) sucrose, 0.65% (w/v) agar, 0.05% (w/v) ascorbic acid, and 0.05% (w/v) polyvinylpyrrolidone-40 (PVP-40) with pH value adjusted to 5.8 before being autoclaved at 121 °C and 15 psi pressure for 20 min (AMSCO®, STERIS Corporation, Mentor, OH). Every 30 mL or 50 mL media were, respectively, distributed per culture jar (CultureJar™ G9, volume 220 mL; inside diameter 43 mm, height 95 mm; PhytoTech Labs, Inc., Lenexa, KS) or culture vessel (PTL-100™, volume 372 mL; length 75 mm, width 75 mm, height 98 mm; PhytoTech Labs, Inc., Lenexa, KS).

All the experiments were completely randomized design with different numbers of replications as mentioned above. The data related to the callus initiation, callus differentiation, shoot proliferation, and rooting experiments were, respectively, analyzed by one-way ANOVA with Tukey's Honestly Significant Difference (HSD) test ($\alpha = 0.05$) to estimate the effects of PGRs on the organogenesis from leaf-derived callus of Queen's crapemyrtle.

RESULTS AND DISCUSSION

Callus initiation. The percentage of response in callus initiation from leaf pieces of Queen's crapemyrtle differed among the eight PGR combinations ($F = 17.9305$; $df = 7, 23$; $p < 0.0001$; **Table 1**). 0.20 mg/L 2,4-D with 1.00 mg/L 6-BA induced the highest percentage of callus initiation (97.94 ± 1.04%), followed by 0.50 mg/L 2,4-D with 2.00 6-BA combination (87.1 ± 4.4%), 0.2 mg/L 2,4-D alone (75.4 ± 12.5%), which was higher than using 0.50 mg/L and 2.00 mg/L NAA alone (7.2 ± 4.1% and 31.00 ± 2.0%, respectively) or 0.10 mg/L NAA with 5.00 mg/L 6-BA (24.1 ± 3.6%).

Together with the growth status of induced callus in each PGR combination (**Fig. 1**), the optimum one was 0.20 mg/L 2,4-D with 1.00 mg/L 6-BA.

Table 1. Effects of different PGR combinations on initiation and growth of callus derived from leaf pieces of Queen's crapemyrtle (*Lagerstroemia speciosa*).

PGR combination (mg/L)			Percentage of successfully initiated callus two weeks after 1st initiation (%)
NAA	2,4-D	6-BA	
0.50	-	-	7.2 ± 4.1[z] e[y]
2.00	-	-	31.0 ± 2.0 d
0.10	-	5.00	24.1 ± 3.6 de
-	0.20	-	75.4 ± 12.5 abc
-	0.20	1.0	97.9 ± 1.0 a
-	0.50	-	70.4 ± 14.8 bc
-	0.50	2.00	87.0 ± 4.4 ab
-	-	2.00	55.0 ± 6.1 d
Statistical analysis			$F_{7, 23} = 17.9305$; $p < 0.0001$

[z] Values represent means ± standard error.

[y] Means followed by different letters within the same column are significantly different as determined by Tukey's Honestly Significant Difference test ($\alpha = 0.05$).

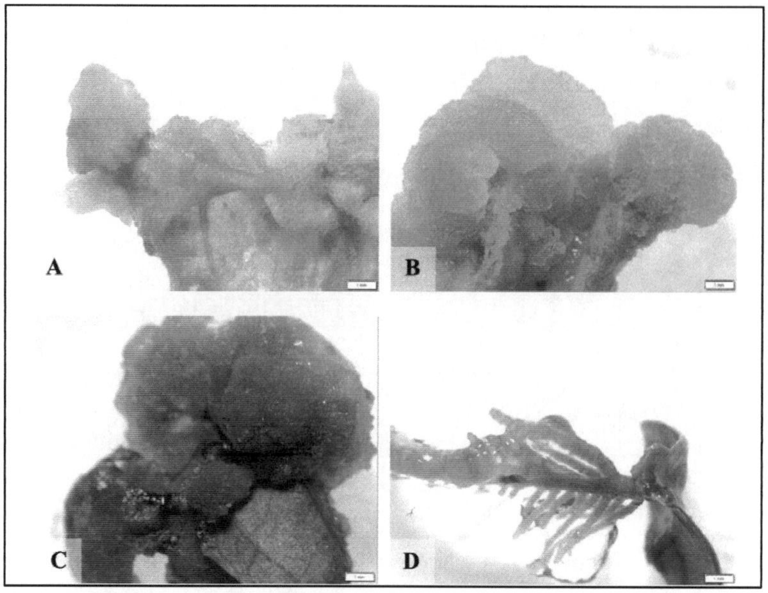

Figure 1. Effects of different PGR combinations on initiation and growth of callus derived from leaf pieces of Queen's crapemyrtle (*Lagerstroemia speciosa*). A: Yellowish, granular, and friable callus along with short flocky roots formed on WPM medium supplemented with 0.20 mg/L 2,4-D. B: Yellowish green, granular, and friable callus formed on WPM medium supplemented with 0.50 mg/L 2,4-D. C: Dark green, poor-granular, and compact callus formed on WPM medium supplemented with 2.00 mg/L 6-BA. D: Leaf pieces were directly rooted on WPM medium supplemented with 0.50 mg/L NAA.

Callus differentiation. The percentage of response in callus differentiation of Queen's crapemyrtle differed among different PGR combinations ($F = 21.609$; $df = 7, 34$; $p < 0.0001$; **Table 2**). 10.00 mg/L 6-BA with 0.50 mg/L NAA initiated the highest percentage of callus differentiation (32.4 ± 3.2%), followed by 8.00 mg/L 6-BA with 0.50 mg/L NAA (16.9 ± 2.7%), which was higher than 10.00 mg/L 6-BA with 0.10 mg/L NAA (5.8 ± 3.4%) or the combinations using ZT and NAA (less than 2.8 ± 1.7). 1.00 mg/L TDZ with 0.1 mg/L NAA did not initiate the callus differentiation.

The number of newly developed adventitious buds ($F = 20.2731$; $df = 7, 34$; $p < 0.0001$; **Table 2**) and the differentiation ratio ($F = 5.9199$; $df = 7, 34$; $p = 0.0003$; Table 2) differed among different PGR combinations. Together with the number of newly differentiated buds and the differentiation ratio, the optimum PGR combination for callus differentiation of Queen's crapemyrtle was 10.00 mg/L 6-BA with 0.50 mg/L NAA, which produced the largest number of newly developed adventitious buds (23.4 ± 3.4) and the highest differentiation ratio (3.9 ± 0.1).

Table 2. Effects of different PGR combinations on callus differentiation of Queen's crapemyrtle (*Lagerstroemia speciosa*).

PGR combination (mg/L)				Repeated number	Percentage of response (%)[z]	Number of newly developed buds	Differentiation ratio
6-BA	ZT	TDZ	NAA				
-	1.00	-	0.10	5	2.8 ± 1.7[y] c[x]	2.4 ± 1.9 c	1.1 ± 0.7 bc
-	1.00	-	0.20	5	2.5 ± 1.6 c	1.4 ± 0.9 c	1.4 ± 0.9 bc
-	1.00	-	0.50	3	2.2 ± 2.2 c	0.7 ± 3.7 c	0.7 ± 0.7 bc
-	1.00	-	1.00	3	2.1 ± 2.1 c	0.3 ± 0.3 c	0.3 ± 0.3 c
8.00	-	-	0.50	6	16.9 ± 2.7 b	7.2 ± 1.4 b	2.2 ± 0.3 b
10.00	-	-	0.10	4	5.8 ± 3.4 c	1.5 ± 1.0 c	0.5 ± 0.5 c
10.00	-	-	0.50	5	32.4 ± 3.2 a	23.4 ± 3.4 a	3.9 ± 0.1 a
-	-	1.00	0.10	4	0.0 ± 0.0 c	0.0 ± 0.0 c	0.0 ± 0.0 c
Statistical analysis					$F_{7,34}$ = 21.609; $p <$ 0.0001	$F_{7,34}$ = 20.2731; $p <$ 0.0001	$F_{7,34}$ = 5.9199; p = 0.0003

[z] The percentage of response for callus differentiation was calculated as (total amount of differentiated callus) ÷ (total amount of inoculated callus) × 100%. The differentiation ratio was calculated as (total amount of newly differentiated adventitious buds) ÷ (total amount of differentiated callus). [y] Values represent means ± standard error. [x] Means followed by different letters within the same column are significantly different as determined by Tukey's Honestly Significant Difference test (α = 0.05).

Micropropagation/shoot proliferation. The percentage of response in micropropagation of Queen's crapemyrtle did not differ among different PGR combinations (F = 0.450; df = 3, 19; p = 0.7209; **Table 3**), which indicated that 6-BA (0.20-5.00 mg/L) combined with NAA (0.02 mg/L) or ZT (0.2 mg/L) initiated the micropropagation with equal effectiveness. However, the number (F = 5.431; df = 3, 19; p = 0.0091; **Table 3**) and the length (F = 4.506; df = 3, 19; p = 0.0179; **Table 3**) of newly developed shoots and the proliferation ratio (F = 6.874; df = 3, 19; p = 0.00351; **Table 3**) differed among the PGR combinations. Together with the growth status of newly shoots in each PGR combination (**Fig. 2**), the optimum PGR combination for the micropropagation of Queen's crapemyrtle was 1.00 mg/L 6-BA with 0.02 mg/L NAA, which induced 94.6 ± 4.0% of nodal segments to produce 80.4 ± 15.2 new shoots (4.1 ± 0.9 cm in length) at the proliferation ratio of 4.5 ± 0.3.

Table 3. Effects of different PGR combinations on micropropagation of Queen's crapemyrtle (*Lagerstroemia speciosa*) at four weeks after inoculation.

PGR combination (mg/L)			Percentage of response (%)[z]	Number of newly developed shoots	Shoot length (cm)	Proliferation ratio
6-BA	ZT	NAA				
0.20	-	0.02	80.8 ± 8.1[y] a[x]	23.4 ± 5.3 b	5.6 ± 1.0 a	1.9 ± 0.1 b
1.00	-	0.02	94.6 ± 4.0 a	80.4 ± 15.2 a	4.1 ± 0.9 ab	4.5 ± 0.3 a
5.00	-	0.02	87.8 ± 6.7 a	84.6 ± 8.3 a	1.6 ± 0.3 b	5.4 ± 0.6 a
1.00	0.2	-	81.0 ± 16.1 a	60.0 ± 15.8 a	3.4 ± 0.8 ab	4.5 ± 0.9 a
Statistical analysis			$F_{3, 19} = 0.450$; $p = 0.7209$	$F_{3, 19} = 5.431$; $p = 0.0091$	$F_{3, 19} = 4.506$; $p = 0.0179$	$F_{3, 19} = 6.874$; $p = 0.0035$

[z] The percentage of response for micropropagation was calculated as (total amount of nodal segments initiated micropropagation) ÷ (total amount of nodal segments inoculated) × 100%. The proliferation ratio was calculated as (total amount of newly developed shoots) ÷ (total amount of inoculated nodal segments). [y] Values represent means ± standard error. [x] Means followed by different letters within the same column are significantly different as determined by Tukey's Honestly Significant Difference test ($\alpha = 0.05$).

Regenerated plantlet rooting and acclimatation. The percentage of response in rooting of Queen's crapemyrtle did not differ among different PGR combinations ($F = 0.4457$; $df = 3, 19$; $p = 0.9372$; **Table 4**), which indicated that IBA (0.00, 0.20, 1.00, and 5.00 mg/L) all initiated the rooting with equal effectiveness. However, the number ($F = 3.8277$; $df = 3, 19$; $p = 0.0305$; **Table 4**) and the length ($F = 4.2890$; $df = 3, 19$; $p = 0.0212$; **Table 4**) of newly developed roots and the survival rate ($F = 5.496$; $df = 3, 19$; $p = 0.0087$; **Table 4**) differed among different IBA concentrations. Together with the growth status of regenerated plantlets in each treatment (**Fig. 3**), the optimum IBA concentration for rooting Queen's crapemyrtle was 0.20 mg/L IBA, which induced 100.0 ± 0.0% shoots to produce 10.4 ± 1.1 roots (3.6 ± 0.7 cm in length) per shoot. After the transplanting, the survival rate of the 0.2 mg/L IBA treated plantlets was 98.3 ± 1.7%, which was significantly higher than the plantlets treated with 0.00 and 5.00 mg/L IBA.

Figure 2. Effects of different PGR combinations micropropagation of Queen's crapemyrtle (*Lagerstroemia speciosa*) at four weeks after inoculation. A: Each nodal segment of Queen's crapemyrtle proliferated into 2-3 new shoots in 7-9 cm length on WPM medium supplemented with 0.20 mg/L 6-BA and 0.02 mg/L NAA. B: Each nodal segment proliferated into 4-5 new shoots in 3-7cm length with small amounts of callus formed on WPM medium supplemented with 1.00 mg/L 6-BA and 0.02 mg/L NAA. C: Each nodal segment proliferated into 3-5 new shoots in 2-4 cm length with relatively large amounts of callus formed on WPM medium supplemented with 5.00 mg/L 6-BA and 0.02 mg/L NAA. D: Each nodal segment proliferated into 3-6 new shoots in 3-5cm length with relatively some callus formed on WPM medium supplemented with 1.00 mg/L 6-BA and 0.20 mg/L ZT.

Table 4. Effects of different PGR combinations on rooting of regenerated plantlets of Queen's crapemyrtle (*Lagerstroemia speciosa*).

IBA (mg/L)	Percentage of response (%)[z]	Number of newly developed roots (>1cm)	Root length (cm)	Survival rate (%)
0	90.0 ± 10.0[y] a[x]	6.3 ± 1.6 b	2.5 ± 0.3 ab	51.9 ± 3.9 c
0.20	100.0 ± 0.0 a	10.4 ± 1.1 a	3.6 ± 0.7 a	98.3 ± 1.7 a
1.00	100.0 ± 0.0 a	7.1 ± 0.9 ab	2.5 ± 0.4 ab	79.7 ± 12.6 ab
5.00	90.7 ± 5.7 a	4.9 ± 1.0 b	1.4 ± 0.2 b	71.4 ± 9.6 bc
Statistical analysis	$F_{3, 19} = 0.9372$; $p = 0.4457$	$F_{3, 19} = 3.8277$; $p = 0.0305$	$F_{3, 19} = 4.2890$; $p = 0.0212$	$F_{3, 19} = 5.496$; $p = 0.0087$

[z] The percentage of response for rooting was calculated as (total amount of rooted shoots) ÷ (total amount of inoculated shoots) × 100%. The survival rate was calculated as (total amount of survived plantlets) ÷ (total amount of transplanted plantlets) × 100%. [y] Values represent means ± standard error. [x] Means followed by different letters within the same column are significantly different as determined by Tukey's Honestly Significant Difference test ($\alpha = 0.05$).

Figure 3. Effects of different PGR combinations rooting of regenerated plantlets of Queen's crapemyrtle (*Lagerstroemia speciosa*). Four weeks after inoculation, A: 4.3 roots in 1.0 cm long developed on half-strength WPM medium supplemented with 0.00 mg/L IBA. B: 7.3 roots in 2.6 cm long developed on half-strength WPM medium supplemented with 0.20 mg/L IBA. C: 7.6 roots in 2.2 cm long developed on half-strength WPM medium supplemented with 1.00 mg/L IBA. D: 3.8 roots in 1.1 cm long developed on half-strength WPM medium supplemented with 5.00 mg/L IBA.

Figure 4. Establishment of callus-induced regeneration system from leaf pieces of Queen's crapemyrtle (*Lagerstroemia speciosa*). A: Initial leaf explants used for callus initiation were collected from Queen's crapemyrtle (L. speciosa) provided by Dr. Gary Knox from UF/IFAS North Florida Research and Education Center (Quincy, FL 32351). B: Leaf-derived callus was initiated on Lloyd & McCown woody plant medium (WPM) supplemented with 0.20 mg/L 2,4-D with 1.00 mg/L 6-BA at 5 days after inoculation. C: Adventitious buds differentiated from the leaf-derived callus on WPM medium supplemented 10.00 mg/L 6-BA with 0.50 mg/L NAA at 4 weeks after inoculation. D: More shoots developed and elongated on WPM medium supplemented with 10.00 mg/L 6-BA with 0.50 mg/L NAA in 14-28 days after inoculation. E: Regenerated shoots were in vitro proliferated on WPM medium supplemented with 1.00 mg/L 6-BA with 0.02 mg/L NAA in 5 weeks after inoculation. F: Strong root system was developed on half-strength WPM supplemented with 0.20 mg/L IBA in 4 weeks after inoculation. G: Rooted plants were transplanted into growth chamber for acclimatization. All the WPM medium (full-strength and half-strength) were also supplemented with 3.0% sucrose, 0.65% agar, 0.05% ascorbic acid, 0.05% PVP-40.

LITERATURE CITED

Ahmad N., Faisal M., Ahmad A., Alatar A.A., Qahtan A.A., Alok A. (2022a). Thidiazuron induced in vitro clonal propagation of Lagerstroemia speciosa (L.) Pers.—an Important avenue tree. Horticulturae 8:359. https://doi.org/10.3390/horticulturae8050359.

Ahmad N., Fatima N., Faisal M., Alatar A.A., Pathirana R. (2022b). Photosynthetic parameters and oxidative stress during acclimation of crepe-myrtle (Lagerstroemia speciosa (L.) Pers.) in a meta-Topolin-based micropropagation system and genetic fidelity of regenerated plants. Plants 11:1163. https://doi.org/10.3390/plants11091163.

Boutigny A.-L., Dohin N., Pornin D., Rolland M. (2020). Overview and detectability of the genetic modifications in ornamental plants. Hortic. Res. 7. https://doi.org/10.1038/s41438-019-0232-5.

Datta S.K. (2021). Breeding of ornamentals: success and technological status. The Nucleus:1-22. https://doi.org/10.1007/s13237-021-00368-x.

Gilman E.F., Watson D.G. (2014). Lagerstroemia speciosa: Queens Crapemyrtle, University of Florida IFAS Extension. pp. 1-3.

Klein G., Kim J., Himmeldirk K., Cao Y., Chen X. (2007). Antidiabetes and anti-obesity activity of Lagerstroemia speciosa. Evid. Based Complement Alternat. Med. 4. https://doi.org/10.1093/ecam/nem013.

Lim-Ho C.L., Lee S.-K. (1985). Micropropagation of Lagerstroemia speciosa (L.) Pers.(Lythraceae). Gardens' Bulletin 38:175-184.

Lloyd G., McCown B. (1980). Commercially-feasible micropropagation of mountain laurel, *Kalmia latifolia*, by use of shoot-tip culture. 30:421-427.

Marwah P., Zhang Y.Y., Gu M. (2021). Investigating Producers' Preferences for Crapemyrtle and Their Perceptions Regarding Crapemyrtle Bark Scale. Horticulturae 7:146. https://doi.org/10.3390/horticulturae7060146.

Merchant M., Garcia J., Gu M. (2018). Longevity of imidacloprid as a soil treatment for crapemyrtle bark scale on crapemyrtle, 2016–2017. AMT 43:tsy012. https://doi.org/10.1093/amt/tsy012.

Rahman M.M., Amin M.N., Rahman M.B., Sultana R.S. (2010). In vitro adventitious shoot organogenesis and plantlet regeneration from leaf-derived callus of *Lagerstroemia speciosa* (L.) Pers. Propag. Ornam. Plants 10:149-155.

Rojas-Sandoval J. (2017). Lagerstroemia speciosa (Pride of India). ISC. http://dx.doi.org/10.1079/ISC.29704.20203482881.

Smith C.M. (2021). Conventional breeding of insect-resistant crop plants: still the best way to feed the world population. Curr. Opin. Insect. Sci. 45:7-13. https://doi.org/10.1016/j.cois.2020.11.008.

Vijayan A., Padmesh Pillai P.,. Hemanthakumar A., Krishnan P. (2015) Improved in vitro propagation, genetic stability and analysis of corosolic acid synthesis in regenerants of Lagerstroemia speciosa (L.) Pers. by HPLC and gene expression profiles. PCTOC 120:1209-1214. https://doi.org/10.1007/s11240-014-0665-3.

Wu B., Xie R., Knox G., Qin H., Gu M. (2021). Host suitability for crapemyrtle bark scale (Acanthococcus lagerstroemiae) differed significantly among crapemyrtle species. Insects 12:6. https://doi.org/10.3390/insects12010006.

Wu B., Xie R., Knox G.W., Qin H., Gu M. (2022). Evaluating Beautyberry and Fig Species as Potential Hosts of Invasive Crapemyrtle Bark Scale in the United States. HortTechnology 32:10-15. https://doi.org/10.21273/HORTTECH04897-21.

Combined Proceedings IPPS 72:79-86. 2022.

Container Grown Plants are Gassy

Forrest J. Brown[1], James S. Owen Jr.[2] and Alex X. Niemiera[1]

[1]School of Plant and Environmental Sciences, Virginia Tech, Blacksburg, VA, USA
[2]U.S. Department of Agriculture, Agriculture Research Service, Application Technology Research Unit, Wooster, OH, USA

fjbrown92@vt.edu

Keywords: Nitrogen, containerized crops, soilless substrates, nursery production, control-release fertilizer, Fourier Transform Infrared Spectroscopy (FTIR)

Summary

Rising mineral nutrient and polymer costs are placing a direct economic burden on nursery crop producers utilizing controlled-release fertilizer. Nitrogen (N) inputs in the containerized crop production have potential inefficiencies under a broad scope of cultural practices. While N gas emissions have been investigated in other forms of crop production, emission research on container-grown nursery production has been limited due to analytical methodology and the complexity of gaseous fate. The objective was to compare two control-release fertilizers (CRFs), ammonium nitrate (AN) vs urea ammonium nitrate (UAN), to determine the gaseous emissions between fertilizer treatment based on time of day using fourier transform infrared spectroscopy (FTIR). These data were then used to estimate seasonal flux loss over a 51-day period. Results showed there was a higher degree of variability of gaseous flux [nitrous oxide (N_2O), nitric oxide (NO), and nitrogen dioxide (NO_2)] in the beginning of the season when CRFs began releasing, more consistent fluxes were exhibited during the mid to late production season. Gaseous fluxes of N species were similar regardless of CRF and time treatments for all N species; only summation of N species ($\sum N$) showed statistical differences. The study of gaseous emissions in nursery production is still in its infancy and more research is necessary to gain a better understanding of gaseous flux and factors influencing flux for container-grown crops.

INTRODUCTION

Daily fertilization is essential in containerized crops and a routine practice in nursery production. Despite modern advancement in mineral nutrient delivery and the science of understating mineral nutrient fate, applied nitrogen (N) remains complex and is poorly understood due to the numerous chemical and biological pathways of the N cycle within the growing media (Creamer et al., 2022). Closing the gap in understanding N fate can help pave the way for a more economically efficient and environmentally friendly nursery industry.

Nitrogen is applied to containerized nursery crops typically as a controlled release fertilizer (CRF) containing a combination of ammonium (NH_4), nitrate (NO_3), or urea [$CO(NH_2)_2$]. Each of these N forms result in different, concurrent microbial and chemical processes. This adds another layer of complexity to the question, "Where is the applied N going?".

Nitrogen budgets quantifying the various pathways of applied N (**Fig. 1**; e.g., plant, substrate, leached) have helped growers and scientists learn and build upon previous estimates of N fate (Navarez et al., 2012; Navarez et al., 2013; Pitton et al., 2022; Warren et al., 1995). However, quantification of N losses vary considerably due to methodology and production practices (Marble et al., 2011; Marble et al., 2012; Pitton et al., 2021).

Nitrogen applied to containerized bark-based substrates undergo several transformations and can ultimately, depending on conditions, form various gaseous species generally termed nitrous oxides (NOx). Specifically, NOx is any form of gaseous N species with an oxide group excluding dinitrogen (N_2) and ammonia (NH_3).

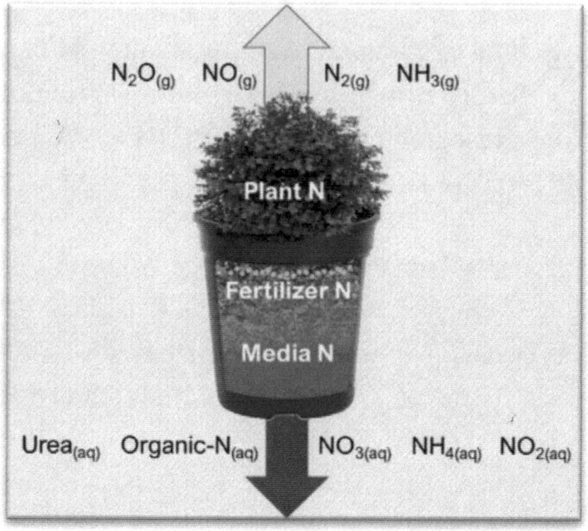

Figure 1. Potential nitrogen pathways and species of a typical containerized crop system.

One of the NOx species receiving significant scrutiny, nitrous oxide (N_2O), is a major climate warming gas of concern due to its long-lived nature and global warming potential 298 times greater than carbon dioxide (CO_2) (Myhre et al., 2013). While gaseous N losses are an environmental concern, there is also an economic concern for growers. The grower's overarching goal is to provide a crop with adequate N throughout the production cycle to produce a saleable product in the shortest time possible. Growers suffer a monetary loss when a large portion of applied N is unused by the plant and specifically lost as a gas from the container.

Fourier Transform Infrared Spectroscopy (FTIR) is an analytical method used for measuring and determining gas emissions and enables individuals to measure NH_3, N_2O, NO, and NO_2 simultaneously, in real-time in field settings with a 'rugged' instrument. By measuring the N gaseous species previously listed we can solve and account for N_2 gas lost from complete denitrification (NO_3 microbially ultimately transformed to N_2 gas in a multi-step reaction). Utilizing this technology, our objective for this portion of the research project was to better understand the gaseous N losses from container-grown plants.

MATERIALS AND METHODS

The experiment took place from 15 July 2020 to 4 September 2020 at the United States Department of Agriculture, Agriculture Research Service (USDA-ARS) Application Technology Research Unit in Wooster, OH.

This experiment was a single factor experiment comparing two N source formulations in a completely randomized block design. Sampling and harvest of three experimental unit replicates were completed on: 1, 2, 3, 6, 9, 12, 15, 19, 23, 27, 31, 35, 39, 43, 47, 51 days after initiation (DAI). At each date, substrate solution extract was collected analyzed for N species using an ion chromograph, fertilizer was collected and digested to determine N content remaining, plant (root and shoot) and substrate were harvested and analyzed for total N content, and N gas flux was measured using FTIR at 1000 hrs. [morning(10:00am)] and 1500 hrs. [afternoon (3:00pm)].

Forsythia ×intermedia Showoff® liners (Baileys Nursery, St. Paul, MN) were transplanted in #2 trade containers (7.33L, C900, Nursery Supplies Inc., Chambersburg, PA) with a pine bark substrate (T.H. Blue Inc., Eagle Springs, NC) amended with 1.51 $kg \cdot m^{-3}$ of ground dolomite lime (95.0% CaCO3 equivalent, 21.6% Ca, 10.0% Mg; Soil Doctor, Atlanta, GA) and 0.89 $kg \cdot m^{-3}$ granular micronutrient fertilizer (Micromax, Everris, Dublin, OH) on July 10, 2020.

The two N source treatments were (3-4 month) control-release fertilizers (CRFs): 42.0 g per container of 15.0N:3.9P:9.9K ammonium-nitrate based CRF (AN) (15-9-12; 8% NH4-N, 7% NO3-N; Osmocote, ICL, Charleston, SC with a water-soluble pre-charge of 0.23 % applied N); and 35.0 g per container of 18.0N:2.2P:6.6K urea ammonium nitrate CRF (UAN) (18-5-8; 6.3% NH4-N, 5.4% NO3-N, 6.3% Urea-N) (Osmocote, ICL, Charleston, SC) Each CRF was top-dressed on July 14, 2020 and surface incorporated into the substrate surface (2-3 cm below the surface by hand). This resulted in 6.3 g of N being supplied per container for CRF treatments based on the fertilizer label.

Containers were then placed in an open-air nursery setting and received 18 mm daily over-head irrigation application (05:00 HR) via upright mini-Wobbler sprinklers (#4 nozzle, 1.6 mm orifice; Senninger, Clermont, FL, U.S.) on 91-cm risers.

Gas analysis occurred on all sampling days using FTIR (Gasmet Terra GT5000, Vantaa, Finland). To measure gas flux from a container, a chamber was made using a plastic 22.7 (L) bucket with Polyurethane Tubing (McMaster-Carr, Cleveland, OH) to make a closed loop with the FTIR (**Fig. 2**). One experimental unit (planted and fertilized container) was placed inside the chamber of the closed loop system, and the gas flux was measured over five minutes. The Gasmet apparatus would then be brought back to ambient gas concentrations; sampling atmospheric air between each experimental sample (3 min) prior to sampling the next replication. Data were used to calculate gas flux by using Equation [1] (Gyawali et al., 2019).

$$f_{NOx} = \frac{P_0 V_c}{R T o A} \times \frac{\Delta C}{\Delta t} [1]$$

The formula is defined by P_o the pressure within the chamber [M L^{-1} t^{-2}]. This is assumed to be equivalent to the atmospheric pressure. V_c is the cumulative volume of the chamber, apparatus internal volume, and tubing volume [L^3], R is the ideal gas law constant [M L^2 N^{-1} K^{-1} t^{-2}], T_0 is the air temperature [K], A is the exposed surface area of an individual container or experimental unit [L^2], ΔC is change in concentration of a given gas on a molar basis [N N^{-1}] which is then compared to change in time in seconds Δt [s]. Three samples per treatment were measured (n = 3). (Gyawali Et al., 2019).

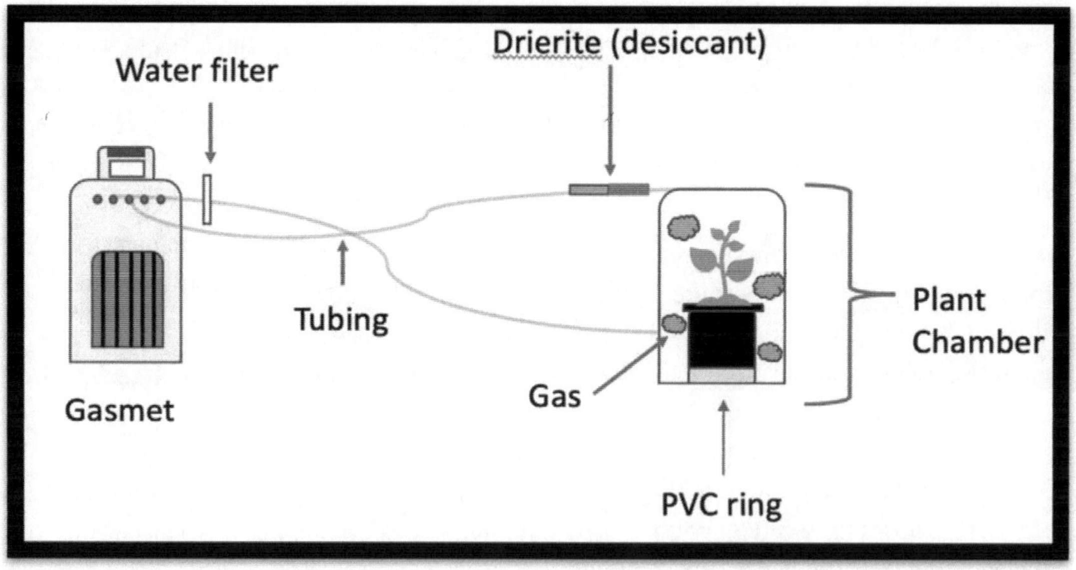

Figure 2. Depiction of gas sampling setup which includes (from left to right) Fourier transform infrared gas analyzer FTIR), 1/4" (6.35 mm) tubing creating a closed loop with a 6-gal (22.7-L) sealed chamber containing #2 containerized *Forsythia* placed upon a 1" (25.4 mm) PVC ring. A desiccant and water filter were added to the chamber return or FTIR intake to control humidity and protect internal mirrors.

Gas measurements occurred during both the morning and the afternoon of sampling dates; this was conducted to see if there was a difference between gaseous flux based on time of the day. Preliminary analysis of N gas measurements was made by graphing ~180s to determine change in concentration over change in time (Equation 1). These data were then graphed and checked by the expected linear increase of CO_2 gas concentration ($R^2 < 0.90$).

Data analysis occurred using JMP Pro (JMP® Pro ver. 14, SAS Institute, Cary, NC). Data was analyzed by morning or afternoon measures (AM or PM); pooling across approximately 14-day segments to represent early, mid, late production period based on observed CRF release and subsequent gas emissions. Additionally, gas emissions occurring in the AM, PM or both were pooled across the entire 51 production period to compare means. All data were not normally distributed, therefore a non-parametric analysis, Wilcoxon / Kruskal-Wallis Test, was used to determine statistically significant differences between means, comparing differences between morning and afternoon or fertilizer formulation.

RESULTS AND DISCUSSION

There was no difference in gaseous species losses during three periods of the growing season: (early, mid, and late production) for fertilizer treatments, however, the sums of all reactive N gaseous species (∑RN) were different between CRF treatment (Fig. 3). For both CRF treatments, ∑RN was highest in the first period (early season) and stabilized thereafter, and gaseous flux was more consistent in the mid to late season periods. Over the entire season time of day data were the same (**Fig. 3**).

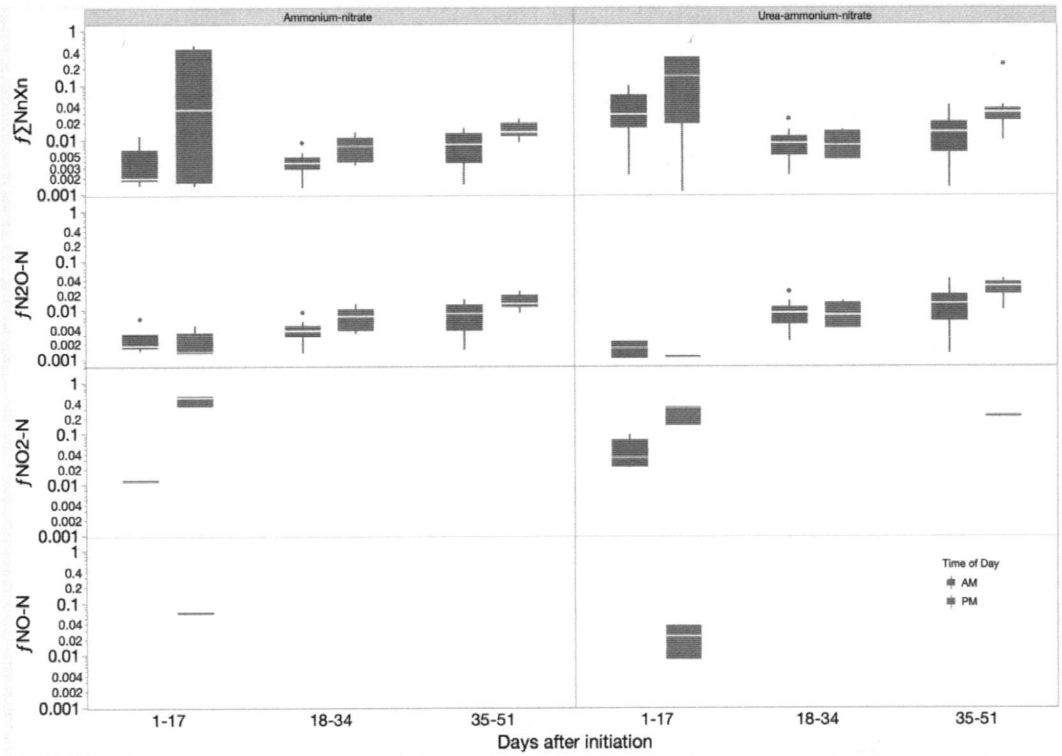

Figure 3. N species gaseous losses based on CRF treatment shown as box plots (AN vs. UAN) during morning (AM) and afternoon (PM) at three stages of the production cycle (early (1-17 DAI), mid (18-34 DAI), late season (35-51 DAI)). Gaseous flux (f_{NO}-N(Eq. 1) is defined as (ug cont-1 sec-1) is shown on a logarithm scale (b=10). $f\sum$NnXn is the sum of all N species measured.

There were no differences in N gaseous species throughout the entire growing season; the only differences are the ∑RN values; PM values were higher than AM values for both fertilizer treatments (**Table 1**).

Table 1. N species gaseous flux of two CRF fertilizer based on time of day (AM/PM) and over entire season. Flux of N species (*f*) and sum of N gaseous species (∑N) Statistically significant (P-value <0.05) values denoted by red text and asterisk (*). Average values and ± standard deviation are shown for all gaseous species.

Fertilizer formulation	Time of Day	N	*f*N2O-N	*f*NH3-N	*f*NO-N (μg cont^{-1} min^{-1})	*f*NO2-N	*f*∑N
Ammonium-nitrate	AM	46	0.26±0.04	0.0±0.0	0.00±0.00	0.01±0.01	0.28±0.04
	PM	36	0.49±0.07	0.0±0.0	0.11±0.11	2.30±1.32	2.90±1.29
	p-value		0.0658	-------	0.2583	0.1888	0.0018
	Mean	82	0.36±0.04	0.0±0.0	0.05±0.05	1.02±0.59	1.43±0.58
Urea-ammonium-nitrate	AM	46	0.45±0.08	0.0±0.0	0.00±0.00	0.29±0.15	0.74±0.15
	PM	34	0.15	0.0±0.0	0.08±0.07	1.75±0.88	2.71±0.85
	p-value		0.1015	-------	0.0979	0.7630	0.0227
	Mean	80	0.63±0.08	0.0±0.0	0.03±0.03	0.91±0.39	1.58±0.38
Fertilizer formulation	p-value		0.2420	-------	0.5564	0.1550	0.0203

Using all the collected RN gaseous and liquid pathways of this study, over half of the applied N was lost via gaseous pathways. We hypothesized that the two major processes influencing gaseous losses are denitrification and nitrification. Denitrification and nitrification are two microbial processes in which there are several gaseous N intermediates or potential fates (NO, N_2O, NO_2, NH_3, N_2). Based on this research, we estimate gasses lost from highest to lowest concentration are: N_2 > NO > N_2O. Since N_2 was not measured, we inferred that N_2 evolved via the microbial oxidation of N_2O (Havlin et al., 2014). Any N not accounted for in plant tissue, fertilizer harvested, substrate, aqueous samples, or RN gas samples was lost as N_2 gas. Unfortunately, there is not currently a preferential methodology to determine N_2 concentrations in gaseous samples (Takaya et al., 2003; Wang et al., 2011).

We maintain that using linear increase of CO_2 gas concentration ($R^2 < 0.90$) to determine gaseous flux of N or other gaseous species is not an appropriate method of analysis. This method would have simplified analysis by using a single gas to determine trends of gaseous species of interest. Each gaseous species exhibits a unique pattern potentially influenced by analytical detection limits and head space saturation that differ on an experimental unit basis. Therefore, analysis of individual RN gaseous species provides more insight into emissions rather than the method examined in this analysis.

Pour-through extractions of the pine bark substrate were conducted prior to gas sampling during this experiment on the same experimental units. This likely led to some of the N cycle products to have been flushed from the system that may have resulted in reduced N gas losses.

CONCLUSION

In conclusion, we found that more than half of applied N exits as a gas or is unaccounted presumably due to denitrification and nitrification. Hence, there is a need to further investigate and understand the various factors affecting gaseous N losses from a container system. Further investigation into the influence of the zone of saturation within the container, irrigation regime, fertilizer type and placement, and substrate physical properties are factors that may influence gaseous losses in nursery production. A comprehensive understanding of N gaseous emissions in nursery production can serve to close the gap on N application inefficiencies presented by gaseous loss of N from the container.

LITERATURE CITED

Creamer, R.E., Barel, J.M., Bongiorno, G., and Zwetsloot, M.J. (2022). The life of soils: Integrating the who and how of multifunctionality. Soil Biol. Biochem. *166*:1-15.

Gyawali, A.J., Lester, B.J., and Stewart R.D. (2019). Talking SMAAC: A new tool to measure soil respiration and microbial activity. Front. Earth Sci. *7*:1-7.

Havlin, J.L., Tisdale, S.L., Nelson, W.L., and Beaton J.D. (2014). Soil Fertility and Fertilizers: An Introduction to Nutrient Management. 8th Edition. Pearson, N.Y., N.Y.

Marble, S.C., Prior, S.A., Runion, G.B., Torbert, H.A., Gilliam, C.H., Fain, G.B., Sibley, J.L., and Knight, P.R. (2011). Determining trace gas flux from container-grown woody ornamentals©. Comb. Procd. Internat. Plant Propagators' Soc. *61*:469-475.

Marble, S.C., Prior, S.A., Runion, G.B., Torbert, H.A., Gilliam, C.H., Fain, G.B., Sibley, J.L., and Knight, P.R. (2012). Effects of fertilizer placement on trace gas emissions from nursery container production. HortScience *47*:1056-1062.

Myhre, G., Shindell, D., Breon, F.M., Collins, W., Fuglestvedt, J., Huang, J., Koch, D., Lamarque, J.F., Lee, D., Mendoza, B., Nakajima, A., Roback, A., Stephens, G., Takemura, T., and Zhang, H. (2013). Anthropogenic and natural radiative forcing, p. 659-740. Stocker, T.F., D. Qin, G.-K. Plattner, M. Tignor, S.K. Allen, J. Boschung, A. Nauels, Y. Xia, V. Bex and P.M. Midgley (eds.). Climate Change 2013: The Physical Science Basis. Contribution of working Group I to the Fifth Assessment Report of the Intergovernmental Panel on Climate Change. Cambridge University Press, Cambridge, United Kingdom and New York, NY, USA.

Narváez, L., Cáceres, and R., Marfá, O. (2012). Effects of climate and fertilization strategy on nitrogen balance in an outdoor potted crop of *Viburnum tinus* L. Span. J. Agric. Res. *10*:471-481.

Narváez, L., Cáceres, and R., Marfá, O. (2013). Effect of different fertilization strategies on nitrogen balance in an outdoor potted crop of *Osteospermum ecklonis* (DC.) Norl. 'Purple Red' under Mediterranean climate condition. Span. J. Agric. Res. *11*:833-841.

Pitton, B.J.L., Evans, R.Y., Zhu-Barker, X., and Oki, L.R. (2021). Greenhouse gas emissions and global warming potential from a woody ornamental production system using a soilless growing substrate. ACS Agric. Sci. *1*:35-43

Pitton, B.J.L., Oki, L.R., Sisneroz, and J. Evans, R.Y. (2022). A nursery system nitrogen balance for production of a containerized woody ornamental plant. Sci. Hortic. *291*:1-7.

Takaya, N., Catalan-Sakairi, M.A., Sakaguchi, Y., Kato, I., Zhou, Z., and Shoun, H., (2003). Aerobic denitrifying bacteria that produce low levels of nitrous oxide. Appl. Environ. Microbiol. *69*:3152-3157.

Wang, R., Willibald, G., Feng, Q., Zheng, X., Liao, T., Brüggemann, N., and Butterbach-Bahl, K. (2011). Measurement of N_2, N_2O, NO, and CO_2 Emissions from soil with the gas-flow-soil-core technique. Environ. Sci. Technol. *45*:6066-6072.

Warren, S. L., Bilderback, and T.E., Tyler, H.H. (1995). Efficacy of three nitrogen and phosphorus sources in container-grown azalea production. J. Environ. Hort. *13*:147-151.

Combined Proceedings IPPS 72:87-93. 2022.

Effect of X-ray Irradiation on Populations of *Pseudomonas amygdali* pv. *loropetali* pv. nov.

Jenny B. Ryals, Patricia R. Knight, Rebecca A. Melanson, Warren E. Copes, Anthony T. Bowden, Christine E. H. Coker, Scott A. Langlois, Haley N. Williams, Jim M. DelPrince, Benny Park, Sam Chang, Kassie Conner, Jeremy M. Pickens, Ronald C. Stephenson, and Gary R. Bachman

Coastal Research and Extension Center, South Mississippi Branch Experiment Station
P.O. Box 193, Poplarville, Mississippi 39470, USA

j.ryals@msstate.edu

Keywords: irradiation, x-ray, loropetalum gall, *Loropetalum chinense*

Summary

Loropetalum, *Loropetalum chinense* (R. Br.) Oliv., is a popular landscape plant, but it can be infected by the gram-negative bacteria *Pseudomonas amygdali* pv *loropetali* pv. nov. Bacterial diseases are difficult to control, and this particular bacterium usually leads to disposal of the plant, resulting in economic losses for the nursery. These bacteria are causing galls on loropetalum which can cause stem girdling leading to reduced growth and possibly death of the plant. The bacteria will infect the plant if it can permeate through a cut or wound in the bark. This creates a major avenue for disease transmission when propagating from cuttings if cuttings are taken from infected plants. It is important that nurseries use proper sanitation steps to reduce the number of infested plants. One of the best ways to begin those sanitation steps is to start with clean cutting material. With growing public concerns on chemical pesticides and their residues, irradiation is becoming a viable alternative and an effective nonchemical treatment for the control of several pathogens. Studies have shown successful results when gamma irradiation was applied to *Pseudomonas* spp., therefore we hypothesize that radiation could eliminate *P. amygdali* pv *loropetali* pv. nov. on loropetalum stock plants. Bacteria were subjected to six levels of x-ray irradiation 0, 0.5, 1, 1.5, 2, 2.5 kGy (0, 500, 1000, 1500, 2000, 2500 Gy). Initial results showed that x-ray treatment to pure bacteria strains resulted in significant bacterial reduction at all levels, with complete inactivity being observed in the 1.5, 2, and 2.5 kGy (1500, 2000, and 2500 Gy) treatments. With these preliminary findings, further studies are being conducted to determine the application of radiation's ability to clean up infected loropetalum plant material.

INTRODUCTION

Loropetalum, *Loropetalum chinense* (R. Br.) Oliv., is a popular landscape plant for USDA Hardiness zones 7 – 10 (Dirr, 2017). The plant is native to eastern Asia and can range in height from dwarf-trailing varieties to upwards of 3.1 m (10 ft). Flower colors can range from white to various shades of pink, purple, or red. The petals are "fringe-like" hence, it is commonly referred to as Chinese fringe flower. In the spring of 2012, in South Alabama, bacterial gall was observed on loropetalum with symptoms similar to observations made on loropetalum in central Alabama, North Carolina and Georgia in nursery and landscape plants (Conner et al., 2013).

Nursery growers reported dieback and plant death and upon further inspection, the plants with dieback had "galling and irregular dark callus formation on the lower stem and lower branches" (Conner et al., 2013). Bacterial colonies were isolated and after testing measures, the bacteria *Pseudomonas savastanoi* was identified as the cause of the galls; this was the first report of *P. savastanoi* causing bacterial gall on loropetalum (Conner et al., 2013). This bacterium will form galls which can cause girdling of the stem leading to reduced growth and eventually death of the plant. In 2018, it was determined, based on pathogenicity assays and molecular tests, that the gram-negative bacteria *P. amygdali* pv *loropetali* pv. nov., are causing the gall on loropetalum (Harmon et al., 2018).

Pseudomonas generally like moist, warm environments (Ramos et al., 2012). The bacteria can be spread by rain or overhead irrigation splashing on to other plants. The bacteria will infect the plant if it can penetrate through a cut or wound in the bark, from there it forms a gall. It is important that nurseries take proper sanitation steps to reduce the number of infested plants. Currently, the best sanitation method is to throw away plants that are contaminated and completely remove them from the nursery. If these plants remain and cuttings are taken from them, it could further spread the bacteria. Also, the areas on the stock plant where the cuts were made are also now open to re-infection. Pruners and other cutting tools should be cleaned and disinfected to prevent the spread of the bacteria from plant to plant. Improper sanitation of these instruments will only amplify the problem. Copes et al., (2019) observed that Clorox at 11% and Virkon S at 1.0% were able to completely eliminate *P. amygdali* pv *loropetali* pv. nov. on stainless steel, and Green-Shield II at 0.5% and KleenGrow at 0.8% nearly eliminated the bacterium on stainless steel. Based on this research, we know that there are "several disinfectants commercially available that can kill *P. amygdali* pv *loropetali* pv. nov. on production surfaces (Copes et al., 2019).

Pickens et al., (2019) reported that in 2013, several large nursery growers disposed over $1,000,000 worth of infected plant material. This disposal was either done by regulatory enforcement or was voluntary (Pickens et al., 2019). Bacterial diseases are difficult to control and currently there are no published recommendations for controlling this disease (Pickens et al., 2019). Auburn University recommends the use of copper treatments in the spring especially after pruning or other events causing damage to the bark (Pickens et al., 2019).

With growing public concerns on environmental and health issues stemming from chemical pesticides and their residues, and also the development of bacterial resistance, irradiation has become a viable alternative and an effective nonchemical treatment for the control of several pathogens (Hallman, 2011; Jeong et al., 2016).

Irradiation causes cellular damage, with direct effects and with indirectly generating reactive oxygen species, which would cease their metabolic functions (Jeong et al., 2016). The unit of irradiation dose is the gray (Gy), which is the energy absorbed in joule of radiation energy per kilogram of matter (Jeong et al., (2016). Gamma irradiation has been successfully observed to inhibit the growth of fungal and bacterial pathogens on fruits or vegetables, such as *Botrytis cinerea* in sweet pepper, *Penicillium purpurogenum* in pineapple, *Rhizopus stolonifer* in sweet potato, *Pseudomonas syringae* pv. *tomato* (*Pst*). in tomato, *Pseudomonas fluorescens* in baby spinach and romaine lettuce and *Monilinia fructicola* in peach (Damayanti et al., 1992; Jeong et al., 2015; Jeong et al., 2016; Kim and Yook, 2009; Olanya et al., 2015; Yoon et al., 2014). Applying gamma irradiation at 25 kGy (25000 Gy) is a common procedure for the sterilization of many food products and medical applications (Jones et al., 2010). McNamara et al., (2003) inferred that most bacteria are eliminated at levels from 15 to 25 kGy (15000 to 25000 Gy). Jeong et al., (2016) observed that gamma irradiation levels of 0.15 kGy (150 Gy) successfully controlled *P. syringae* pv. *tomato* (*Pst*). in tomato seedlings with no phytotoxicity observed. *Pseudomonas fluorescens* was controlled when levels were 0.04 to 0.05 kGy (40 to 50 Gy) and 0.05 to 0.06 kGy (50 to 60 Gy) in baby spinach and romaine lettuce respectively (Olanya et al., 2015). The same species of *Pseudomonas* was also controlled in radiated beef steaks at levels of 1.5 and 3.0 kGy (1,500 and 3,000 Gy) (Chung et al., 2000).

Based on the results of gamma irradiation applied to *Pseudomonas* in these studies, we hypothesize that radiation could eliminate *P. amygdali* pv *loropetali* pv. nov. on loropetalum stock plants. Successfully eliminating plant pathogens using radiation could save nurseries substantial money.

MATERIALS AND METHODS

Standard operating procedures for a sanitary microbiology laboratory were followed in order to encourage aseptic processes

(Plant Bacteriology Specialist, 2022). Suppression of bacterial colony formation could be observed if levels of contamination are high. Pure colonies of *Pseudomonas amygdali* pv *loropetali* pv. nov. were obtained from Dr. Kassie Conner, Auburn University, in March and June 2022. Dr. Conner verified the strain of bacteria using PCR. Colonies were maintained on Pseudomonas F (BD Diagnostic Systems) nutrient agar filled petri dishes. Liquid culture, utilizing Tryptic Soy Broth (TSB), were inoculated with colonies of *P. amygdali pv loropetali pv. nov.* and placed in a 25°C incubator 72 hours prior to irradiation.

The day of irradiation, the concentration of bacteria in the liquid culture was determined using a spectrophotometer (Evolution 60S UV-Visible Spectrophotometer). TSB was used as the blank in the spectrophotometer and the bacteria suspended in TSB were adjusted to an optical density of 0.302 at 420nm (OD_{420}). Based on calculations by Dr. Shien Lu, Mississippi State University, this would result in approximately $2x10^8$ cells per ml (Plant Bacteriology Specialist, 2022). Next, 10mL of the liquid culture were placed into petri dishes for irradiation in the Kimtron 350 X-ray (Kimtron Inc, Oxford, CT, USA). The Kimtron 350 utilized is a customized X-ray irradiator for food irradiation research (Wu and Chang, 2020).

Petri dishes were exposed to 6 levels of irradiation: 0, 0.5, 1, 1.5, 2, or 2.5 kGy (0, 500, 1000, 1500, 2000, or 2500 Gy). Each treatment contained three petri dishes of bacteria, with a total of 18 dishes irradiated (**Fig. 1**). Along with the three petri dishes, three alanine markers (dosimeters) were also placed with the samples to record the amount of radiation to which the samples were exposed (**Fig. 2**).

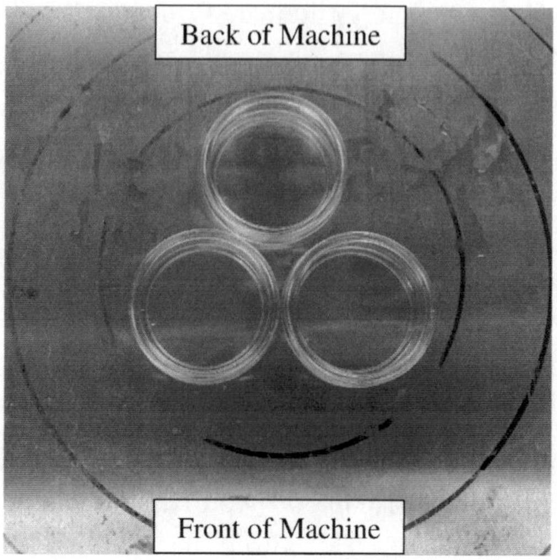

Figure 1. Plate orientation in Kimtron IC 350 X-ray machine.

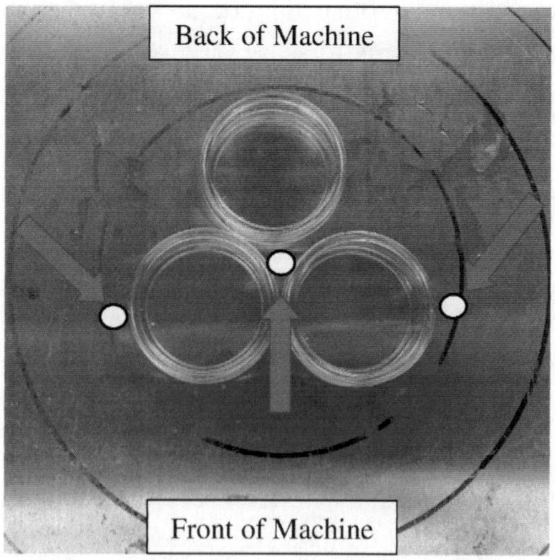

Figure 2. Representation of placement of dosimeters in Kimtron IC 350 X-ray machine.

After irradiation treatements, bacteria were transferred to 15mL tubes and placed in a 2°C (37°F) refrigerator for 24 hours. Tubes were then removed and six serial dilutions and six platings were done for each tube containing the irradiated liquid culture. Plates were then placed back in the incubator for cultures to grow for 72 hours at 25°C. Plates were removed from the incubator and colony forming units (CFUs) were counted using a Reichert Darkfield Quebec Colony Counter (RE-3325). Data were analyzed using SAS 9.4 PROC GLIMMIX and pairwise comparisons were done using Tukey's HSD where our alpha was 0.05.

RESULTS AND DISCUSSION

The levels chosen for this experiment were based on previous research. *Pseudomonas fluorescens* was controlled when levels were 0.04 to 0.05 kGy (40 to 50 Gy) and 0.05 to 0.06 kGy (50 to 60 Gy) in baby spinach and romaine lettuce respectively (Olanya et al., 2015). *Pseudomonas fluorescens* was also controlled in radiated beef steaks at levels of 1.5 and 3.0 kGy (1500 and 3000 Gy) (Chung et al., 2000). It was also reported to be controlled at irradiation levels of 1.2 - 2.3 kGy (1200 – 2300 Gy) (Frazier and Westhoff, 1988). In our experiment, we observed significant control of *P. amygdali* pv *loropetali* pv. nov. in all treatments compared to our non-irradiated control (**Table 1**; *P*<0.0001).

Our preliminary results are similar to reports by Jeong et al., (2016) who observed that CO_{60} gamma irradiation levels of 0.1 – 0.2 kGy (100 - 200 Gy) successfully controlled *P. syringae* pv. *tomato* (*Pst*). in tomato seedlings with no phytotoxicity observed. They observed that their treatment of 0.1 kGy (100 Gy) resulted in an approximate 5-log reduction of the viable count compared to the initial counts and complete inactivity at 0.2 kGy (200 Gy), which was their lethal dose (Jeong et al., 2016).

Using X-ray irradiation, we observed a 2.38-log reduction from our control colony counts compared to the counts of our first treatment of 0.5 kGy (500 Gy). At 1.5 kGy (1500 Gy) all bacterial colonies were rendered completely inactive. The differences in levels needed to inactivate our respective bacteria could be due to we were each experimenting with a different *Pseudomonas* sp. and sources of radiation, while similar, utilized different radiation sources.

Table 1. X-ray irradiation effect on average log of colony forming units.

Irradiation Level (kGy)	Mean Log CFU/mL^{z}
0 (Non-irradiated Control)	3.02 a
0.5	0.64 b
1.0	0.41 b
1.5	0.0 b
2.0	0.0 b
2.5	0.0 b
P-value^{y}	<0.0001

^{z}Analysis of variance was performed using PROC GLIMMIX (SAS 9.4). Means followed by the same letter are similar and not significantly different (α = 0.05). Each irradiation level had three replications. ^{y}P values for differences between means were obtained using Tukey's honest significant difference (HSD) at $P \leq 0.05$.

According to the EPA (2022), X-rays and gamma rays have the same basic properties, however, they are generated from different parts of the atom with X-rays being emitted from processes outside the nucleus, and gamma rays originating inside the nucleus. X-rays also are generally lower in energy and can be less penetrating than gamma rays, explaining why our required doses to reach bacterial inactivity were so much higher than Jeong et al., (2016) observed (EPA, 2022).

Knowing that we can irradiate and effectively control *P. amygdali* pv *loropetali* pv. nov. provides a foundation for future work examining radiation as a source of plant pathogen control. Further bacterial biological replications are currently being irradiated, as well as loropetalum cuttings, to provide us with further confirmation on optimal irradiation levels. If treatment with irradiation can provide growers with clean stock plants for cuttings, it is possible this treatment could provide a viable control option for loropetalum gall to limit economic losses for nurseries.

ACKNOWLEDGEMENTS

The authors thank Dr. Shin Lu, Mississippi State University, for providing guidance for plant bacteria laboratory procedures. The project was founded through the Mississippi State University William White Grant, supported by the Mississippi State University Division of Agriculture, Forestry, & Veterinary Medicine. This material is based upon work that is supported by the National Institute of Food and Agriculture, U.S. Department of Agriculture, Hatch project under accession number MIS-149190.

LITERATURE CITED

Chung, M.S., Ko, Y.T. and Kim, W.S. (2000). Survival of *Pseudomonas fluorescens* and *Salmonella typhimurium* after electron beam and gamma irradiation of refrigerated beef.

J. Food Protec. *63*: 162-166.

Conner, K.N., Olive, J., Zhang, L., Jacobi, J. and Putnam, M.L. (2013). First report of bacterial gall on *Loropetalum chinense* caused by *Pseudomonas savastanoi* in the United States. Plant Dis. *97*: 835-835.

Copes, W.E., Mavrodi, O.V. and Mavrodi, D.V. (2019). Control of *Pseudomonas amygdali* pv. *loropetali* on metal, wood, and *Loropetalum chinense* stem surfaces. Plant Health Prog. *20*: 270-277.

Damayanti, M., Susheela, K.H. and Sharma, G.J. (1996). Effect of plant extracts and systemic fungicide on the pineapple fruit-rotting fungus, *Ceratocystis paradoxa*. Cytobios-Cambridge. 155-166.

Dirr, M.A. (2016). Dirr's Encyclopedia of Trees and Shrubs. Timber Press, Portland, OR.

Environmental Protection Agency (EPA). 2022. Radiation Protection. (June 24, 2022). *Radiation Basics.* https://www.epa.gov/radiation/radiation-basics.

Frazier W.C. and Westhoff, D.C. (1988). Preservation by radiation. *Food Microbiology*. 4th ed. McGraw-Hill; New York, NY.

Hallman G. (2011). Phytosanitary applications of irradiation. Comprehensive Reviews in Food Science and Food Safety. *10*: 143–151.

Harmon, C.L., Timilsina, S., Bonkowski, J., Jones, D.D., Sun, X., Vallad, G.E., Sepulveda, L.R., Bull, C. and Jones, J.B. (2018). Bacterial gall of *Loropetalum chinense* caused by *Pseudomonas amygdali* pv. *loropetali* pv. nov. Plant Dis. *102*: 799-806.

Jeong, R.D., Chu, E.H., Shin, E.J., Lee, E.S., Kwak, Y.S. and Park, H.J. (2015). Antifungal effect of gamma irradiation and sodium dichloroisocyanurate against *Penicillium expansum* on pears. Letters in Appl. Microbio. *61*: 437-445.

Jeong, R.D., Chu, E.H., Lee, G.W., Park, J.M. and Park, H.J. (2016). Effect of gamma irradiation on *Pseudomonas syringae* pv. *tomato* DC3000-short communication. Plant Protection Sci. *52*: 107-112.

Jones, L.M., Grice, K.R.E. and Davis, R.I. (2010). Gamma irradiation of dried plant material: implications for the identification of plant pathogens employing molecular techniques. Australasian Plant Path. *39*: 358-362.

Kim K.H., and Yook H.S. (2009). Effect of gamma irradiation on quality of kiwifruit (*Actinidia deliciosa* var. *deliciosa* cv. Hayward). Radiation Phys. Chem. *78*: 414–421.

McNamara, N.P., Black, H.I.J., Beresford, N.A. and Parekh, N.R. (2003). Effects of acute gamma irradiation on chemical, physical and biological properties of soils. Appl. Soil Ecol. *24:* 117-132.

Olanya, O.M., Niemira, B.A. and Phillips, J.G. (2015). Effects of gamma irradiation on the survival of *Pseudomonas fluorescens* inoculated on romaine lettuce and baby spinach. LWT-Food Sci. Tech. *62*: 55-61.

J.W. (2019). Research updates on bacterial gall of loropetalum. Proc. South. Nurs. Assoc. Res. Conf. 63rd Annual Rpt.110-114.

Plant Bacteriology Specialist, (2022, August 17). Personal communication.

Ramos, C., Matas, I.M., Bardaji, L., Aragón, I.M. and Murillo, J. (2012). *Pseudomonas savastanoi* pv. *savastanoi*: some like it knot. Molec. Plant Path. *13*: 998-1009.

Wu, Y., and Chang, S. K. (2020). The efficacy of X-ray doses on *Vibrio vulnificus* in pure culture and *Vibrio parahaemolyticus* in pure culture and inoculated farm raised live oysters (*Crassostrea virginica*) with different acceleration voltages. Food Control *115*: 107277.

Yoon M., Jung K., Lee K.Y., Jeong J.Y., Lee J.W., and Park H.J. 2014. Synergistic effect of the combined treatment with gamma irradiation and sodium dichloroisocyanurate to control gray mold (*Botrytis cinerea*) on paprika. Rad. Physics Chem. *98*: 103–108.

Combined Proceedings IPPS 72:94-101. 2022.

Physiological Response of Wax Begonia to Heat and Light Stress

Julian Ginori[1], Heqiang Huo[1], Sandra Wilson[2]

[1]University of Florida, Department of Environmental Horticulture, Mid-Florida Research and Education Center, University of Florida, 2725 S Binion Road. Apopka, FL 32703, USA. [2]Department of Environmental Horticulture, P.O. Box 110670, Gainesville, FL 32611,USA

julian08@ufl.edu

Keywords: Stomatal conductance, carbon assimilation, transpiration, ion leakage, fluorescence

Summary

Wax begonia (*Begonia ×semperflorens-cultorum*) is a common ornamental plant used in flower beds for its diverse flower coverage to beautify public spaces. The intense Florida summers can increase its greenhouse production costs and hinder its year-round landscape potential, especially in full sun conditions. The physiological response of wax begonia to stress associated with heat, drought and light is not well understood but necessary for plant selection and improvement. Experiments were conducted to compare physiological plant responses (i.e., photosynthesis, fluorescence, and ion leakage) of four different wax begonia genotypes (FB08-059, OPGC 5104, 'Sprint White' and 'Cocktail Vodka') grown under light 35/22.5 °C or shade (30/22.5°C) conditions for 41 d. Results showed that when stressed (nonshaded and hot) FB08-059 (a noncommercial red genotype) had greater stomatal conductance (0.23 compared to 0.12-0.16 mol m^{-2} s^{-1}), greater chlorophyll fluorescence (0.7-0.75 compared to 0.45-0.64), and less ion leakage (11.91% compared to 20.34%-33.72%) than the other three genotypes. Results of this study combined with subsequent morphological data are foundational for breeding enhanced abiotic stress tolerance in wax begonia. Additional studies are in place to measure physiological stress and severe drought tolerance of these same genotypes.

INTRODUCTION

Commonly called wax begonia, *Begonia ×semperflorens-cultorum* refers to a group of cultivated hybrid begonias derived from *Begonia cucullata* and *Begonia schmidtiana* (Hvoslef-Eide and Munster, 2007; Neale et al., 2006). These fibrous-rooted begonias are primarily used as bedding plants for lining walkways, roads, and flowerbeds. The market for begonias has increased to over $130 million with production primarily in Florida (USDA, 2020). Plants thrive in cooler parts of the year but decline noticeably in the summer months of southern states due to heat stress. The development of new cultivars withstanding high temperatures and intense sunlight is much needed to expand the year-round utilization of wax begonia in warm climates of the U.S. and beyond.

With global temperatures expected to increase between 1.8 and 4 °C in the next eighty years (Hasanuzzaman et al., 2013), it is essential to better understand plant response to heat and intense UV light as a consequence of climate change. The negative effects of heat stress and elevated UV light on agricultural crop production are already evident, with significant yield losses that may lead to global food insecurity (Christensen and Christensen, 2007). Despite their economic and ecological importance, how heat stress and/or elevated UV light physiologically affect ornamental plants is less studied compared to row crops. Under stress conditions, physiological responses in plants are triggered to protect against stress-induced damage (Abdelmageed and Gruda, 2009). Carbon assimilation, indicative of photosynthetic efficiency, is highly sensitive to various environmental stresses. For example, Urban et al. (2017b) found that carbon assimilation of poplar (*Populus deltoides × nigra*) and loblolly pine (*Pinus taeda*) was significantly reduced under high temperatures (40 °C), a response due to increased stomata closure and photorespiration (Farquhar and Sharkey, 1982; Urban et al., 2017a). While heat stress responses of agronomic crops have been extensively studied, research on the effect of such environmental stress on ornamentals including the top bedding plant, wax begonia, is lacking. As such, the following study was conducted to understand heat stress responses in wax begonia. Specific objectives were to compare photosynthesis, chlorophyll fluorescence and ion leakage among four wax begonia genotypes grown under shade and non-shaded conditions.

MATERIALS AND METHODS

Plant Material and Light Treatments. Four different wax begonia genotypes, two with red foliage and two with green foliage, were propagated from cuttings and grown for ten weeks in a low light greenhouse (PAR value of 750 μmol m^{-2} s^{-1}) with day/night temperatures of 28/23 °C (**Fig. 1**). FB08-059 is a putatively heat-tolerant begonia with dark green/red foliage in low light environments (Pounders et al., 2015); OPGC 5104 is a bright-green wild begonia that was collected from Hawaii. 'Sprint White' and 'Cocktail Vodka' are two commercial cultivars with green and red foliage, respectively. Plugs were transplanted to 1-gal. pots filled with a bark and peat- based soilless media (Premium Nursery and Veg Mix, Reliable Peat Company, Leesburg, FL) supplemented with 5gm/qt 14N-14P-14K Osmocote slow- release fertilizer two weeks prior to the commencement of the experiment. The plants were randomly assorted into two experimental groups, non-shaded and shaded.

The non-shaded treatment had a peak PAR value of 2100 μmol m^{-2} s^{-1} and day/night temperatures of 35/22.5 °C while the shaded treatment PAR value peaked at 750 μmol m^{-2} s^{-1} with day/night temperatures of 30/22.5 °C. Relative humidity fluctuated between 65% and 100% depending on time of day in both treatment groups. Water was consistently available through either rain or above ground sprinklers to eliminate the effect of drought conditions on plant response.

Figure 1. Top and side views of wax begonia genotypes FB08-059 (A,B), 'Cocktail Vodka' (C,D), OPGC 5104 (E,F), and 'Sprint White' (G,H) utilized to examine stress tolerance under shaded and non-shade conditions. Images were taken ten weeks after initial cutting, prior to commencement of the study.

Measurements of Stress Response. Carbon assimilation, stomatal conductance, and transpiration were measured throughout the 41-d study using a LiCOR-6800 Portable Photosynthesis System (Lincoln, NE). After stabilization, three readings were taken (30 sec. apart) for each plant at peak daylight hours, between 12pm and 4pm, to ensure stability in the chamber. The Fv/Fm measurements were taken with an OS30p chlorophyll fluorometer (Opti-Sciences, Hudson, NH). All plants were measured two hours prior to dawn to allow the plant to enter a dormant state overnight prior to taking measurements. Readings were taken from three fully expanded leaves per plant, with which mean Fv/Fm was recorded. Leaves were handled gently to avoid any damage. For measurement of ion leakage, three leaf disks from each leaf sample were placed each in 5mL of deionized water for four hours. Initial readings for each leaf disc were taken using an Orion Star A215 conductivity meter (Waltham, MA). Second measurements were performed with the same samples after being autoclaved for 20 min. and allowing to cool for 15 min.

Statistical Analysis. A completely random block design with three replications for each block and four plants per genotype for each replication were utilized for each light treatment. Mean value with the standard error (n=12) was calculated for each measured trait at each time interval. Final data collected at day 41, were subjected to a two-way analysis of variance (genotype x light) and when appropriate significant means were separated using Tukey's honestly significant difference test at $P \leq 0.05$ (Agricolae package in R studio, Boston, MA). Images for leaf folding were taken with a Nikon and leaf angle determined using ImageJ. ROS accumulation leaf images were modified using GIMP (version 2.10) followed by quantification in ImageJ (Bethesda, MD).

RESULTS

Photosynthetic Parameters. The interaction between genotype x light was nonsignificant for carbon assimilation (P=0.1772), and significant for both transpiration (P=0.0025) and stomatal conductance (P=0.0006). For the non-shaded plants, response in photosynthesis was nonsignificant among genotypes, remaining steady at 9-10 µmol m^{-2} s^{-1} for much of the experiment (**Fig. 2A**).

Transpiration of non-shaded genotypes was also nonsignificant (**Fig. 2B**), yet stomatal conductance of FB08-059 (red, noncommercial genotype) was 1.6 times greater compared to all other genotypes (**Fig. 2C**).

When shaded, commercial genotypes had noticeably higher carbon assimilation by day 12 than noncommercial genotypes; and at 41 days 'Sprint White' assimilated 1.1 times more carbon than OPGC 5104 (Fig. 2A). Under the same shaded conditions, stomatal conductance was similarly high among the commercial genotypes ('Sprint White and 'Cocktail Vodka'), being 43% greater than the OPGC 5104 genotype (green, noncommercial) (**Fig. 2C**).

Chlorophyll Fluorescence. There was a nearly significant genotype effect (P=0.0519) and a strong significant light effect (P=0.0012) on Fv/Fm, with a nonsignificant genotype x light interaction (P=0.2512). At each time interval (days after treatment), chlorophyll fluorescence (Fv/Fm) was greater among genotypes when shaded compared to non-shaded treatments (**Fig. 2B**). At day 41, Fv/Fm of shaded genotypes was 1.08 times greater than non-shaded genotypes. For the duration of the experiment (days 3-41), Fv/Fm values were between 0.45 and 0.71 when non-shaded and 0.68 and 0.76 when shaded. Under direct sunlight and heat, the effect of these stressors on Fv/Fm was much more prominent, resulting in very low Fv/Fm at days 3-6 of the treatment (**Fig. 2B**). However, the Fv/Fm in non-shaded FB08-059 and OPGC 5104 plants gradually increased to levels comparable to the shaded treatment after 20 d.

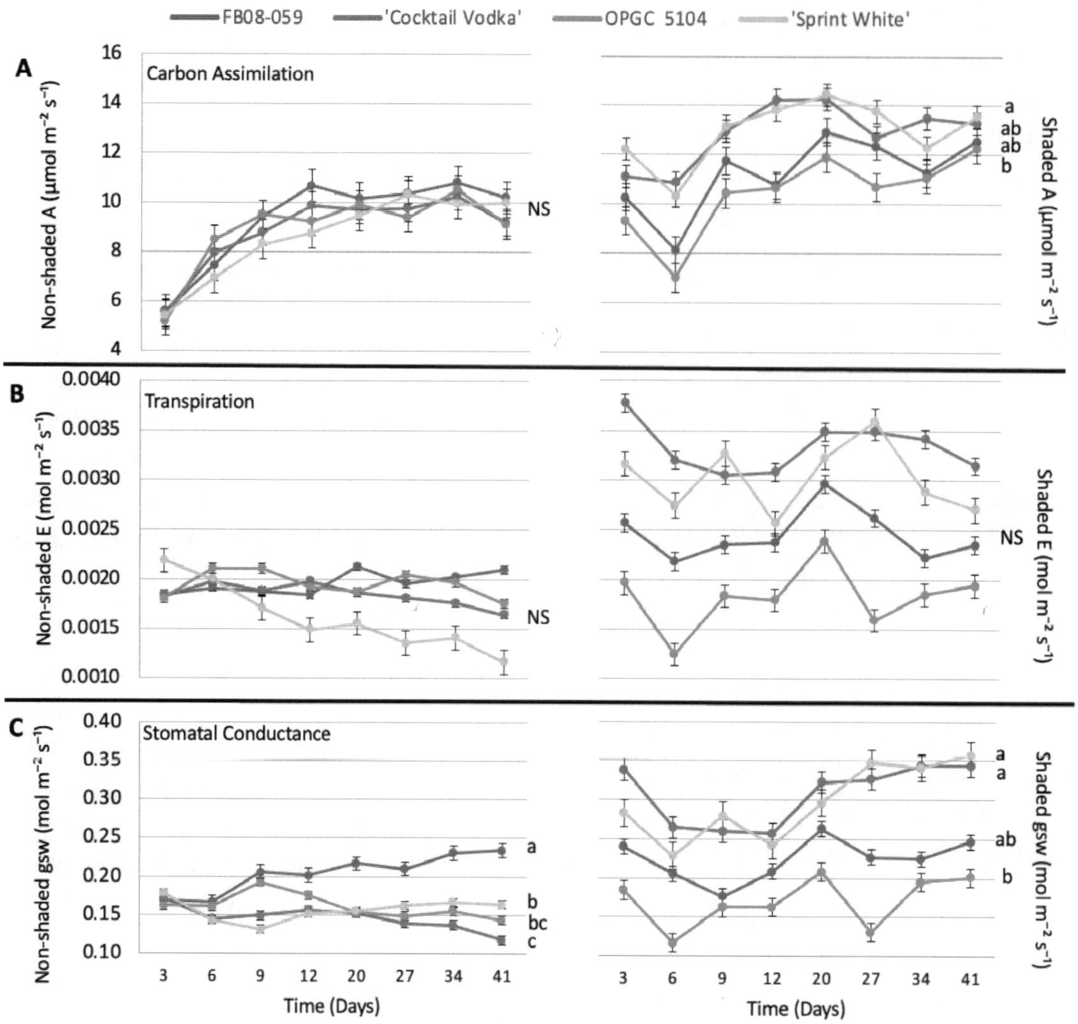

Figure 2. Carbon assimilation (A), transpiration (B), and stomatal conductance (C) responses of wax begonia genotypes (FB08-059, OPGC 5104, 'Sprint White' and 'Cocktail Vodka') grown under non-shaded (graphs to the left) or shaded (graphs to the right) conditions for 41 d. Plants were measured during the peak light and heat intensity between 12pm and 3pm. Bars represent mean ± standard error (n=12). Means followed by a different letter are significantly different according to a Tukey's HSD test, at $P \leq 0.05$.

Ion Leakage. Both light ($P<0.0001$) and genotype ($P<0.0001$) affected plant response to stress indicated by K+ leakage with a non-significant interaction (**Fig. 4**; P=0.1859). Percentages of ion leakage in non-shaded genotypes FB08-059 (11.91%) and 'Sprint White' (33.72%) were higher than their corresponding shaded treatments (9.65% and 21.49%, respectively), revealing the detrimental effect inflicted by direct sunlight and heat (**Fig. 4**). In general, non-commercial genotypes had less K+ leakage than commercial genotypes. In particular, FB08-059 had 2.32 times less K+ leakage when non-shaded, and 1.89 times less K+ leakage when shaded compared to all other genotypes in respective treatments.

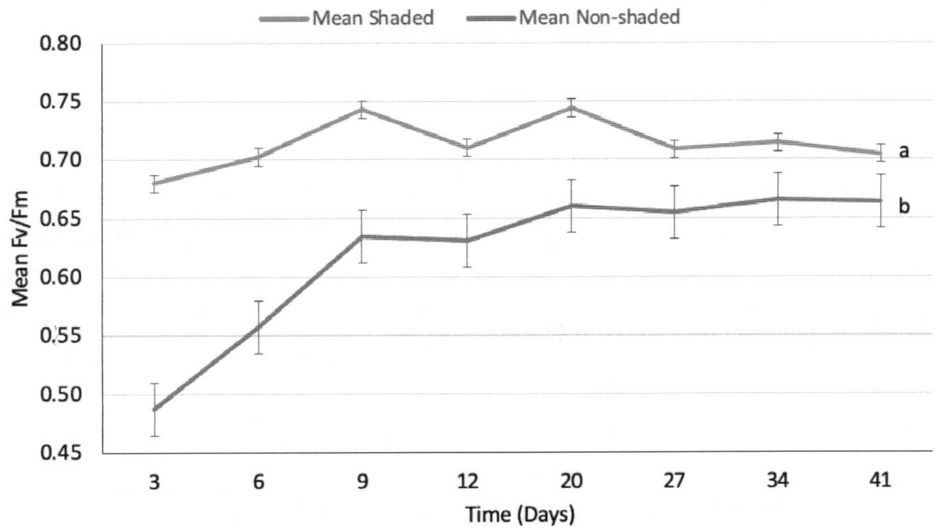

Figure 3. Mean chlorophyll fluorescence (Fv/Fm) responses of four wax begonia genotypes (FB08-059, OPGC 5104, 'Sprint White' and 'Cocktail Vodka') grown under shaded or non-shaded conditions for 41 d. Fv/Fm values at 41d did not differ by genotype (P=0.0519) and there was no genotype × light interaction (P=0.2512). Therefore, main treatment effects of light (P=0.0012) are reported for the combined genotypes. As a reference, Fv/Fm values above 0.70 are indicative of plants not under stress. Bars represent mean ± standard error (n=12). Means followed by a different letter are significantly different according to a Tukey's HSD test, at $P \leq 0.05$ for the day 41 measurements.

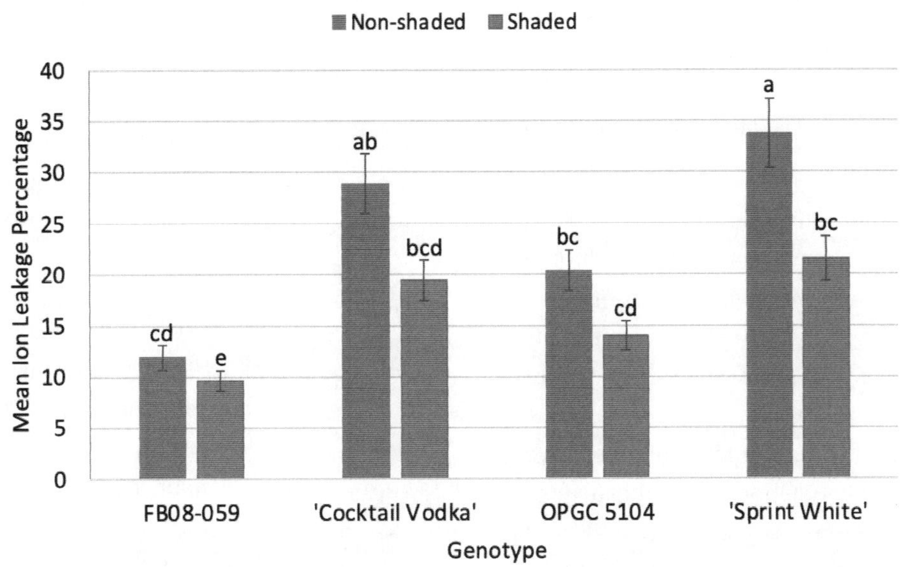

Figure 4. Ion leakage of four wax begonia genotypes (FB08-059, OPGC 5104, 'Sprint White' and 'Cocktail Vodka') grown under shaded or non-shaded conditions for 41 d prior to sample collection. Means followed by the same letter are not significantly different according to Tukey's HSD at $P \leq 0.05$. Bars represent mean ± standard error (n=12).

DISCUSSION

This study compared heat stress tolerance of four wax begonia genotypes grown under shaded and non-shaded conditions and found the non-commercial genotypes to be overall more tolerant to stress (light and heat) than the commercial genotypes. In particular, the FB08-059 genotype grown in full sun had higher stomatal conductance and less K+ leakage than other genotypes grown in the same conditions revealing its pronounced ability to tolerate stress. Comparatively, when shaded (non-stressed), 'Sprint White' (green commercial genotype) had higher carbon assimilation and stomatal conductance than OPGC 5104 (green, non-commercial genotype). It is of interest to consider the effect of foliage color on physiological responses in plants such as anthocyanin and chlorophyll content, cuticle thickness and leaf folding angle. Darker colored leaves of begonia have been shown to have enhanced photoprotective properties, enabling them to better withstand negative environmental stressors (Zhang et al., 2010).

Chlorophyll fluorescence is an important indicator of plant stress well before there are morphological responses. A Fv/Fm between 0.75-0.80 implies a plant is functioning at optimal performance, implying slight stress was present for the shaded treatment but not at a level that negatively impacted the functioning of the plants. It should be noted that, although Fv/Fm was nearly significant among genotypes (P=.0519) there was a general trend where the commercial red and green genotypes ('Sprint White' and 'Cocktail Vodka') had relatively low Fv/Fm values, suggesting they were less tolerant to stress than the red and green non-commercial genotypes (FB08-059 and OPGC 5104).

Ion leakage caused by irreversible membrane damage is another commonly used method to examine plant response to various stresses. In this study, genotypes responded similarly to light treatments where ion leakage was greater for non-shaded plants compared to shaded plants. Also, the non-commercial red and green genotypes (FB08-059 and OPGC 5104, respectively) generally had less ion leakage than the commercial red and green genotypes ('Sprint White' and 'Cocktail Vodka') indicating they were able to tolerate more stress. Interestingly, OPGC 5104 originated from the Hawaiian Islands where it grows in full sunlight and reproduces by seed freely. It's success in a similar environment could explain why this cultivar does not succumb to the same levels of stress response as its commercial counterpart.

CONCLUSION

The results presented herein show genotypic responses to abiotic stress under different environmental conditions. This data along with ongoing studies provide the necessary framework to uncover the physiological and morphological basis of abiotic stress tolerance in begonia, knowledge that is essential to plant breeding programs.

LITERATURE CITED

Abdelmageed, A.H.A. and Gruda, N. (2009). Influence of high temperatures on gas exchange rate and growth of eight tomato cultivars under controlled heat stress conditions. European J. Hort. Sci. *74*:152–159.

Christensen, J.H. and Christensen, O.B. (2007). A summary of the PRUDENCE model projections of changes in European climate by the end of this century. Climat. Chang. *81*:7–30.

Farquhar, G.D. and Sharkey, T.D. (1982). Stomatal conductance and photosynthesis. Annu. Rev. Plant Physiol. *33*:317–345.

Hasanuzzaman, M., Nahar, K., Alam, Md., Roychowdhury, R., and Fujita, M. (2013). Physiological, biochemical, and molecular mechanisms of heat stress tolerance in plants. Intl. J. Mol. Sci.*14*: 9643–9684.

Hvoslef-Eide, A.K. and Munster, C. (2007). Begonia: history and breeding. Flower Breeding Genetics 241–275.

Neale, S., Goodall-Copestake, W., Kidner, C.A. (2006). The evolution of diversity in begonia. Flor. Ornam. Plant Biotechnol. *4*:606–611.

Pounders, C.T., Sakhanokho, H.F., and Nyochembeng, L.M. (2015). *Begonia ×semperflorens* FB08-59 and FB08-163 clonal germplasm. HortScience *50*:145–146.

Urban, L., Aarrouf, J., and Bidel, L.P.R. (2017a). Assessing the effects of water deficit on photosynthesis using parameters derived from measurements of leaf gas exchange and of chlorophyll a fluorescence. Front. Plant Sci. doi: 10.3389/fpls.2017.02068.

Urban, J., Ingwers, M., McGuire, M.A., Teskey, R.O. (2017b). Stomatal conductance increases with rising temperature. Plant Sig. Behavior. doi:10.1080/15592324.2017.1356534.

[USDA] United States Department of Agriculture. (2020). Floriculture crops 2019 census data summary. <https://www.nass.usda.gov/Publications/Todays_Reports/reports/floran20.pdf> (Accessed 5 Aug. 2022).

Zhang, K.-M., Yu, H.-J., Shi, K., Zhou, Y.-H., Yu, J.-Q., Xia, X.-J. (2010). Photoprotective roles of anthocyanins in *Begonia semperflorens*. Plat Sci. 179:202–208.

Combined Proceedings IPPS 72:102-110. 2022.

Quantifying the Influence of Moisture Content on Bark Screening for Improved Particle Separation

Kristopher S. Criscione, Jeb S. Fields, Amanda Mizell

Louisiana State University (LSU) AgCenter Hammond Research Station, 21549 Old Covington Highway, Hammond, Louisiana 70403, USA

Kcrisc2@lsu.edu

Keywords: Greenhouse growing, growing media, material processing, nursery crops, pine bark, screening, soilless substrates

Summary

Historically, tree bark was regarded as a waste product of the timber industry for decades. After lumber harvesting and debarking, bark is often hammer-milled and screened to decrease its particle size for further use. This bark processing is impacted by many variables such as moisture content, which can influence the manufacturing and alter the final product. However, little work has been conducted to quantify moisture contents effect on bark screening. Thus, this study consisted of bark screened at five different moisture contents (50, 55, 60, 65, and 70%) and its yield were quantified and analysed. In general, as moisture content increased, bark that was processed through the screen (unders) had a decrease in yield on a volume and mass basis, whereas bark that did not process through the screen (overs), increased in volume and mass. More fine particles attached to the overs bark; however, this did not largely influence container capacity or air-filled porosity values. Hence, the drier the bark prior to screening resulted in more balanced particle separation.

INTRODUCTION

The timber industry has been heavily relied upon for fuel, building components, and energy, especially prior to the 1970's (Raviv and Lieth, 2008). Formerly, softwood tree species such as pine or fir, would be harvested and debarked, where the xylem wood would be used for lumber or pulp, and the bark discarded. However, bark accounts for approximately 10% of the tree volume (Bunt, 1988) and is often buried or burned (Naasz et al., 2009). Nevertheless, advances in research discovered more sustainable uses of lumber harvests, particularly in the bark sector. Bark has been known to be used for a variety of commodities, including wood planks and pulpwood for paper products (Harkin and Rowe, 1971), biofuels (Nosek et al., 2016), cork (wine bottles), and in the horticulture industry as a mulch and growing medium (Baker, 1957; Pokorny, 1979; Bunt, 1988).

Once the log is debarked, the bark initially is not suitable for use due to large particle sizes (Pokorny, 1983). Therefore, the bark requires further processing such as aging, hammer-milling, and screening. Previously, Pokorny and Delaney (1975) demonstrated that hammer-milled/screened bark-based substrates that contain a majority of bark particles > 0.60 mm provide suitable horticultural media. Currently, the stripped bark is hammer-milled through screen apertures that often range from 4.0 – 9.5 mm (Fain et al., 2008; Jackson et al., 2009).

With regards to horticultural bark-based substrates, particle size has a tremendous influence on the physical/hydraulic properties, which subsequently impacts plant performance (Fields et al., 2017). A bark with an unbalanced proportion of coarse particles results in insufficient water-holding characteristics, whereas increased percentages of fine bark in a substrate material leads to poor aeration; both of which can be deleterious to containerized crop growth and development (Mathers et al., 2007).

Successful and continued use of screened pine bark depends on consistency, reproducibility, and predicting the proportions of partitioned bark particles on each screen after processing (Pokorny and Henry, 1984). Though biological and mechanical factors can influence the efficiency of bark hammer-milling and screening (Solbraa, 1979), moisture content may have the largest impact on the proportions of bark particle separation (Stewart et al., 2019). Water has high surface tension which frequently results in water remaining adsorbed to bark particle surface areas and internal porosities (Raviv and Lieth, 2008). However, greater moisture contents enable bark particles to more readily "stick" to each other via adhesion/cohesion. Jackson et al. (2010) concluded that the bark moisture content at the time of hammer-milling/screening can influence particle partitioning, where increasing bark moisture contents can decrease the quantity of bark particles screened.

Research is sparse in quantifying particle separation via screening under different initial screening moisture contents. This presents opportunities to 1) identify suitable moisture contents for bark screening and 2) further understand how moisture content at the time of screening influences bark processing yield. We hypothesize that the particle separation efficiency will decrease as moisture content increases.

MATERIALS AND METHODS

Bark Moisture Content Preparation. Fifteen plastic bags were each filled with exactly 0.03 m^3 of aged loblolly pine (*Pinus taeda*) bark (Phillips Bark Processing Co; Brookhaven, MS, U.S.) and sealed shut to ensure no moisture loss. Thereafter, the moisture content (MC) of the collected bark was gravimetrically determined prior to bark screening by weighing, drying (105°C for 48 h), and reweighing four samples, resulting in a MC of 55% ± 0.01 SD. Therefore, five MC treatments were chosen: 50-, 55-, 60-, 65-, and 70% MC.

To estimate the dry weight of 0.03 m^3 of the pine bark samples, a porometer analysis (Fonteno and Harden, 2010) was conducted on three unscreened bark replicate samples. The total dry weight of the 0.03 m^3 samples were estimated by using bulk density (D_b) values (0.17 $g{\cdot}cm^{-3}$ ± 0.00 SD) and were calculated (4,919 g ± 59 SD). Subsequently, the quantity of water for each MC treatment required to be lost via evaporation (50% MC) or added (55, 60, 65, and 70% MC) was calculated and gravimetrically measured. Treatments contained target weights of 9,838 (50% MC), 10,931 (55% MC), 12,298 (60% MC), 14,055 (65% MC), and 16,397 g (70% MC). For the 50% MC treatment, the bark remained within the plastic bag and the bag was left open for evaporative demand to reduce the MC. The bags were continuously mixed and weighed until the desired weight was reached. For all other treatments, water was added to each bag and was equilibrated 72 h prior to screening. Each MC treatment contained three replicates ($5_{\text{MC treatments}}$ x $3_{\text{replicates}}$ = $15_{\text{total bags}}$). Once all MC treatments attained the desired weights, the treatments were prepared for screening.

Bark Processing. Once equilibrated and ready for processing, three small (~50 g) samples were randomly collected from each bag to measure the MC immediately before screening. The actual MC of the 50, 55, 60, 65, and 70% MC treatments were 52% ± 2 SD, 58% ± 2 SD, 61% ± 5 SD, 65% ± 1 SD, 68% ± 1 SD, respectively.

From each bag (replicate), 0.014 m^3 of bark was removed and placed in a 0.03 m^3 container and was immediately processed. The bark passed through a continuous flow screen (CF-1; Gilson Company Inc. Model; Lewis Center, OH, U.S.) fitted with a 6.3 mm aperture screen, set to 569 revolutions per minute, and screen level was maintained at 5° inclined slope. The bark particles passed through the screen at a rate of 56.21 $cm^3\,min^{-1}$.

Measurements. All bark that was passed through the screen will be referred to as 'unders' and all bark that did not pass through the screen will be referred to as 'overs' for the remainder of this paper. Multiple measurements were assessed during and immediately after each replicate of 0.014 m^3 of bark was processed in MC treatments: The time it took for all the bark to be fed through the screen, the mass (g) of the overs and unders, and the volume (m^3) of the overs and unders. After the bark replicate within each MC treatment was processed, the bark was placed in a plastic bag and sealed to prevent moisture loss.

Physical Properties. Substrate physical properties, [container capacity (CC), air space (AS), total porosity (TP), and D_b] were measured via porometers of each over and under replicate within all MC treatments after screening. Thereafter, each replicate within the MC treatments was measured for its particle size distribution by sieving three, 100 g dry substrate replicates through a Ro-Tap shaker (Rx-29; W.S. Tyler, Mentor, OH, U.S.) for five min with a column of stacked sieves with aperture sizes of 6.3, 2.0, 0.7, 0.5, 0.3, and 0.1 mm, with a catch pan at the bottom.

Data Analysis. All data presented in tables and figures with corresponding statistical analysis was analysed in JMP Pro (16.2.0; SAS Institute, Inc.; Cary, NC, U.S.) utilizing Analysis of variance (ANOVA) and Tukey's Honestly Significant Difference at the $\alpha = 0.05$ significance level. Pearson correlation coefficient values were also calculated in JMP Pro (16.2.0) to correlate screening parameters across different types of measured yield.

RESULTS AND DISCUSSION

Physical Properties. There were slight differences observed across CC values within unders, where 55% unders held more than 65% unders (**Table 1**; p = 0.0239). However, there were no differences observed in overs CC values (**Table 1**; p = 0.4820). All unders ranged within recommended nursery substrate standards for water holding capacities (0.45 – 0.65 cm^3 cm^3; Bilderback et al., 2013). However, overs were far below recommendations (< 0.40 cm^3 cm^3). This trend continued for AS, where recommended values range within 0.10 – 0.30 cm^3 cm^3 and all overs and the 65% unders exceeded suggested air-filled porosity values (Bilderback et al., 2013).

There were slight differences examined in unders TP values (p = 0.0079) and no differences in overs (**Table 1**; p = 0.2375;). However, a t-test of summarized overs against unders showed no differences between total porosity values (**Table 1**; p = 0.3947). Screening bark substrates can have strong impacts on air-filled porosity (p <0.0001) and water holding characteristics (p <0.0001; Table 1) simply by shifting the AS:CC ratio due to alterations in particle arrangement and surface area proportions, while typically having negligible effect on TP (Altland et al., 2011). This follows the fundamental geometric principle that a group of uniform spherical objects will always occupy 66.7% (vol.) of a cylindrical container (Jury and Horton, 2004), regardless of sphere volume. Though bark particles are relatively platy, this principle more-or-less follows the results herein, where similar findings were also observed by Fields et al. (2021). The 70% unders had the greatest D_b values (p = 0.0003), and after 60% MC in overs, D_b began to increase with increasing moisture (**Table 1**; p = 0.0162).

In this study, moisture content had relatively little influence on resultant bark physical properties (Table 1). However, it has been demonstrated that increasing proportions of fine particles can increase CC and decrease AS (Altland et al., 2018).

Among the PSD analysis, unders had no differences (p = 0.0646); however, in overs, as moisture content increased, extra-large particle percentages decreased (**Table 1**; p = 0.0007). These results were inverted for unders large (p < 0.0001) particles, and overs medium (p < 0.0001) and fine (p < 0.0001) particle diameter proportions, where, as moisture content increased, particle proportions also increased (**Table 1**). This is likely due to fine particles adhesively attaching to the bark particles at time of screening (Jackson et al., 2010). In bark fines, MC played key roles in both overs and unders as MC increased, where less fine particles were present in unders and more fine particles were present in overs (**Table 1**; p < 0.0001).

Table 1. Static physical properties and particle size distribution of screened pine bark substrates under different moisture contents.

Substrate	Partitioned Particles	Static Physical Properties [a] Container capacity $cm^3\ cm^{-3}$	Air space $cm^3\ cm^{-3}$	Total porosity $cm^3\ cm^{-3}$	Bulk density $g\ cm^{-3}$
Unscreened	-	0.32	0.35	0.66	0.17
50%	Overs	0.29 a [c]	0.43 a	0.73 a	0.15 ab
55%	Overs	0.32 a	0.47 a	0.79 a	0.16 ab
60%	Overs	0.34 a	0.49 a	0.82 a	0.15 b
65%	Overs	0.28 a	0.44 a	0.72 a	0.16 a
70%	Overs	0.40 a	0.44 a	0.84 a	0.16 a
P-value overs [d]	-	0.4820	0.1211	0.2375	0.0162
50%	Unders	0.55 ab	0.17 b	0.72 b	0.19 a
55%	Unders	0.57 a	0.15 b	0.72 b	0.19 a
60%	Unders	0.55 ab	0.22 b	0.77 ab	0.19 a
65%	Unders	0.47 b	0.35 a	0.82 a	0.17 b
70%	Unders	0.54 ab	0.22 b	0.76 ab	0.20 a
P-value unders	-	0.0239	<0.0001	0.0079	0.0003
P-value overs vs unders [e]		<0.0001	<0.0001	0.3947	<0.0001

Substrate	Partitioned Particles	Particle Size Distribution [b] Extra Large (>6.3 mm) %	Large (6.3–2.00 mm) %	Medium (2.00-0.71 mm) %	Fines (<0.71 mm) %
Unscreened	-	38.2	43.2	12.7	7.0
50%	Overs	59.3 a	37.3 a	1.3 b	2.5 c
55%	Overs	55.4 a	40.1 a	2.1 b	3.2 bc
60%	Overs	51.9 a	40.2 a	3.7 b	4.2 bc
65%	Overs	52.1 a	40.9 a	3.5 b	4.4 b
70%	Overs	38.9 b	41.7 a	11.9 a	9.0 a
P-value overs		0.0007	0.3908	<0.0001	<0.0001
50%	Unders	0.0 a	40.7 b	31.8 bc	27.0 a
55%	Unders	0.0 a	40.4 b	34.1 ab	25.2 a
60%	Unders	0.0 a	40.7 b	35.1 a	25.0 a
65%	Unders	0.4 a	53.3 a	33.2 ab	14.2 b
70%	Unders	0.6 a	50.4 a	29.1 c	21.0 a
P-value unders		0.0646	<0.0001	0.0007	0.0004
P-value overs vs unders [e]		<0.0001	0.0068	<0.0001	<0.0001

[a] Measured via porometer analysis. Total porosity = air space (minimum air-filled porosity after free drainage) + container capacity (maximum water holding capacity after free drainage). [b] Percent of total sample dry mass within the particle size range. [c] Letters denote detected differences amongst means separately (overs within overs; unders within unders) utilizing Tukey's HSD (α = 0.05). [d] Measures of overall treatment effects utilizing ANOVA analysis with a significance value of (α = 0.05) separately (overs within overs; unders within unders) utilizing Tukey's HSD (α = 0.05). [e] Measures of overall treatment effects utilizing ANOVA analysis with a significance value of (α = 0.05) separately (overs against unders) utilizing Tukey's HSD (α = 0.05).

The particle size distribution curve demonstrates that overs bark contains significantly fewer fine bark particles (> 0.71 mm) than unders bark proportions (**Fig. 1**). Moreover, a large gap exists between the two screened barks as % of bark particles below a particular diameter increases (**Fig. 1**).

Figure 1. Particle size distribution curve of screened bark at different initial moisture contents. Each error bar is constructed using a 95% confidence interval of the mean.

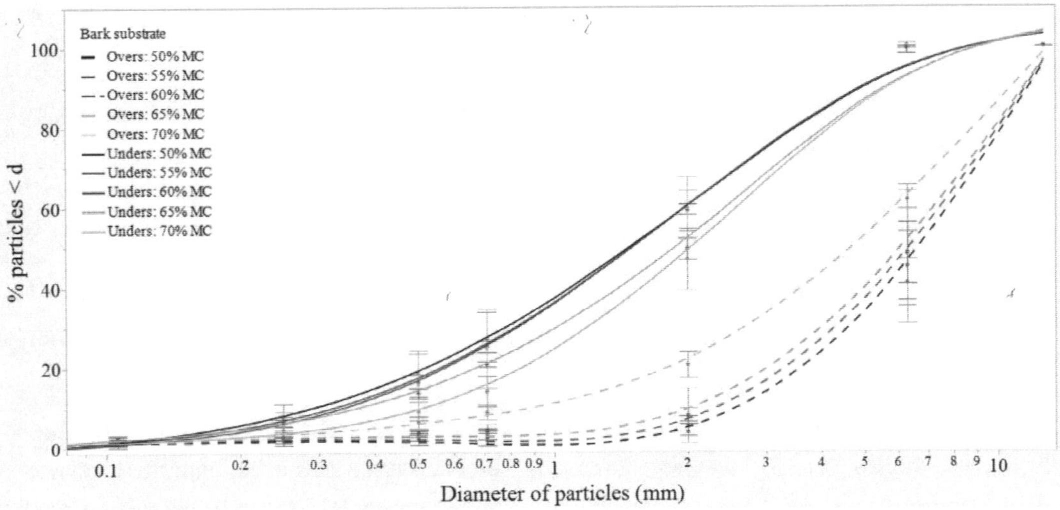

Uniquely, there was a concave down arrangement observed in the PSD bark fines unders, where particle proportions decreased with increasing moisture content, and then inverted at 65% MC (**Table 1**). A comparison of summarized unders against summarized overs values show significant differences, regardless of moisture content, of all particle diameter classifications (**Table 1**).

Screening. There was a strong correlation ($r = 0.7624$) between moisture content and the time it took to clear the screen after the final feed. Screening bark at 55% MC resulted among the fastest time to clear the aperture, while screening bark at 70% MC took the longest ($p < 0.0001$; Table 2). This is likely due to the increased proportions of medium and fine over particles blocking screen apertures (**Table 1**; Jackson et al., 2010).

Moisture content prior to screening played significant roles in processed bark output (**Table 2**). Generally, as moisture content increased, bark volume ($p < 0.0001$) and mass ($p < 0.0001$) decreased for unders (**Table 2**). These results were inverted for overs (**Table 2**; $p = 0.0049$). Furthermore, there were strong negative correlations between MC and volume ($r = -0.9171$) and mass ($r = -0.9386$) in under particles. Conversely, there were strong positive correlations between MC and volume ($r = 0.7941$) and mass ($r = 0.9383$) in over particles.

The more balanced bark separation on a volume or mass basis decreased as moisture content increased (**Table 2**), alluding that the drier the bark prior to screening will result in more particle separation.

Table 2. Screening parameters

Moisture Content	Partitioned Particles	Time to clear screen after final feed (sec)[a]	Volume (cm^3)	Mass (g)	Particle separation ratio (%; volume basis)[c]	Particle separation ratio (%; mass basis)[b]	Particle mass remaining on screen (g)
50%	Overs	15 bc	60.1 bc	2886.3 c	55 c	55 c	NA
55%	Overs	13 c	56.0 c	3106.0 c	57 c	59 bc	NA
60%	Overs	15 bc	65.6 abc	3805.0 c	64 bc	67 b	NA
65%	Overs	18 b	76.5 ab	5061.7 b	76 b	80 a	88.3 b
70%	Overs	43 a	79.2 a	6278.0 a	92 a	87 a	759.3 a
P-value	-	<0.0001	0.0049	<0.0001	<0.0001	<0.0001	<0.0001
50%	Unders	15 bc [d]	49.2 a	2339.0 a	45 a	45 a	-
55%	Unders	13 c	43.7 a	2182.5 a	43 a	41 ab	-
60%	Unders	15 bc	38.2 ab	1913.3 a	36 ab	34 b	-
65%	Unders	18 b	24.6 b	1240.0 b	24 b	20 c	-
70%	Unders	43 a	6.8 c	1043.7 b	8 c	13 c	-
P-value [e]	-	<0.0001	<0.0001	<0.0001	<0.0001	<0.0001	-
P-value overs vs unders [f]	-	-	<0.0001	<0.0001	<0.0001	<0.0001	-

The particle separation ratio (% of partitioned particles relative to the total mass or volume) based off both volume ($p < 0.0001$) and mass ($p < 0.0001$) was lowest in 70% MC unders and conversely greatest in 70% MC overs (**Table 2**). Jackson et al. (2010) reported that the greater the moisture content results in decreased screened bark proportions, which is parallel to the results herein (**Table 2**). In opposition to the results evidenced in this study, Fields et al. (2017) screened bark with a 4-mm aperture at 66.4% MC and received practically an equal (~50%) partition by volume. This is further validation that there are several variations in bark that can affect processing (Kaderabek et al., 2016; Stewart et al., 2019).

After the bark processing was completed, the remaining particles on the screen were collected and weighed. No particles remained on the screen in overs ≤ 60% MC treatments (**Table 2**). However, in both the 65% and 70% MC treatments, 2% ± 0 SD and 10% ± 0 SD of the total bark mass screened remained on the aperture, respectively (Table 2). This is likely due to more fine particles are adhered to larger coarse bark (Table 1), which creates a bark particle obstruction, blocking other bark particles from being partitioned. This phenomenon was demonstrated to be more pronounced as the bark had greater MC (**Table 2**).

CONCLUSION

It is evident that bark moisture content prior to screening is a key factor in bark processing output, affecting the final product on both a volume or mass basis. While initial moisture content had minimal impacts on the physical properties of the growing substrates; there were significant effects on particle separation ratios. The greatest partition of particles occurred when the bark was processed as lower moisture contents (i.e., 50%). Future research should identify an optimal MC range for bark particle processing with minimal hydrophobicity concerns.

LITERATURE CITED

Altland, J.E., Owen, J.S., and Gabriel, M.Z. (2011). Influence of pumice and plant roots on substrate physical properties over time. HortTech. *21*:554–557.

Baker, K.F. (1957). The U.C. System for producing healthy container-grown plants. Ed. California Agric. Expt. Sta. Extension Service Manual 23, Univ. California, College of Agriculture Publication, Berkeley, California U.S.A.

Bilderback, T.E., Owen, J.S., Altland, J.E., Fain, G.B, Jackson, B.E. Riley, E.D., Kraus, H.T., Fonteno, W.C. (2013). Strategies for developing sustainable substrates in nursery crop production. Acta. Hort. *1013*:43-56.

Bunt, A.C. (1988). Media and mixes for container-grown plants. 2nd ed. Unwin Hyman Ltd, London, England.

Harkin, J.M. and Rowe, J.W. (1971). Bark and its possible uses. (Research note) Forest Products Lab. 091:56p.

Fain, G.B., Gilliam, C.H., Sibley, J.L., and Boyer, C.R. (2008). Wholetree substrates derived from three species of pine in production of annual vinca. HortTech. *18*:13–17.

Fields, J.S., Owen, J.S., and Scoggins, H.L. (2017). The influence of substrate hydraulic conductivity on plant water status of an ornamental container crop grown in suboptimal substrate water potentials. HortSci. *52*:1419-1428.

Fields, J.S., Owen, J.S., and Altland, J.E. (2021). Substrate stratification: layering unique substrates within a container increases resource efficiency without impacting growth of shrub rose. Agronomy 11, 1454. https://doi.org/10.3390/agronomy11081454

Fonteno, W.C. and Harden, C.T. (2010). North Carolina State University Horticultural Substrates Lab Manual. North Carolina State University.

Jackson, B.E., Wright, R.D., and Seiler, J.R. (2009). Changes in chemical and physical properties of pine tree substrate and pine bark during long-term nursery crop production. HortSci. *44*:791-799.

Jackson, B.E., Wright, R.D., and Barnes, M.C. (2010). Methods of constructing a pine tree substrate from various wood particle sizes, organic amendments, and sand for desired physical properties and plant growth. HortSci. *45*:103-112.

Jury, W.A. and Horton, R. (2004). Soil Physics; Wiley: Hoboken, NJ, USA.

Kaderabek, L., Jackson, B.E., and Fonteno, W. (2016). Changes in the physical, chemical, and hydrologic properties of pine bark over twelve months of aging Comb. Proc. Intl. Plant Prop. Soc. *41*:313-317.

Mathers, H.M., Lowe, S.G., Scagel, C., Struve, D.K.. and Case. L.T. (2007). Abiotic factors influencing root growth of woody nursery plants in containers. HortTech. *17*:151-162.

Naasz, R., Caron,J., Legault, J., and Pichette, A. (2009). Efficiency factors for bark substrates: biostability, aeration or phytotoxicity. Soil Sci. Soc. Am. J. *73*:780-791.

Nosek, R., Holubcik, M., and Jandack. J. (2016). The impact of bark content of wood biomass on biofuel properties. BioRes. *1*:44-53.

Pokorny, F.A. (1979). Pine bark container media - an overview. Comb. Proc. Inter. Plant Prop. Soc. *29*:484-494.

Pokorny, F.A. (1983). Pine bark as a soil amendment [Chemical properties, advantages, use in forest tree nurseries]. Technical publication R8 TP USDA Forest Service, Southern Region.

Pokorny, F.A. and Delaney,S. (1975). Synthesizing a pine bark potting substrate from component particles. Proc. SNA Res. Conf. 20th Annu. Rpt. 24-25.

Pokorny, F.A., and Henny, B.K. (1984). Construction of a milled pine bark and sand potting medium from component particles ii. Medium synthesis. J. Amer. Soc. Hort. Sci. *109*:774-776.

Raviv, M. and Leith, J.H. (2008). Soilless culture theory and practice. Elsevier, San Diego, CA.

Solbraa, K. (1979). Composting of Bark. I. Different bark qualities and their uses in plant production. Med. Fra Norsk Instituut for. Skogforskning *34*:285-323.

Stewart, C.J., Marble, S.C., Jackson, B., Pearson, B.J., Wilson, P.C., and Lauer, D.K. (2019). Influence of pine bark substrate age on performance and leaching of nursery preemergence herbicides. HortSci. *54*:896-902.

Combined Proceedings IPPS 72:111-118. 2022.

Seed Germination and Cryopreservation of Wild Lime (*Zanthoxylum fagara*)

Lindsay Mikell[1], Sandra B. Wilson[1], S. Christopher Marble [2], Wagner Vendrame[1] and Edzard van Santen[3]

[1]University of Florida, Department of Environmental Horticulture, 2550 Hull Road, Fifield Hall PO Box 110670, Gainesville, FL 32611; [2]University of Florida, Department of Environmental Horticulture, Mid-Florida Research & Education Center, 2725 S Binion Rd, Apopka, FL 32703; [3]IFAS Statistical Consulting Unit and Department of Agronomy, University of Florida, P.O. Box 110500, Gainesville, FL 32611

mikell.lindsay@ufl.edu

Keywords: Rutaceae, native plants, plant propagation

Summary

Wild lime *Zanthoxylum fagara* (L.) Sarg., is a small tree having attractive evergreen foliage and fragrant, yellow flowers. Native to Florida and Texas, this drought tolerant species is sought after for use in butterfly gardens in warm climates or conservatories. Yet, its nursery availability and ornamental use remain limited, as propagation protocols are largely unknown. To determine if seed propagation is a reliable and efficient means of producing wild lime, a series of studies were conducted in incubators to evaluate the effects of seed origin, temperature, dormancy, and cryopreservation on germination. Initial x-ray analysis and seed viability tetrazolium (TZ) tests determined the majority of seeds were filled (90-98%) and moderately viable (86-87%), regardless of the location they were collected from. However, seeds collected from northcentral Florida (Gainesville) and south Florida (Miami) responded differently to temperature treatments deployed to mimic spring (29/19 °C), summer (33/24 °C), fall (27/15 °C) and winter (22/11 °C) conditions. After 8 weeks, northcentral FL seeds had 10.7-41.1% germination, with seeds in the coolest temperature (winter) having the lowest germination, whereas south FL seed germination ranged from 30.2-71.2% across temperatures with the lowest germination occurring in the warmest temperature (summer). Additionally, seeds were found to imbibe regularly (do not require scarification) and tolerate cryopreservation procedures for long-term storage but possess physiological dormancy that must be overcome prior to germination.

INTRODUCTION

Wild lime (*Zanthoxylum fagara* (L.) Sarg.,) exhibits a suite of characteristics that may warrant its widened use for ornamental and ecologically friendly landscaping (**Fig. 1A-H**). Native to Florida and Texas in the U.S., this small-sized tree belongs to the citrus family (*Rutaceae*) and is a host plant to several swallowtail butterfly species (FNPS, 2022). Wild lime is adaptable to different landscape conditions tolerating partial shade to sun, moderate salt spray, periods of drought, and cold hardiness in zones 8b-11. In the spring, this species boasts fragrant, yellow flowers occurring in the leaf axils (**Fig. 1C**) of separate male and female plants that are proceeded by one-seeded, long-stalked follicles (**Fig. 1D-G**) attractive to birds.

Figure 1. Ornamental and reproductive traits of wild lime featuring: (A) glossy, odd pinnately compound leaf with winged rachis, (B) woody growth habit, (C) inflorescences in yellow axillary panicles and stems with short, recurved stipular spines, (D) immature, glandulous fruit, (E) mature seeds in dehiscent capsules, (F) x-ray at 2000x showing embryo presence or absence, (G) globose, black seeds, and (H) swallowtail butterfly visiting host plant.

Consistent with poor germination of other closely related *Zanthoxylum* species (USDA Forestry Services, 2008), the germination of wild lime is irregular and sporadic. Seeds are likely to possess physiological dormancy (Datt et al., 2017; Patade et al., 2020), with unknown ability to tolerate storage. Physiological dormancy is described as an inhibiting mechanism in the embryo that prevents radicle emergence (Baskin and Baskin, 2014). Practices to alleviate physiological dormancy include moist-chilling stratification, after-ripening, or application of gibberellic acid to replace a chilling requirement (Davies et al., 2018). Physical dormancy in comparison, is a condition when seeds fail to germinate because they are impermeable to water. To relieve this type of dormancy, scarification practices must be employed to mechanically or chemically abrade the seed coverings to allow for water uptake (imbibition) (Davies et al., 2018). In addition to dormancy, seed longevity (storability) is yet another factor that may influence germination response over time. Seeds that can tolerate drying may be ideal candidates for cryopreservation, a long-known technique used to maintain germplasm at ultra-low temperatures (Vendrame and Faria, 2011). With appropriate protocol development, cryopreservation could be a potential means of preserving seeds of wild lime for long periods of time and ultimately broadening its year-round availability for nursery production.

The overall goal of this study was to develop efficient and reliable methods to sexually propagate wild lime and subsequently increase its availability and use. A series of three studies was conducted with specific objectives to determine if: 1) germination differs by seed origin and temperature, 2) dormancy can be overcome by seed treatments and 3) seeds can be cryopreserved for long term storage.

MATERIALS AND METHODS

Effects of origin and temperature on seed germination. On 27 Sept. 2021 and 31 Aug. 2021 wild lime seeds were harvested from two locations, the first from a cultivated population in northcentral Florida (Gainesville, FL) and the second from a natural population in south Florida (Miami, FL). Seeds were stored in paper bags at room temperature (22-25 °C) for 2-3 weeks prior to experimentation. A subsample of seeds from each location was examined using an Ultra Focus x-ray system with embryo fill calculated using Faxitron Vision software (US Forest Service National Seed Laboratory, Dry Branch, GA). Two replicates of 100 seeds were then cut laterally and stained overnight at 37 °C in a 1.0% TZ (2,3,5-triphenyl-2H-tetrazolium chloride) solution in accordance with the Association of Official Seeds Analysts (AOSA) rules for TZ testing (Iseldy, 2016). Seeds were considered viable when firm embryos stained evenly red under 10× magnification. From each location, an additional four replicates of 50 seed (experimental unit) were subjected to germination tests conducted in Percival I30VL light and temperature-controlled incubators set to mimic spring (29/19 °C), summer (33/24 °C), fall (27/15 °C) and winter (22/11 °C) alternating temperatures in Florida with a 12 hr photoperiod. Seeds were surface sterilized with a 10% bleach solution (0.75% a.i. NaClO) for 10 min., triple rinsed, and soaked overnight in sterile deionized (DI) water prior to placing in 10.9 x 10.9 cm transparent polystyrene germination boxes lined with two sheets of moistened white blotter paper beneath one sheet of unbleached crepe germination paper. Germination was checked three times a week and recorded as the first sign of radical emergence for a period of 6 months.

Effects of seed scarification and pre-hormone treatment on imbibition and germination. Using the seeds from northcentral FL a seed imbibition study was conducted using four replicates of 25 seeds subjected to one of three treatments. The first subsample of seeds was mechanically scarified using coarse grit sandpaper and then soaked overnight in DI water. The second set of seeds was also mechanically scarified but soaked in 500 mg/L gibberellic acid (GA_3) for 24 hr. The third set of seeds served as the control and was only soaked overnight in DI water for 24 hr. Dry weight (W_0) of each replicate was recorded prior to soaking and then wet weight (W_1) was recorded in 12-hr increments until the experiment ended after 336 hr (2 wk). Increase in fresh weight was calculated using the following equation: $[(W_1 - W_0)/ W_0] \times 100$.

A second study was conducted using four replications of 50 seeds that were either scarified with coarse grit sandpaper or treated with GA_3 and kinetin at two different rates (200 mg/L GA_3 + 100 mg/L kinetin or 400 mg/L GA_3 + 200 mg/L kinetin). Seeds were soaked overnight prior to placement in germination boxes arranged in incubators set at 29/19 °C with a 12-hr photoperiod. Germination was observed three times a week and recorded for 6 months.

Effects of cryopreservation on seed germination. Seeds from the northcentral FL origin (7 wk post collection) were placed in a desiccator to determine initial moisture content. Four replications of 50 seeds (northcentral FL origin) were subjected to five different cryopreservation treatments and included a non-treated control. The control was not immersed in a plant vitrification solution 2 (PVS2) nor liquid nitrogen (LN). Cryopreservation treatments consisted of seed 1) immersed in PVS2 but not LN, and 2) immersed in LN but not PVS2, 3) immersed in PVS2 followed by LN, 4) pre-cooled with ice prior to immersion in PVS2 and then LN, and 5) pre-cooled with ice, immersed in a solution of PVS2 plus 1.0% phloroglucinol (PG) followed by LN. Both the control and the PVS2 only (treatment 1) were held at room temperature for 72 hours. After 72 hr in LN (treatments 2-5) cryotubes were removed and rapidly reheated in a water bath held at 39.7 °C. Seeds were rinsed and soaked overnight prior to placing in incubators set to 29/19° C. Germinates were recorded weekly for 6 months and presented at 1-month intervals.

Statistical analysis. Germination data were analyzed using Generalized Non-Linear Models procedures as implemented in SAS PROC NLMIXED (SAS/STAT 14.1; SAS Institute, Cary, NC) through a 3-parameter logistic growth model

$$Proportion\ germinated = \frac{c}{(1 + Exp(-a \cdot (DAS - b)))}$$

where a = growth rate, b = inflection point, c = asymptotic final germination, and DAS = Days after the start of the experiment. Monthly means were predicted from the fitted curve and treatments and or parameters compared using pairwise t-tests.

RESULTS AND DISCUSSION

Seed viability and germination. Seeds from both origins (northcentral and south FL) had similarly high embryo fill (90-98%) and pre-germination viability (86-87%) revealing that both seed populations were of similar quality (P =0.5425) at the start of experimental treatments (**Table 1**). When seeds were placed in incubators set to mimic seasons in Florida (29/19, 33/24,

27/15 and 22/11 °C), a significant origin × season interaction occurred ($P \leq 0.0001$) revealing that seeds from each origin did not respond similarly to temperature. As such, seeds from northcentral FL reached 10.7-41.1% germination among all temperatures with winter showing the lowest germination, and nonsignificant differences observed between all other seasons (**Table 1**).

Table 1. Initial embryo fill was calculated using x-ray software to non-destructively determine presence or absence of embryos in seed, viability was based on 24-hr tetrazolium (0.1%) staining of laterally cut seeds was cut laterally of a subsample of wild lime seeds collected from northcentral and south Florida prior to germination test. Least squares germination means from a generalized linear mixed model analysis 35, 45 and 55 Days after the start of the experiment conducted in incubators set to mimic summer (33/24 °C), spring (29/19 °C), fall (27/15 °C) and winter (22/11 °C) seasons.

	Days after start (DAS)		
Origin/Season	35	45	55
North Central FL			
Spring	23.9 (16.9, 32.8) a[z]	28.2 (19.2, 39.3) a	28.9 (20.9, 38.4) a
Summer	26.0 (18.0, 35.9) a	37.6 (26.5, 50.2) a	32.9 (23.8, 43.5) a
Fall	24.0 (16.7, 33.3) a	34.1 (23.8, 46.2) a	41.1 (31.4, 51.6) a
Winter	7.6 (4.5, 12.7) b	8.9 (4.9, 15.7) b	10.7 (6.5, 17.0) b
	Embryo fill = 90%	Initial Viability = 86%	
South FL			
Spring	56.0 (50.7, 61.2) b	61.3 (56.6, 65.7) b	62.0 (56.7, 67.0) b
Summer	21.3 (17.1, 26.2) c	26.5 (22.3, 31.1) c	30.2 (25.1, 35.9) c
Fall	67.5 (61.4, 73.0) a	70.3 (65.0, 75.0) a	71.2 (65.4, 76.4) a
Winter	52.4 (47.2, 57.6) b	56.2 (51.6, 60.7) b	59.3 (54.1, 64.2) b
	Embryo fill = 98%	Initial Viability = 87%	

Source	P > F
Origin	≤ 0.0001
Season	≤ 0.0001
Origin*Season	≤ 0.0001
DAS	≤ 0.0001
Origin*DAS	0.3891
Season*DAS	0.692
Origin*Season*DAS	0.2954

[z] Means within origin and time (DAS) followed by the same letter are not statistically different at $\alpha = 0.05$.

Whereas seeds from south FL reached similar germination percentages between 59.3-71.2% at 29/19, 27/15 and 22/11 °C. While germination in the fall temperature (27/15°C) was significantly higher than spring and winter (29/19 and 22/11 °C), they were on average 2.0-2.4 times greater than the 33/24 °C (summer) treatment (**Table 1**). This suggests that seeds possess some level of dormancy that cannot be overcome passively by these fixed temperatures in an 8-wk study. Further, seeds collected from warmer climates of south FL preferred cooler temperatures of spring, winter and fall for optimal germination.

Results from the imbibition study showed that sandpaper scarified seeds with or without GA_3 imbibed similarly to non-scarified seeds (P=0.6574) with fresh mass increases ranging from 11.7 to 13.8% among treatments (Data not presented). This suggests seeds of wild lime can absorb water normally and do not have physical dormancy. Further, when seeds were subjected to scarification or hormonal pretreatments, after one month germination was 4.9 times greater when seeds were treated with 400 mg/L GA_3 and 200 mg/L kinetin compared to seeds treated with sandpaper (**Table 2**).

Table 2. Mean germination percentage and rate (T50FG) of mechanically scarified (seeds were placed between coarse grit sandpaper sheets and gently rubbed until visibly abraded), hormonally pretreated [seeds were soaked for 24 hr in gibberellic acid (GA_3) and kinetin at two rates] and a non-treated control for wild lime seeds placed in incubators set at 29/19 °C with a 12-hr photoperiod for 6 months.

Month after treatment	Sandpaper scarification	200 mg/L GA3 + 100 mg/L kinetin	400 mg/L GA3 + 200 mg/L kinetin	Control (non-treated)
	Germination percent (%)			
1	5.5 (4.6, 6.4) d[z]	13.5 (11.5, 15.6) b	26.7 (24.1, 29.3) a	10.1 (8.6, 11.5) c
2	24.3 (22.0, 26.6) b	31.1 (29.6, 32.5) a	30.4 (29.2, 31.5) a	23.2 (21.2, 25.1) b
3	34.1 (32.7, 35.4) a	32.8 (31.4, 34.2) a	30.4 (29.2, 31.6) a	26.7 (25.4, 28.0) b
4	35.2 (33.6, 36.9) a	32.9 (31.4, 34.4) a	30.4 (29.2, 31.6) a	27.1 (25.5, 28.7) b
5	35.3 (33.6, 37.1) a	32.9 (31.4, 34.4) a	30.4 (29.2, 31.6) a	27.1 (25.5, 28.8) b
6	35.3 (33.6, 37.1) a	32.9 (31.4, 34.4) a	30.4 (29.2, 31.6) a	27.2 (25.5, 28.8) b
	Number of days until 50% of final germination (T50FG) was reached			
	51 (47, 55) a	34 (30, 37) b	21 (18, 23) c	38 (32, 43) b

[z] Means within a row followed by the same letter are not statistically different at $\alpha = 0.05$.

However, this effect diminished as germination time increased. After 6 months, germination of non-treated control seeds was the lowest at 27.2%, while germination amongst seed treatments was non-significant and ranged from 30.4-35.3% (treated with sandpaper or GA + kinetin). More pronounced was the effect of treatments on germination rate. Seeds treated with the higher 400 mg/L GA_3 + 200 mg/L kinetin solution took the least amount of time (21 d) to reach 50% of their final germination compared to 34 d for seeds that were treated with half the concentration (200 mg/L GA_3 + 100 mg/L kinetin) and 51 days for seeds that were scarified. Gibberellins have long been known to stimulate germination by increasing hydrolytic enzyme activity in seeds and embryo growth potential (Baskin and Baskin, 2014). The positive effect of GA_3 and kinetin on germination percentage and speed is consistent with Datt et al. (2017) who found that seeds from the closely related Timroo (*Zanthoxylum armatum*) responded better to this combination of hormones than any other pre-treatment.

Seed cryopreservation. Initial percent moisture of seeds was 6.7%, considered acceptable for cryopreservation (Davies et al., 2018). Seeds that germinated after precooling on ice before submersion into solutions of PVS2, or PVS2 and 1% Phloroglucinol, then immersed into LN were statistically significant from the control (**Fig. 2**), revealing that seeds can be cryopreserved successfully using either method. This also reveals that seeds are tolerant of maturation drying and therefore considered orthodox, a benefit to commercial propagation. It is plausible that collection time and storage length influenced seed deterioration of wild lime.

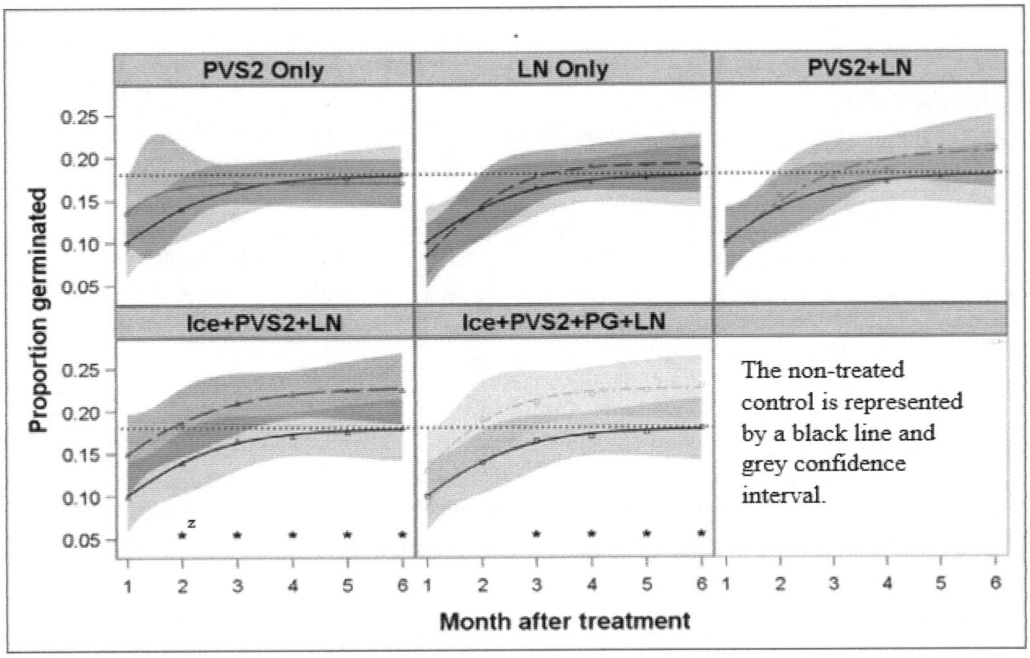

Figure 2. Mean germination over a 6-month period and approximate 95% prediction intervals estimated from a 3-parameter logistic growth mixed model as implemented in SAS® PROC NLMIXED (SAS/STAT 14.1; SAS Institute, Cary, NC). The experiment consisted of a non-treated control and five treatments. The control had no immersion in plant vitrification solution 2 (PVS2) nor liquid nitrogen (LN), and treated seeds were either immersed in PVS2 but not LN, immersed in LN but not PVS2, immersed in PVS2 then LN, pre-cooled on ice for one hour, immersed in PVS2 and then LN, and pre-cooled on ice for one hour, immersed in a solution of PVS2 plus 1.0% phloroglucinol (PG) then LN. The control and PVS2 only (treatment 1) were held at room temperature for 72 hrs. Rinsed seeds were then placed in germination boxes for 6 months at 29/19 °C. The dotted horizontal reference line indicated the maximum germination for the non-treated control.

CONCLUSION

Results presented herein suggest that wild lime seeds are not physically dormant but do possess physiological dormancy. Seed source influences germination of this species and should be considered when propagating. Seeds are tolerant of desiccation and cryopreservation making them an ideal candidate for long term storage. Cold/warm stratification through move-along experiments where seeds are manually shifted through mimicked seasons during a single year are currently in progress to ascertain the depth of physiological dormancy and reach germination at levels high enough for commercial application. In addition, promising experiments are in progress to explore asexual propagation of this species by cuttings and tissue culture.

LITERATURE CITED

Baskin, C. and Baskin, J. (2014). Seeds: Ecology, Biogeography and Evolution of Dormancy and Germination. Academic Press. Cambridge, MA.

Datt, G., Chauhan, J., and Ballabha, R. (2017). Influence of pre-sowing treatments on seed germination of various accessions of timroo (*Zanthoxylum armatum* DC.) in the garhwal himalaya. J. Appl. Res. Med. Aromat. Plants. 7:89-94.

Davies, F.T., Geneve, R.L., and Wilson, S.B. (2018). Hartmann and Kester's Plant Propagation. Principles and Practices. 9th ed. Pearson Education, NY.

[FNPS] Florida Native Plant Society. (2022). *Zanthoxylum fagara*. (Accessed 10 July 2022).
https://www.fnps.org/plant/zanthoxylum-fagara.

Iseldy, D. (2016). Rules for Testing Seeds. Assoc. of Official Seed Analysts. Las Cruces, NM.

Patade, V.Y., Kumar, K., Gupta, A.K., Grover, A., Negi, P.S., and Dwivedi, S.K. (2020). Improvement in seed germination through pre-treatments in timur (*Zanthoxylum armatum* DC.): a plant with high medicinal, economical and ecological importance. Natl. Acad. Sci. Lett. 43:295-297, doi:10.1007/s40009-019-00851-9.

[USDA] United States Department of Agriculture Forestry Services. (2008). The Woody Plant Seed Manual. (Accessed 15 March 2022).
https://www.fs/fed/is/r,/pubs_series/wo/wo_ah727.pdf.

Vendrame, W.A. and Faria, R.T. (2011). Phloroglucinol enhances recovery and survival of cryopreserved *Dendrobium nobile* protocorms. Scientia Hort. 128:31-135. doi:10.1016/j.scienta.2010.12.018.

Vendrame, W.A. and Faria, R.T. (2011). Phloroglucinol enhances recovery and survival of cryopreserved *Dendrobium nobile* protocorms. Sci. Hort. *128*:31-135.
doi:10.1016/j.scienta.2010.12.018.

Combined Proceedings IPPS 72:119-126. 2022.

The Effects of Microalgae as a Biostimulant on Seed Germination

Runshi Xie[1,3], Kevin Dang[2], Sonya Kan[3], Jakob Sauve[3], Riley Johnson[3], Meredith Clay[1], Russell Jessup[4], Hongmin Qin[3*]

[1]Department of Horticultural Sciences, Texas A&M University, College Station TX 77843, USA. [2]Artie McFerrin Department of Chemical Engineering, Texas A&M University, College Station, TX 77843, USA. [3]Department of Biology, Texas A&M University, College Station 77843, USA. [4]Department of Soil and Crop Sciences, Texas A&M University, College Station, TX 77843, USA

Fushe001@tamu.edu

Keywords: Organic farming, basil, tomato, hemp, soil amendment, biofertilizer

Summary

Microalgae have been considered the safe and sustainable new source of biostimulant or soil amendment in organic plant production. As an emerging concept, further research on the effect of different microalgae strains on the production of different crops is needed to develop successful algal biostimulant products. In this study, the effects of microalgae (under different culturing conditions) on seed germination were investigated by treating different plant seeds with microalgae extractions or deionized water. This study included two horticultural crops, basil and tomato, for their fast-growing cycle and evaluation of their nutritional values. Industrial hemp was also included in this study. Seed germination parameters, including daily germination rate, germination index, and seedling growths (root length and shoot length), were evaluated. The results show that the microalgae treatments positively affected the initial seed radicle emergence and final germination percentage of hemp and tomato, respectively. All microalgae treatments had increased the seedling vigor of basil by positively influencing root growth. The results suggest that microalgae have the potential to be used as biostimulants in different crop productions, and further research is required.

INTRODUCTION

Modern agriculture faces numerous challenges, including increasing global food demand and loss of productivity due to climate change, soil erosion, and other environmental issues such as lack of biodiversity. Meanwhile, better awareness of public food safety has led to the rising demand for high-quality and organically produced agricultural products (Devlet, 2021). To this end, the agriculture industry has been adopting novel and environmentally friendly approaches, such as using plant biostimulants (Colla and Rouphael, 2020). Plant biostimulants are considered as substance(s) or microorganisms that, when applied to plants, can benefit the process of nutrient uptake, nutrient efficiency, and tolerance to abiotic stress, leading to better crop performance independently of its nutrient content (Ricci et al., 2019).

Microalgae, which comprise eukaryotic green microalgae and prokaryotic cyanobacteria (blue-green algae), are praised for their extraordinary capability of production of biomass and various value-added products (Hadipoor et al., 2021), and have been increasingly explored recently as the safe and sustainable alternative source of biostimulant or soil amendment in organic plant production (Colla and Rouphael, 2020). Studies have shown numerous benefits of microalgae, including better seed germination, seedling growth, increased yield, and enhanced tolerance to diseases and environmental stresses (Kim et al., 2018; Martini et al., 2021; Supraja et al., 2020). Although studies have shown that microalgae produce bioactive and signaling molecules such as phytohormones that have biostimulant effects on horticultural and agronomic crops, their targeted applications (e.g., microalgae strains and plant species) and specific mechanism remains unknown or unevaluated (Colla and Rouphael, 2020).

Several important phytohormones in higher plants have also been found within microalgae (Stirk et al., 2014; Stirk et al., 2013). Thus, the biostimulant effects associated with this phytohormone presence detected within microalgae were hypothesized to positively influence the overall crop yields, seed germination and seedling growth, and reduced seedlings diseases such as damping off (commonly found in hemp). Furthermore, the gibberellins (GA) amount was reported to be lower in actively growing cultures compared to slow growing cultures (Stirk et al., 2014). Therefore, this study aims to investigate the differences in effects light conditions have on the endogenous phytohormones within different eukaryotic microalgae strains (*Chlorella* and *Chlamydomonas*) to improve seed germination and seedling growths in different plant species. The objectives of this study are (1) to study microalgae as a source of biostimulant in the seed germination of tomato, basil, and hemp, and (2) to identify the effect of different light conditions on microalgae biostimulant activities by monitoring the seed germination and seedling growth.

MATERIALS AND METHODS

Plant material and microalgae strains. Basil (*Ocimum basilicum* 'Genovese') and tomato (*Solanum lycopersicum*, Homestead 24 red tomato) were included in this study for their fast-growing cycle and evaluation as candidates for organic farming (**Fig. 1 A, B**). Industrial hemp (*Cannabis sativa*) (**Fig. 1C**) was also included in this study due to its recent legalization and its increased use by Texan farmers. The microalgae strains [*Chlorella Vulgaris* (**Fig. 1D**) and *Chlamydomonas Reinhardtii* (**Fig. 1E**)] used in this study were sourced from the Department of Plant Pathology and Microbiology and the Department of Biology at Texas A&M University.

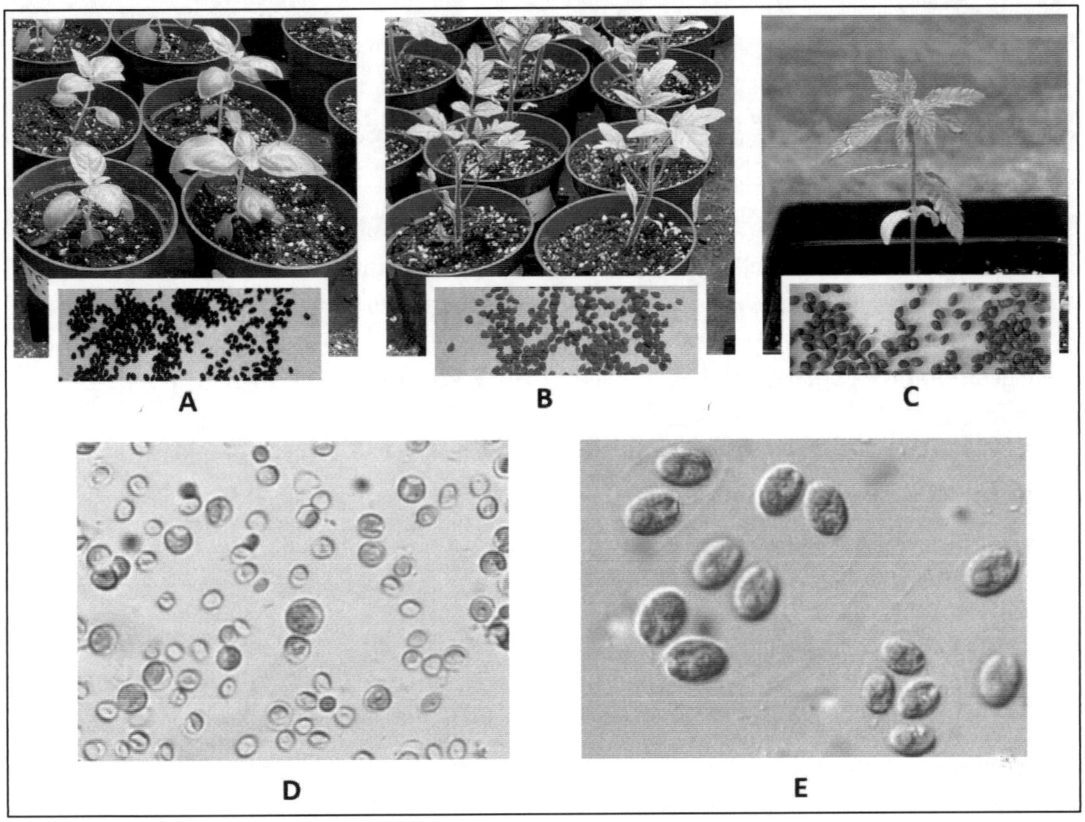

Figure 1. Plant material (seeds): (A) basil, (B) tomato, and (C) hemp; and microalgae strains: (D) *Chlorella Vulgaris* and (E) *Chlamydomonas Reinhardtii* used in this study.

Microalgae culturing and biomass harvesting. Stock microalgal cell cultures were grown in 1 L TAP (Tris Acetate Phosphate) solutions in a climatic chamber with the following variables controlled: continuous light with a light intensity of (120 μmol m^{-2} s^{-1} PPFD), a temperature of 22 °C and a cell concentration of 1 x $10^{\wedge 7}$ cells/M (**Fig. 2 A, B**). The cultures were then subcultured into 4 L flasks after 96 hours following a one to ten dilutions (**Fig. 2C**). Once the subculture reaches a concentration of 1 x $10^{\wedge 7}$ cells/mL (after 96 hours), they are moved into two lighting conditions: (1) a continuous light (CL) condition (same as described above) and (2) a continuous dark (CD) condition for two days before harvesting.

Microalgal culture (1 x 10^7 cells/mL or more) was harvested by centrifugation at 2000×g (**Fig. 2**); the collected biomass was washed and freeze-dried (**Fig. 2E**). The dried powder (algae biomass) can be used for long-term storage and re-suspended in DI water (0.5 mg/mL) right before use. The algae suspension was treated with sonication for 3 min (e.g., Branson sonicator 150, amplitude 40%, 3 min) to disrupt the cell walls.

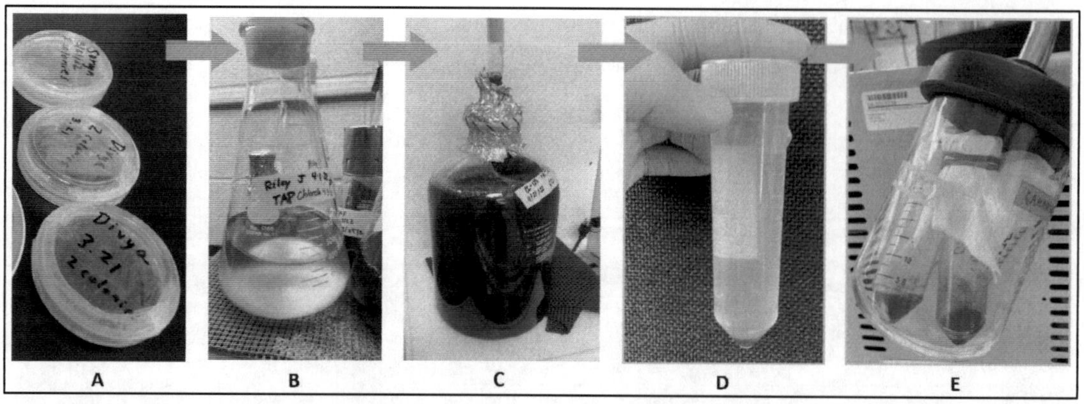

Figure 2. Procedure of microalgae culturing and harvesting of biomass: (A) microalgae colonies developed from TAP plates, (B) inoculation in liquid medium, (C) increasing cell density in 4 L container, (D) harvesting by centrifuging, and (E) dried-freezing.

Seed germination experiments. Seeds of different plant species were soaked for 3 min in 5% aqueous NaClO for sterilization and thereafter washed three times with DI water. The sterilized seeds [10 seeds per plate (5 replications)] were placed on filter paper in Petri dishes and soaked with 4 mL of microalgae solution or deionized water, respectively. The Petri dishes with treated seeds were then sealed with parafilm and placed in a growth chamber under a temperature of 25 °C. The number of germinated seeds was monitored and recorded daily to determine the germination percentage, and the root and shoot length were recorded at the end of the experiment.

Seed germination data were analyzed using JMP software (JMP Pro16, Statistical Analysis System, Cary, NC, USA). One-way analysis of variance (ANOVA) was used to analyze the number of germinated seeds, root length, and shoot length under different microalgae treatments. The difference in germination percentages was tested using the Adjusted Wald Test for comparing proportions. The multiple means under the treatment groups were compared to the control using Dunnett's test.

RESULTS AND DISCUSSION

The effect of microalgae on seed germination. The germination of seeds in different plant species showed different responses to the microalgae treatments. For basil, the radicle emergence of seeds was observed on day two of the experiments, and the seeds under Chlamydomonas-CD treatment showed a higher response/emergence compared to the control group ($p = 0.0023$) (**Fig. 3A**). Overall, the germination rate of basil seeds was high (>90%) across all treatments and the control group, and most seeds germinated after day three of the experiment (**Fig. 3A**). Microalgae treatments did not show any adverse effects on basil seed germination.

For the tomato seed germination, the highest germination rate was reached approximately after day six of the experiment, while significant differences in germination rates between microalgae and the control group (according to Dunnett's test) were detected on days three and four (**Fig. 3B**).

The germinated seeds in the control group were initially more than the ones under microalgae treatment on day three ($p < 0.001$). However, the final germination rate of the control group (70%) was lower than the *Chlorella*-CD ($p = 0.0067$) and *Chlamydomonas*-CD ($p = 0.0291$) according to the Adjusted Wald Test.

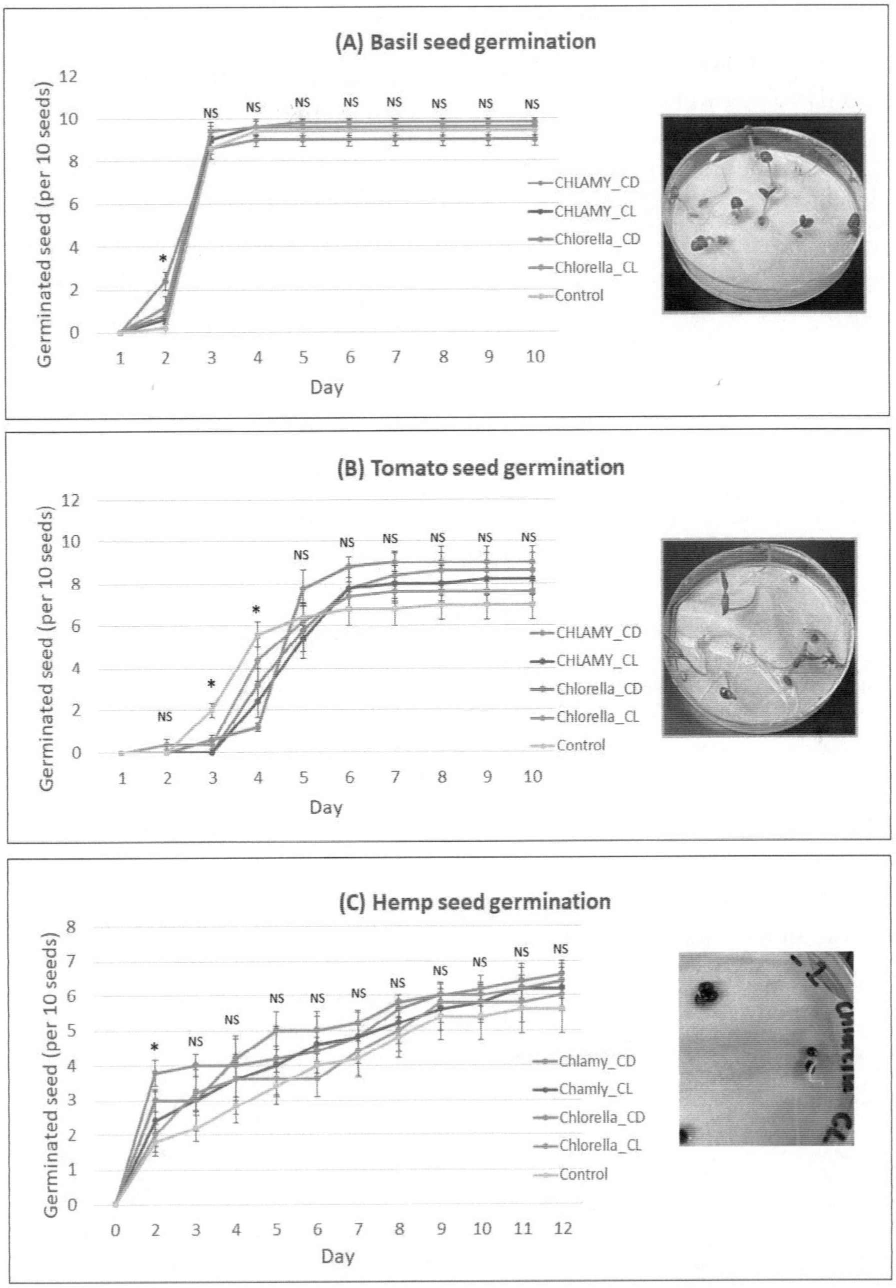

Figure 3. Average germinated seed (per 10 seeds) of (A) basil, (B) tomato, and (C) hemp, during the experiment; an asterisk (*) denotes significant differences detected, and 'NS' means no significant difference detected between microalgae treatments and controls at the same day according to the Dunnett's test.

For the hemp seed germination, the effects of microalgae on the radicle emergence were significant, while the difference between treatments and control decreased as time progressed (**Fig. 3C**). On day three, the seeds treated with *Chlamydomonas*-CD had a significantly higher germination rate compared to the control group ($p = 0.0337$). The hemp seeds treated with deionized water (control) are among the lowest germination levels throughout the entire experiment; however, the effects were not statistically significant at the end of the data collection.

The effect of microalgae on seedling growth. Different effects of microalgae treatments on seedling growth were observed and specific to different plant species (**Table 1**).

Table 1. The effect of microalgae on seedling growth; the same letter in each column denotes no significance detected according to the student's t-test at 0.05 level.

	Mean length (mm)					
	Tomato		Basil		Hemp	
	Root	Shoot	Root	Shoot	Root	Shoot
*Chlorella*_CD	46.56[ab]	23.98[abc]	28.69[ab]	10.96[b]	11.3[a]	6.13[b]
*Chlorella*_CL	58.15[a]	26.32[a]	26.60[b]	10.92[b]	22.21[a]	5.94[b]
*Chlamy*_CD	43.53[b]	20.86[c]	31.02[a]	12.13[a]	13.92[a]	7.94[ab]
*Chlamy*_CL	57.03[a]	21.38[bc]	29.98[a]	9.79[c]	29.8[a]	13.21[a]
Control	50.41[ab]	24.22[ab]	21.68[c]	7.09[d]	19.61[a]	5.46[b]

Among the three tested plant species in this study, basil showed the most apparent and positive responses to all microalgae treatments in terms of higher root and shoot growth. For instance, basil seeds treated with the Chlorella-CD or -CL solution had roots grow 32% or 22% in length, respectively, more than the control. Similarly, basil seeds treated with the Chlamydomonas-CD or -CL solution had a more than 30% increase in root growth compared to the control. The basil seedlings treated with deionized water (control) also exhibited the lowest shoot growth (7.09 mm), while the highest shoot length was found in the Chlamydomonas-CD treatment (**Table 1**).

In the tomato experiment, despite *Chlamydomonas*-CL or *Chlorella*-CL solutions resulting in an increased root length growth of 13% or 15% compared to control, the effects of microalgae on root growth were not statistically significant between treatments and the control. On the other hand, tomato seedlings under microalgae treatments and the control group showed similar root and shoot growth, with the only exception of *Chlamydomonas*-CD treatment, which had a shorter shoot length compared to the control ($p = 0.0487$).

For hemp seedlings' growth, there were no statistical effects detected among the treatments and the control in terms of root development. In contrast, microalgae treatments had equivalent or positive effects on the shoot development compared to the control. For instance, the average shoot length of hemp seedlings was more than doubled when treated with *Chlamydomonas*-CL compared to the control ($p = 0.01$). Overall, no negative effects of microalgae treatments were detected on hemp seed growth.

CONCLUSION

In this study, the effects of microalgae on seed germination were investigated by treating different plant seeds with microalgae extractions or deionized water. The microalgae treatments positively affected the initial seed radicle emergence and final germination percentage of hemp and tomato, respectively. All microalgae treatments had increased the seedling vigor of basil by positively influencing root growth. Overall, the microalgae treatments were an equivalent or positive influence on seed germination and seedling growth in all tested plant species, except for tomato shoot growth when treated with *Chlamydomonas*-CD.

The results suggest that microalgae have the potential to be used as biostimulants in selected crop production; however, further research is required. Irrigation experiments will be needed to further evaluate microalgae treatments on plant growth and yields. In order to gain insight into how microalgae influence plant growth, it is also crucial to conduct different assays to quantify the factors, such as the phytohormone compounds of microalgae under different conditions.

Acknowledgments. The authors thank Dr. Mengmeng Gu at Colorado State University for providing the tomato and basil seeds and suggestions for the experiments. Special thanks to the Aggie Research Program at Texas A&M for providing the platform for team member recruitment in this project.

LITERATURE CITED

Colla, G., and Rouphael, Y. (2020). Microalgae: new source of plant biostimulants, MDPI. pp. 1240.

Devlet, A. (2021). Modern agriculture and challenges. Frontiers in Life Sciences and Related Technolog. *2*:21-29.

Hadipoor, M., Keivanimehr, F., Baghban, A., Ganjali, M.R., and Habibzadeh, S. (2021). Chapter 24 - Carbon dioxide as a main source of air pollution: Prospective and current trends to control, in: A. Núñez-Delgado (Ed.), Sorbents Materials for Controlling Environmental Pollution, Elsevier. pp. 623-688.

Kim, M., J., Shim, C.K., Kim, Y.K., Ko, B.G., Park J.H., Hwang S.G., and Kim B.H. (2018). Effect of biostimulator *Chlorella fusca* on improving growth and qualities of chinese chives and spinach in organic farm. Plant Path. J. *34*:567.

Martini, F., Beghini, G., Zanin, L., Varanini, Z., Zamboni, A., and Ballottari, M. (2021). The potential use of *Chlamydomonas reinhardtii* and *Chlorella sorokiniana* as biostimulants on maize plants. Algal Res. *60*:102515.

Ricci, M., Tilbury, L., Daridon, B., and Sukalac, K. (2019). General principles to justify plant biostimulant claims. Front. Plant Sci. *10:*494.

Stirk, W.A., Bálint, P., Tarkowská, D., Novák, O., Maróti, G., Ljung, K., Turečková, V., Strnad, M., Ördög, V., and Van Staden, J. (2014). Effect of light on growth and endogenous hormones in *Chlorella minutissima* (Trebouxiophyceae). Plant Phys. Bioch. *79*:66-76.

Stirk, W.A., Ördög, V., Novák, O., Rolčík, J., Strnad, M., Bálint, P., and van Staden, J. (2013). Auxin and cytokinin relationships in 24 microalgal strains1. J. Phycol. *49*:459-467.

Supraja, K.V., Behera, B., and Balasubramanian, P. (2020). Efficacy of microalgal extracts as biostimulants through seed treatment and foliar spray for tomato cultivation. Industrial crops and products *151*:112453.

Combined Proceedings IPPS 72:127-132. 2022.

Germination of *Viola odorata*, a Genetic Resource for Fragrance in *Viola* Breeding

Shea A. Keene and Thomas A. Colquhoun

Environmental Horticulture Department, UF/IFAS, 1523 Fifield Hall, Gainesville, FL 32611, USA

keene284@ufl.edu

Keywords: Seed germination, non-deep physiological dormancy, cold stratification

Summary

The highly scented *Viola odorata* is a potential genetic resource for fragrance-focused *Viola* breeding; however, the species is thought to exhibit seed dormancy and germination of the seeds is difficult. The objective of this study was to investigate two methods to break dormancy and promote germination in *V. odorata*, including different concentrations of gibberellic acid in the culture media, and different durations of cold stratification. Ultimately, the highest germination percentage was achieved via cold stratification at 4 °C for 8 and 12 weeks.

INTRODUCTION

Viola odorata, also known as sweet violet, is a cleistogamous, temperate, perennial species in the Violaceae family that is famous for its fragrance (Marcussen, 2006). As genetic resources of aroma traits, sweet violets could be used in a fragrance-focused breeding program to develop fragrant *Viola* hybrids that can tolerate the intense heat and humidity common in the southeastern US. However, like other *Viola* species, *V. odorata* is thought to exhibit seed dormancy as its seeds are notoriously difficult to germinate and overall germination rates are low (Banasinksa and Kuta, 1996; Barekat et al., 2013). Achieving consistent, high rates of germination is very important in a breeding program that utilizes interspecific hybridization, as poor germination would result in smaller parental and hybrid populations, and thus less genetic variance from which to select (Brown et al., 2014). Seed dormancy may be defined as a barrier that prevents the complete germination of a viable seed under otherwise favorable environmental conditions (Baskin and Baskin, 2004). Dormancy evolved across species as an adaptation to different environments—resulting in a diverse range of dormancy mechanisms—and serves to prevent germination until certain conditions have been met and the environment is favorable to the establishment of a new generation (Finch Savage and Leubner-Metzger, 2006). The requirements to break dormancy vary just as widely and can be highly specific to different taxa (Willis et al., 2014). The most common form of dormancy exhibited by *Viola* species—non-deep physiological dormancy (PD)—may be broken by stratification or treatment with gibberellic acid (GA_3) (Franklin et al., 2017; Gehring et al., 2013). Thus, the objectives of this research were to evaluate two methods to break dormancy and achieve germination of *V. odorata* seeds, including different concentrations of GA_3 in the culture media and different durations of cold stratification.

MATERIALS AND METHODS

Experiment 1: Different GA_3 concentrations in the culture media. Seeds of *V. odorata* 'Reine de Neiges' were purchased from Jelitto Seeds (Schwarmstedt, Germany) and stored in their original packaging before use. The culture media was prepared with Murashige and Skoog (MS) basal salt mixture, sucrose, and agar. After the media was autoclaved, a filter-sterilized GA_3 stock solution was added to three of the bottles to achieve GA_3 concentrations of 2 mg/L, 6 mg/L, and 250 mg/L. Media with 0 mg/L GA_3 was also prepared. The media was then poured to solidify in petri dishes with an 8 cm diameter. Prior to sterilization, 240 seeds were mechanically scarified by gently rubbing them between two pieces of 120-grit sandpaper for 1 min. All seeds were surface sterilized in a 4.125% sodium hypochlorite solution with two drops of Tween-20 for 15 min on a gyratory shaker, then rinsed three times with sterile distilled water. Twenty seeds were transferred onto each petri dish using sterile forceps in a transfer hood. The petri dishes were sealed with a single layer of sealing film and placed on a shelf under cool white fluorescent tube lights with a 16-h light/8-h dark photoperiod and maintained at 25 °C.

Experiment 2: Different durations of cold stratification. Seeds of *V. odorata* 'Rubra' were purchased from Jelitto Seeds (Schwarmstedt, Germany) and stored in their original packaging before use. Media for stratification was prepared as described in Experiment 1, with the following changes. Media for a given treatment group was prepared on the day the stratification treatment was initiated, and no GA_3 was added. No seeds were scarified before sterilization. Seeds were sterilized as described in Experiment 1. Once seeds were sown on the media (20 seeds per dish), the sealed petri dishes were stored in a laboratory refrigerator at 4 °C for the appropriate duration. Stratification was staggered so that all seeds were moved to germination conditions at the same time. Seeds in the 0-week stratification treatments were sterilized before they were sown on the germination media. For germination, each petri dish of seeds was transferred to a 20-cell insert containing high-porosity peat-based media (PRO-MIX HP; Premier Horticulture Ltd., Rivière-du-Loup, Quebec, CA), with one seed per cell. These inserts were placed in solid-bottom plastic trays without holes and covered with clear plastic humidity domes. The covered trays were placed on a shelf under cool white fluorescent tube lights with a 16-h light/8-h dark photoperiod and maintained at 25 °C. Deionized (DI) water was added as needed to keep the seeds moist.

Experimental design, data collection, and analysis. The experimental design of Experiment 1 was a single factor (GA_3 concentration), completely randomized design (CRD) with five treatments: no scarification and 0 mg/L GA_3 (control); 0 mg/L GA_3; 2 mg/L GA_3; 6 mg/L GA_3; and 250 mg/L GA_3. All seeds were scarified, except the seeds in the control group. Each treatment was replicated three times and the experimental unit was an individual petri dish. Seeds were considered to have germinated when the radicle emerged. The experimental design of Experiment 2 was a single factor (stratification duration at 4 °C) CRD with four treatments: 0 weeks, 4 weeks, 8 weeks, and 12 weeks. Each treatment was replicated four times, and the experimental unit was an individual petri dish/20-cell insert. Seeds were considered to have germinated when the shoot/cotyledons emerged. For both experiments, a CRD was chosen because of the homogenous nature of the experimental units, and because the experiments were conducted in a laboratory with stable environmental conditions. For both experiments, the final germination percentage (FGP) was calculated for each experimental unit as the number of seeds germinated divided by the total number of seeds in each unit, then multiplied by 100. The FGP of each experiment was analyzed using analysis of variance (ANOVA) based on the model y = overall mean + treatment effect + residual. The ANOVA assumptions of homogeneity of variance and normally distributed residuals were checked with Levene's test and the Shapiro-Wilk test, respectively. If the F-statistic of the ANOVA indicated significant treatment effects ($P < 0.05$), the data were further analyzed by Tukey's range test with a P-value of 0.05.

RESULTS

Experiment 1. The FGP of 'Reine de Neiges' seeds was not significantly different among the GA_3 treatments (**Table 1**). Overall, the FGP was very low, reaching a maximum of only 10% to 13.33%.

Experiment 2. The FGP of 'Rubra' seeds was significantly different among the stratification duration treatments, with the highest FGP (approximately 70%) achieved for seeds stratified for 8 weeks and 12 weeks (**Table 2**). The FGP of seeds stratified for 8 weeks was not significantly different than the FGP of seeds stratified for 12 weeks.

Table 1. Effect of GA_3 concentration in the media on the germination of *V. odorata* 'Reine de Neiges' seeds.

Treatment	Cumulative number of seeds germinated[1]	Final germination percentage (%)[2]
Control	0	0.00 ns
0 mg/L GA_3	4	6.7 ns
2 mg/L GA_3	6	10.0 ns
6 mg/L GA_3	6	10.0 ns
250 mg/L GA_3	8	13.3 ns

[1] For a given treatment, the combined total number of seeds that germinated in replicates 1-3. [2] For a given treatment, the final germination percentage (FGP) is the average of the FGP of replicates 1-3. Within this column, ns indicates the ANOVA was not significant with $\alpha = 0.05$.

Table 2. Effect of cold stratification duration on the germination of *V. odorata* 'Rubra' seeds.

Treatment	Cumulative number of seeds germinated[1]	Final germination percentage (%)[2]
12 weeks	55	68.8 A
8 weeks	56	70.0 A
4 weeks	24	30.0 B
0 weeks	9	11.3 C

[1] For a given treatment, the combined total number of seeds that germinated in replicates 1-4. [2] For a given treatment, the final germination percentage (FGP) is the average of the FGP of replicates 1-4. Within this column, values followed by different letters were significantly different as determined by Tukey's HSD ($P < 0.05$).

DISCUSSION

In Experiment 1, the FGP of 'Reine de Neiges' seeds was very low, and it was lower than the FGP achieved under similar conditions in other studies. For example, in the 2 mg/L GA_3 treatment, only 10% of 'Reine de Neiges' seeds germinated. In contrast, Banasinska and Kuta (1996) achieved approximately 60% germination for *V. odorata* seeds that were mechanically scarified and sown on MS media with 2 mg/L GA_3. Similarly, only 10% of 'Reine de Neiges' seeds germinated on MS media containing 6 mg/L GA_3, compared to 29.99% of *V. odorata* seeds sown on media with the same GA_3 concentration (Barekat et al., 2013). The failure to find significant differences ($P < 0.05$) may be due in part to inadequate replication of the treatments, as the residual degrees of freedom fell short of the rule-of-thumb value, 12 (Clewer and Scarisbrick, 2001). To a certain point, increasing the number of replicates can increase the precision of treatment comparisons and allow for the detection of smaller differences among those treatments (Casler et al.,

2015). The amount of available seed material constrained the number of replicates in this experiment. In subsequent experiments, additional seed was acquired so that more replicates could be done. Interestingly though, a percentage of violet seeds germinated at all treatment levels, except one (**Table 1**). In the control condition, non-scarified seeds on GA_3-free media did not germinate; however, scarified seeds on GA_3-free media did germinate. These results suggest that sweet violet seeds exhibit non-deep PD, as scarification can help promote germination in water-permeable seeds with non-deep PD (Baskin and Baskin, 2014). The results may also suggest that more GA_3 is not necessarily better at promoting germination. From a practical standpoint, using less (or no) GA_3 would be better, as it would save time and money since GA_3 is an expensive and heat-labile compound that must be filter-sterilized and added to the media after the media has been autoclaved. Ultimately, the results of this experiment demonstrated that seeds of a *V. odorata* cultivar can be successfully germinated under *in vitro* conditions; however, high rates of germination were not achieved.

In Experiment 2, both the 8-week and 12-week stratification durations resulted in high germination success for 'Rubra' seeds, with FGPs of approximately 70%. The germination rates achieved here are comparably high when compared to the germination rates achieved in other studies of *Viola*. For *V. odorata*, Barekat et al. (2013) achieved a maximum FGP of 74.58% and 54.31% with endosperm culture and embryo culture, respectively. However, these percentages were based on the germination of four treatment replicates with only five seeds per replicate. Moreover, such methods are expensive, as well as time- and labor-intensive, as they require aseptic conditions and *in vitro* culturing. Banasinska and Kuta (1996) achieved germination for 37 out of 61 seeds of *V. odorata*, a germination rate of approximately 60%, with mechanical scarification and *in vitro* culture on MS media supplemented with 2 mg/L GA_3. In research on other *Viola* species, Gehring et al. (2013) achieved maximum germination of approximately 50-60% for seeds of *V. pedata* that were exposed to a 12-week warm-dry treatment, followed by cold stratification for 12 weeks. Franklin et al. (2017) achieved maximum germination of 90-100% for seeds of *V. pedunculata* and *V. purpurea* that were exposed to several months of dry storage and several weeks of cold stratification at 4 °C. Like Franklin et al. (2017), the FGPs achieved here suggest that experimenting with dormancy-breaking conditions that mimic natural conditions—in this case, extended cool-wet conditions—can increase germination success.

CONCLUSION

Of the two methods evaluated in this study, the highest FGPs (approximately 70%) of seeds of a *V. odorata* cultivar were achieved via cold stratification at 4 °C for 8 and 12 weeks. Thus, based on these results, it is recommended that seeds of *V. odorata* and its cultivars be cold stratified at 4 °C for at least 8 weeks to break dormancy and promote germination.

LITERATURE CITED

Baskin, J.M. and Baskin, C.C. (2004). A classification system for seed dormancy. Seed Sci. Res. *14*:1-16.

Brown, J., Caligari, P.D.S., and Campos, H.A. (2014). Plant Breeding. 2nd ed. John Wiley & Sons Ltd. West Sussex, UK.

Casler, M.D., Vermerris, W., Dixon, R.A. (2015). Replication concepts for bioenergy research experiments. Bioenerg. Res. *8*:1-16.

Clewer, A.G. and Scarisbrick, D.H. (2001). Practical statistics and experimental design for plant and crop science. John Wiley & Sons, Ltd.

Gehring, J.L., Cusac, T., Shaw, C., and Timian, A. (2013). Seed germination of *Viola pedata*, a key larval host of a rare butterfly. Nat. Plants J. *14*:205-212.

Finch-Savage, W.E. and Leubner-Metzger, G. (2006). Seed dormancy and the control of germination. New Phytol. *171*:501-523.

Franklin, S., Tran, L.B., Farzad, D., and Hill, R.I. (2017). Seed germination in *Viola pedunculata* and *Viola purpurea* subsp. *Quercetorum* (Violaceae), critical food plants for two rare butterflies. Madrono *64*:43-50.

Marcussen, T. (2006). Allozymic variation in the widespread and cultivated *Viola odorata* (Violaceae) in western Eurasia. Bot. J. Linnean Soc. *151*:563-571.

Willis, C.G., Baskin, C.C., Baskin, J.M., Auld, J.R., Venable, D.L., Cavender-Bares, J., Donohue, K., and Rubio de Casas, R. (2014). The evolution of seed dormancy: environmental cues, evolutionary hubs, and diversification of the seed plants. New Phytol. *203*:300-309.

Combined Proceedings IPPS 72:133-141. 2022.

Groundcover Type and Irrigation Delivery Affect Soil Moisture Dynamics in the Landscape

Max McKeown[1], Jeb S. Fields[1], Jeff Kuehny[2], and Heather Kirk Ballard[3]

[1]Hammond Research Station 21549 Old Covington Hwy. Hammond, LA 70403, USA [2]LSU AgCenter Burden Museum & Gardens 4560 Essen Lane Baton Rouge, LA 70809, USA. [3]Louisiana State University School of Plant, Environmental, and Soil Sciences137 J.C. Miller Hall Baton Rouge, LA 70803, USA

tmckeo2@lsu.edu

Keywords: *Liriope*, mulch, *Sphagneticola trilobata,* wedelia, soil temperature, volumetric water

Summary

There is little research published on the effect ornamentals groundcovers have on soil health. Soil properties can be manipulated by groundcover growth habit and irrigation type. This research was designed to evaluate the effects groundcover form and habit have on soil moisture and temperature under different irrigation regimes. A bunching (*Liriope muscari* 'Big Blue') and matting groundcover (*Sphagneticola trilobata*) were planted in individual plots that were irrigated by either overhead or micro-irrigation. Soil volumetric water content (VWC) and temperature were monitored by soil sensors buried 15cm deep in each plot. Overhead and micro spray irrigation, along with groundcover growth habit, affected soil temperature and soil VWC. Plots with Wedelia had the largest increase in VWC during irrigation events, regardless of irrigation type. Soil VWC was found to be lower in planted treatments than fallow treatments. At each irrigation event, micro spray showed a greater increase in VWC when compared to overhead irrigation across all treatments. However micro spray irrigation soil VWC decreased at the same rate for the overhead irrigation. Soil temperatures fluctuations were reduced under both groundcover species, when compared to fallow plots. Irrigation delivery method was also found to influence soil temperatures. Micro spray irrigation caused a slight increase in temperature at each irrigation event, while there was no temperature increase with overhead irrigation events. Ornamental groundcovers can lower soil VWC and temperature through increased transpiration and shielding solar radiation. Furthermore, groundcovers mitigate the rapid fluctuations in temperature creating a more normalized soil dynamic.

INTRODUCTION

Sustainable practices are growing in popularity for the horticulture industry. Sustainable practices should focus on the financial gains, environmental advantages, and human enrichment (Doxon, 1996). However, landscapes are often slower to adopt sustainable practices than production agriculture (Doxon, 1996). With landscapes occupying millions of acres of land in the United States (Steinberg, 2005), it is critical that we not only continue to develop sustainable landscape practices, but we must implement more of these practices. The lawn is the ubiquitous in the American landscape, making up approximately 25 to 40 million acres of land (Steinberg, 2005). While there are many practices (reduced pesticide use, planting native species, water management) to include in sustainable landscapes - one is to include is planting and maintaining ground covers, which support reduced labor cost and maintenance, lowers water and fertilizer usage, and reduces landscape runoff. Encouraging the installation of landscapes that require fewer inputs (e.g., irrigation, fertilizer, and maintenance) may decrease negative environmental outcomes (Khachatryan, 2020). Fertilizing lawns can contribute to non-point pollution, produce algae blooms, and cause waterway degradation (Campbell et al., 2020). Fertilization mismanagement of urban vegetation represents a potential source of nutrients that may contribute to water quality impairment (Carey et al., 2012).

Water movement under different groundcover management systems (GMSs) has been well-studied under orchards. Several comprehensive reviews assessing the relative advantages and disadvantages of various GMSs have emphasized the need for additional information on the physiological, economic, and edaphic impacts of alternative orchard GMSs (Merwin et al., 1994). These are systems where various material or vegetation is used to cover bare soil to prevent erosion, add nutrients to the soil or cool soil temperature. Many studies cite groundcovers increasing water infiltration rates of soil (Folorunso et al., 1992, Krohn et al., 2005).

Groundcovers have been shown to reduce high soil temperatures, which factor into the rates of biochemical reactions and have strong influence on plant and root growth (Song et al. 2013). Vegetation cover has proven to have significant effects on soil temperature (Michelsen-Correa and Scull, 2005). Effects of groundcover canopies have been studied widely in vineyard management systems and orchards. Temperatures were found to be consistently cooler under a living groundcover system, Wimmera ryegrass (*Lolium multiflorum*), and vetch (*Vicia sativa*) in vineyards in South Africa (Van Huyssteen et al., 2017). Temperatures were also found to be lower under living mulch systems in vineyards than under conventional mulch systems. The groundcover treatments may have reduced soil temperatures because of the evaporative demand of the vegetation (Bavougian and Read, 2018). One study found that vegetation heights have an inverse relationship to soil temperatures (Song et al., 2013). Soil temperatures are lower under grass groundcover systems than bare soil (Wu et al., 2014).

Cover crops are well studied in production agriculture, with vast research quantifying their benefits on crop productivity and soil health. However, there has been little research in documenting the benefits of ornamental groundcover systems beyond aesthetics and other ecosystem services such as pollinator and wildlife support. However, as landscapes cover such a vast quantity of land, it is important to quantify the benefits of ornamental groundcovers on soil health. Thus, the objective of this experiment was to study the influence of ornamental groundcover on soil moisture and temperature. Additionally, this research aims to understand how the various groundcover growth habits (matting vs bunching) interact with different irrigation (overhead vs. micro spray) on these dynamic soil properties to quantify the benefits and develop best practices.

MATERIALS AND METHODS

This research was conducted at the Louisiana State University Agricultural Center Hammond Research Station located in Hammond, LA. A 68 m^2 area plot (4 x 17m) was prepared for this research. Wherein, the soil was tilled to a depth of 4 cm and amended with a locally sourced landscape mix consisting of pine bark, sand, and dolomitic lime. The plot was divided into 18 individual 1 m^2 plots, with half being irrigated by overhead sprinklers (Model 15 UH; U15Q, Rainbird, Azusa, CA) on 1 m risers, and the other half irrigated via micro sprayers (Model XS360TS Adj True Spray, Rainbird, Azusa, CA) on 30 cm straws. The plots were irrigated every three days with overhead irrigation plots receiving 15 min and micro spray plots receiving 22 min at each irrigation event. The difference was determined by calculating the total quantity of water applied and adjusting irrigation timing, so each section received the same volume of water per irrigation event. A VWC sensor (Teros 12; METERGROUP, Pullman, WA) was buried in the center of each plot at a depth of 15 cm to monitor soil volumetric water content and temperature. The sensors were attached to a data logger (CR1000x; Campbell Scientific, Logan, UT) along with a tipping bucket rain gauge (TR-525I; Texas Electronics, Dallas, TX). Data was collected every 10 minutes and hourly averages were recorded.

The entire research plot was mulched with pine straw at a depth of 7.5cm. Within each irrigation system, three randomly selected plots were planted with wedelia (*Sphagneticola trilobata*), three were planted with Big Blue liriope (*Liriope muscari* 'Big Blue'), and the remaining three were left fallow. The wedelia was selected as a quick growing groundcover that would spread and entirely cover the surface, potentially uniformly dispersing water, while the liriope was selected as a bunching groundcover that would potentially channel water. Each plot was fertilized with 100 g-controlled release fertilizer (Osmocote Plus 15-9-12, 5-6 months; ICL Specialty Fertilizers, Dublin, OH) spread uniformly across the entire 1 m^2 plot. Overhead photos were collected using a bracket (1 m x 1m) and stand that ensured the camera was positioned 150 cm high above the center of the plots so each photo was taken from the same height with the entire 1 m^2 plot within the frame.

RESULTS AND DISCUSSION

Soil Moisture. Data was collected over a period of one week, 7/12/2022-7/18/2022. Soil volumetric water content (VWC) started rising approximately 45 min after irrigation and reached a maximum approximately 2 hrs. after irrigation. The VWC gradually declined over the week with all treatments (Fig. 1). Although the decline of the liriope and fallow was relatively uniform across the week, the wedelia experienced more pronounced daily moisture depletion, indicating a greater transpirational reduction of moisture from the soil **(Fig. 1)**.

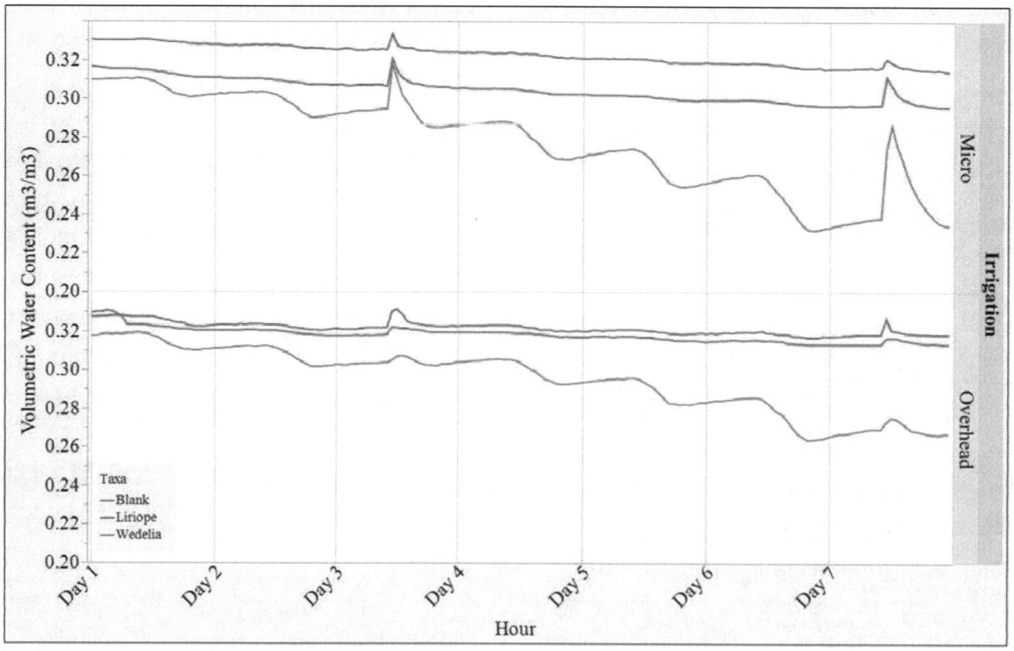

Figure 1. The change in volumetric water content over a 7-day period (7/12/22 to 7/18/22) under wedelia and Big Blue liriope with micro spray and overhead irrigation treatments with two irrigation events.

The wedelia has a faster rate of growth and greater canopy coverage than the liriope (**Fig. 2**), and thus the wedelia has more biomass to uptake water.

In cropping systems, cover crops reduce excess soil moisture, and work with the main crop to uptake more water from within the root zone (Kahimba et al., 2008).

Figure 2. Comparing the different growth habits of wedelia and Big Blue liriope under overhead micro spray irrigation.

Overall, the wedelia has the lowest VWC value on average of all three treatments. The fallow plots had the greatest VWC value. After the irrigation events, these wedelia plots experienced the greatest increase in VWC. Cover crops have been known to increase water infiltration rates (Kahimba et al., 2008). The VWC started rising in the overhead irrigation plots approximately 30 minutes after the irrigation event and reaches its maximum within 2 hours (**Fig. 1**). The irrigation spikes in VWC were considerably smaller in the overhead irrigation treatments than the micro irrigation treatments. This is likely due to the uniform wetting of overhead irrigation in the soil profile. The micro spray irrigation coverage extended only to the edges of each plot. Even though both overhead and micro spay irrigation received the same volume of water, the irrigation distribution had an impact on VWC. Sprinkler irrigation is less efficient, and more water is placed where it is not needed by the plant (Wang, 2000). In the overhead irrigation treatments, the greatest spike was in the fallow plots. The fallow plots have no vegetative canopies to deflect the irrigation water, and thus all water enters the soil profile. Conversely, the planted plots will not only deflect the irrigation and retard its entry into the soil, but also allow for evaporation of moisture remaining on the foliage. Like the micro-irrigation treatment, the wedelia treatment had the most observable daily reduction in soil VWC, further indicating increased plant-water uptake (**Fig. 1**). The liriope treatments consistently decreased in VWC over the 7-day period similarly to the fallow plots.

Soil Temperature. Temperatures in the micro spray irrigation experienced more variation between treatments than the overhead irrigation (**Fig. 3**). Similar to the soil moisture values, plots with wedelia have the lowest temperature overall, while the fallow plots had the highest soil temperature. Holmes et al. (2008) showed that maximum soil temperature occurs shortly after solar noon at the soil surface, but lags in time with increasing depth (Holmes et al., 2008). In this research, the peak temperatures were consistently measured at 19:00. Soil temperatures were lowest at around 10:00. During the irrigation events temperature increased slightly and then decreased soon after (Fig. 3). Water is known to transfer solar heat from the surface as it infiltrates the soil profile, increasing subsurface temperatures in response to irrigation events (Hillel, 2004).

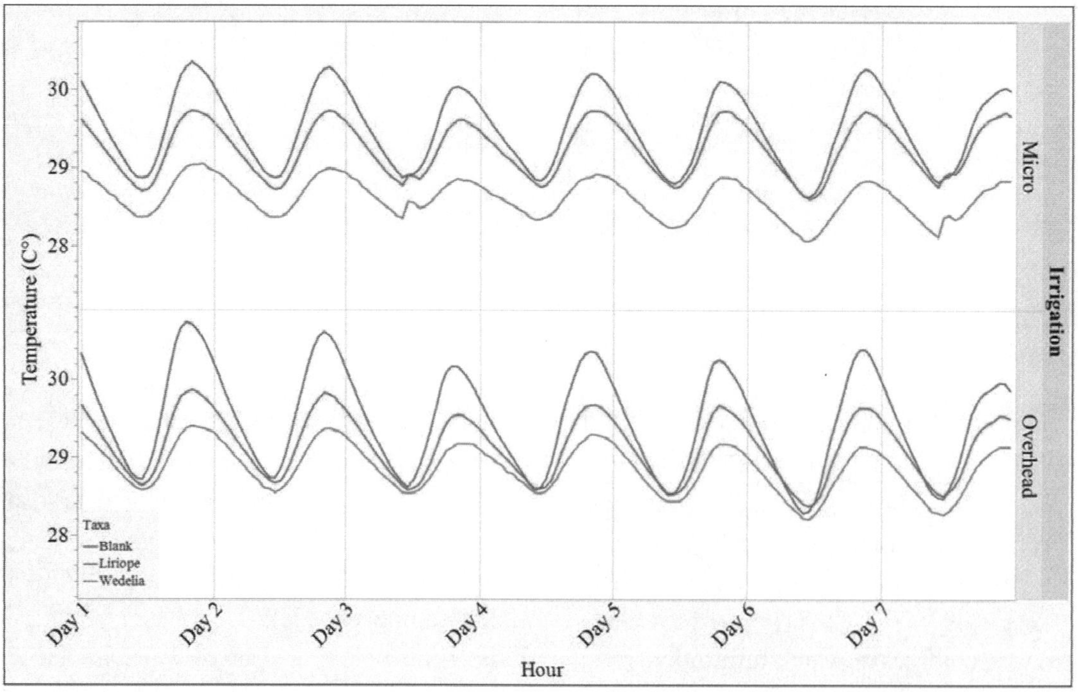

Figure 3. The change in temperature 15 cm below the soil surface over a 7-day period (7/12/22 to 7/18/22) under wedelia and Big Blue liriope with micro spray and overhead irrigation treatments with two irrigation events.

The more rapid soil temperature fluctuations in the overhead irrigated plots are likely a result of the increased water infiltration rate in the overhead irrigation systems. The overhead irrigation is wetting a larger area than micro irrigation. The temperature mitigating effect of the micro spray irrigation was likely due to the smaller irrigation area. Thus, more water entered the profile under the groundcover, raising the specific heat of the soil. Soil temperature is greatly affected by soil water content (Zhang et al., 2022).

Wedelia had the lowest temperature overall in the overhead irrigation system. Wedelia had a denser, closed canopy (**Fig. 2**) which blocked solar radiation reaching the soil surface. The fallow treatment had the highest temperature. The temperature of the wedelia treatment was similar to the liriope and fallow treatments in the overhead irrigation plot, than in the micro irrigated plot (**Fig. 3**). There may be due to the smaller droplet size of the overhead irrigation system – and evaporative cooling. Wherein the dense canopy likely deflected the water preventing some from entering the soil. But in addition to wedelia deflecting the water, the temperature was still the lowest because of the shade of the dense canopy. Unlike the micro spray irrigation system, the overhead plots had no observable soil temperature increases with each irrigation event. The soil profile was wetted over a larger area, but the wetting front most likely penetrated less than the micro irrigation. There was not a large enough influx of water to carry heat below the surface. In both irrigation systems, soil temperature increased, and decreases were less extreme in the planted treatments (**Fig. 3**). Vegetative canopies cool the environment by providing shade (i.e., reducing solar radiation) and by transpiration of water through leaves (Wu et al., 2014). Furthermore, the presence of the plants may provide small breaks in the mulch layer where evaporation (and subsequent evaporative cooling) may occur. This shading and the increased plot coverage in the wedelia is likely the reason be why the wedelia plots had the lowest temperature in both irrigation systems.

CONCLUSION

The objective of this study was to determine if groundcover growth habit and irrigation delivery method would affect soil moisture and temperature. Soil moisture and temperature were found to be lower in planted treatments versus fallow treatments. The irrigation delivery method also influenced soil temperature, with micro spray irrigation resulting in a more gradual daily flux than overhead. Groundcover habit also affected both soil temperature and VWC, with the matting groundcover (wedelia) shielding the plots from more solar radiation and deflecting more water than bunching (liriope) groundcovers. Our results can help support more informed decisions in residential and commercial landscapes through improving soil health. Finally, by incorporating ornamental groundcovers, there is increased sustainability of landscape systems -enhancing the ecosystem of urban areas.

LITERATURE CITED

Bavougian, CM., and, Read, P.E. (2018). Mulch and groundcover effects on soil temperature and moisture, surface reflectance, grapevine water potential, and vineyard weed management. Peer 6:e5082. doi: 10.7717/peerj.5082.

Campbell, J., Rihn, A., and Khachatryan, H. (2020). Factors influencing home lawn fertilizer choice in the United States. HortTech. *30*:296-305.

Carey, R. O., Hochmuth, G. J., Martinez, C.J., Boyer, T. H., Nair, V. D., Dukes, M.D., Toor, G.S., Shober, A.L., Cisar, J. L., Trenholm, L.E., and Sartain, J. B. (2012). A review of turfgrass fertilizer management practices: implications for urban water quality. HortTech.*22*: 280-291. https://doi.org/10.21273/HORTTECH.22.3.280>.

Doxon, L. E. (1996). Landscape sustainability: environmental, human, and financial factors. HortTech. 6: 362-365.

Folorunso, O., Rolston, D., Prichard, P., and Louie, D. (1992). Cover crops lower soil surface strength, may improve soil permeability. Hilgardia 46:6:26-27. DOI:10.3733/ca. v046n06p26

Hillel, D. (2004). Soil temperature and heat flow, p. 215-233. Introduction to Environmental Soil Physics. Elsevier Academic Press, San Diego, CA.

Holmes, T. R. H., Owe, M., De Jeu, R. A. M., and Kooi. H. (2008). Estimating the soil temperature profile from a single depth observation: a simple empirical heat flow solution, Water Resour. Res., *44*: W02412, doi:10.1029/2007WR005994.

Kahimba, F.C., Sri Ranjan, R., Froese, J., Entz, M., and Nason, R. (2008). Cover crop effects on infiltration, soil temperature, and soil moisture distribution in the Canadian prairies. Appl. Engineer. Agricult. *24*: 321-333.

Khachatryan, H. (2020). Landscape aesthetics and maintenance perceptions: assessing the relationship between homeowners' visual attention and landscape care knowledge. Land Use Policy *95*:104645.

Krohn, N.G., and Ferree, D.C. (2005). Effects of low-growing perennial ornamental groundcovers on the growth and fruiting of `Seyval Blanc' grapevines. HortSci. *40*:561-568.

Merwin, I. A., Stiles, W.C., and van Es, H. M. (1994). Groundcover management impacts on soil physical properties. J. Amer. Soc. Hort. Sci. *119*:216-222.

Michelsen-Correa, S., and Scull, P. (2005). The impact of reforestation on soil temperature. Middl. States Geograp. *38*:39-44.

Song Y. T., Zhou, D. W., and Zhang, H. X. (2013). Effects of vegetation height and density on soil temperature variations. Chin. Sci. Bull. *58*:907–912. doi: 10.1007/s11434-012-5596-y

Steinberg T. (2005). Lawn and landscape in world context, 1945-2000. OAH Maga. Hist. *19*:62-68. doi:10.1093/maghis/19.6.62

Van Huyssteen, L., Van Zyl, J.L., and Koen, A. P. (2017). The effect of cover crop management on soil conditions and weed control in a Colombar vineyard in Oudtshoorn. S. Afri. J. Enolog. Viticul. *5*:1. https://doi.org/10.21548/5-1-2364

Wang, D., Shannon, M.C, Grieve, C.M., Yates, S.R. (2000). Soil water and temperature regimes in drip and sprinkler irrigation, and implications to soybean emergence. Ag. Water Manage. *43*:15-28.

Wu, J., Tang, C. S., Shi, B., Gao, L., Jiang, H. T., and Daniels, J. (2014). Effect of ground covers on soil temperature in urban and rural areas. Environ. Engineer. Geosci. *20*:225-237. 10.2113/gseegeosci.20.3.225.

Zhang, M., Li, Y., Liu, J., Wang, J., Zhang, Z., and Xiao, N. (2022). Changes of soil water and heat transport and yield of tomato (*Solanum lycopersicum*) in greenhouses with micro-sprinkler irrigation under plastic film. Agron. *12*:664. https://doi.org/10.3390/agroomy12030664

Combined Proceedings IPPS 72:142-150. 2022.

Comparison of Auxin Formulations and Concentrations on Rooting Woody Softwood Cuttings

Jerry H. Yu, Andra W. Nus, and Thomas G. Ranney

Mountain Horticultural Crops Research and Extension Center, Department of Horticultural Science, North Carolina State University, 455 Research Drive, Mills River, NC 28759, USA

Hyu23@ncsu.edu

Keywords: Asexual propagation, stem cuttings, *Acer palmatum*, *Cornus* [(*Cornus florida* x *kousa*) x *florida*], *Metasequoia glyptostroboides* 'Ogon', *Zelkova serrata* 'Goshiki'.

Summary

This study evaluated a range of auxin formulations and concentrations for their effects on root initiation on difficult-to-root, softwood cuttings including *Acer palmatum*, *Cornus* [(*Cornus florida* x *kousa*) x *florida*], *Metasequoia glyptostroboides* 'Ogon', and *Zelkova serrata* 'Goshiki'. Auxin formulations included gels, powders, and liquid quick-dips. Auxin concentrations ranged from 0 to 16,000 ppm as either IBA or a mixture of KIBA and KNAA. Results varied by taxa. The highest rooting percentages for *Cornus* (75.3 – 92.9%) were achieved with rooting gels containing 3,100-5,500 ppm IBA, a powder containing 16,000 ppm IBA, and liquid quick-dips containing 5,000 – 10,000 ppm auxin (mixture of 2/3 KIBA and 1/3 KNAA). For *Metasequoia glyptostroboides* 'Ogon', the treatments with the highest rooting percentages (19.1 – 37.2%) included two gels, powders ranging from 1,000 to 16,000 ppm IBA, and liquid quick-dip containing 5,000-10,000 ppm auxin (mixture of 2/3 KIBA and 1/3 KNAA), plus the control. Rooting percentages for *Acer palmatum* and *Zelkova serrata* 'Goshiki' ranged from 2.4 to 23.5% and 0 to 9.5%, respectively, with no significant treatment effects. These results identify numerous treatments that were effective for rooting selected cultivars of *Cornus* and *Metasequoia*; the effectiveness of rooting treatments varied considerably among taxa and environment.

INTRODUCTION

The age-old practice of propagating plants by stem cuttings has progressed by leaps and bounds with the widespread adoption of the nursery industry in applying synthetic auxins to cuttings (Mendel, 1992). Of the many methods of applying auxin to cuttings, liquid quick-dip and powder applications remain the industry standard (Blythe et al., 2007). Even so, some taxa of plants remain difficult to root by liquid and powder means (Dirr and Heuser, Jr. 2006).

Since the 1980s, several commercial rooting gels have been introduced that promise another method of delivering auxin to cuttings. The gels are marketed as more effective and safer to use than other auxin delivery mechanisms because of their ability to adhere to where they are applied, thus delivering auxin for a longer period while reducing the risk of inhalation (Hydrodynamics International, 2022; Technaflora Plant Products, 2022). Producers of other types of rooting products have countered by claiming gels produce a hypoxic environment at the cutting base which leads to root rot (Hormex, 2020). Despite these claims, commercial rooting gels have seen widespread adoption by hydroponic growers and home propagators but are less common in the industry (Amazon, 2022). However, if a commercial rooting gel were to significantly outperform other auxin delivery methods in harder to root cuttings, it could be an economically desirable alternative for commercial propagators.

Previous research into rooting gels has been paltry and mixed. Ismaili (2016) reported that Clonex® Rooting Gel outperformed a 3,000 ppm liquid treatment of IBA (acid form) on softwood cuttings of seven varieties of olive (*Olea europaea*). Hepburn and Matthews (1985) did not observe a significant difference in roots per cutting between an IBA-based powder and an NAA-based gel in the potato (*Solanum tuberosum*) cultivar 'Désirée'. Caplan et. al, (2018) found that 2,000 ppm IBA gel outperformed a 2,000 ppm *Salix alba* extract gel for *Cannabis sativa*, but did not evaluate against more conventional powder or liquid formulations of IBA.

For this experiment, gels were selected based on Google and Amazon rankings, which approximate sales numbers (Amazon, 2022; Google, 2022). Therefore, these gels represent what a typical consumer might find when searching for a rooting gel in the USA. The selection of the powder and liquid products was based on the standard operating procedure of the Mountain Crop Improvement (MCI) Lab and common nursery practice (Dirr and Heuser, Jr., 2006).

Taxa were selected for propagation difficulty, diversity, prevalence in the trade, and availability. Taxa that were moderately difficult-to-root from softwood cuttings were chosen, based on industry consensus (Pers. Communication) and comments by Dirr and Heuser (2006). Botanical diversity was also emphasized, with taxa from 4 families and both gymnosperms and angiosperms selected. The 4 taxa selected were: 1) *Acer palmatum* (unnamed selection), 2) *Cornus* H2016-019-001 [(*Cornus florida* x *kousa*) x *florida*], 3) *Metasequoia glyptostroboides* 'Ogon', and 4) *Zelkova serrata* 'Goshiki'.

The objective of this study was to evaluate the effectiveness of a diverse collection of rooting gels, powders, and liquids on softwood cuttings of four moderately-difficult-to-root taxa of woody plants.

MATERIALS AND METHODS

The studies were conducted as a randomized complete block design (RCBD). There were 12 treatments plus one control. Four proprietary cloning gels were selected including Clonex® Rooting Gel (3,100 ppm IBA; Hydrodynamics International, Lansing, MI), Midas Hydro Rooting Gel (3,500 ppm IBA, Midas Hydro, Southfield MI), Rootech Cloning Gel™ (5,500 ppm IBA, Technaflora Plant Products, Mission, B.C.), and FOOP Clone Gel (natural ingredients including fish excrement, mineralized aquaculture water, mycorrhizal fungi, willow water, aloe vera, hydrolyzed fish, sea kelp, volcanic silica, peppermint oil, and xanthan gum; FOOP Organic Biosciences, Silver Spring, Maryland), along with 4 concentrations of Hormex Rooting Hormone Powder (1000, 3000, 8000, and 16000 ppm IBA; Maia Products, Westlake Village, CA), and 4 concentrations of a liquid 5-sec. quick-dip of ⅔ K-IBA and ⅓ K-NAA (2500, 5000, 7500, and 10000 ppm total auxin concentration, Sigma-Aldrich, Saint Louis, Mo.). The control consisted of a quick dip in distilled water. A single rep consisted of 7 cuttings (subsamples), with 6 reps (complete blocks) for a total of n = 6 per taxa. Taxa were treated as separate experiments.

Softwood cuttings were taken between 8:30 and 14:00 pm on 26-27 May 26, 2022, from stock plants grown in full sun between 6 and 30 years old, depending on the taxa. All stock plants were established in the landscape of the Mountain Horticultural Crops Research and Extension Center in Mills River, NC (USA zone 7a). Cuttings were taken from the portions of the stock plants which were in full sun. Once cut, cuttings were transported and stored in a moistened cooler and placed in a 4 °C (40 °F) refrigerator until sticking time.

At sticking time, cuttings were re-trimmed and leaves from the bottom half of the cuttings were removed. The bottom 2.5 cm (1 in) of the cutting was moistened before being dipped into either the liquid quick-dip, gel, or powder treatments. Cuttings for the control and liquid quick-dip sections were kept in the auxin solution for 5 seconds. Excess powder and gel were shaken off. All cuttings were then stuck in a 48 x 51 x 11 cm (18.9 x 20 x 4.3 in) flat to a depth of 8 - 10 cm (3.2 - 3.9 in), with each flat treated as a complete block. All cuttings in a given flat/block were as similar in length and node number as practical.

The rooting media consisted of a mixture of 66% horticultural perlite (Krum, Carolina Perlite Company, Gold Hill, NC) with 33% ground Canadian sphagnum peat moss (Sunshine®, Sun Gro® Horticulture, Agawam, MA; ground with a W.W. grinder, Wichita, KS). Stuck flats were placed under intermittent mist at a rate of 10 seconds every 10 minutes, with flats containing the same taxa randomly grouped together on a bench. No bottom heat was provided. Cuttings were harvested 52 to 91 days after sticking, when most cuttings of the taxa showed some sign of rooting or callusing.

Data collected included the percentage of cuttings in a rep that had rooted, the average number of primary roots per rooted cutting per rep and the average length of the primary roots per rooted cutting per rep. Data were analyzed using a one-way ANOVA (Proc GLM, SAS Institute, NC) with significant difference between means determined by Fisher's protected least significant difference. In addition, a two-way ANCOVA was performed for all 3 dependent variables (Proc GLM, SAS). Additional regression analyses (Proc GLM, SAS) were performed for auxin concentration for liquid and powder treatments based on the ANCOVA (Proc GLM, SAS). Taxa specific differences in method are recounted below:

***Cornus* H2016-019-001.** Terminal cuttings were taken from six-year-old stock plants between 8:30 and 9:30 am on May 27 in partially cloudy conditions. These 3 node cuttings were stuck at 9:40 am on May 27 after being recut to between 8 cm (3.5 in) to 16 cm (6.3 in). They were harvested 52 days after sticking on July 18, 2022.

***Metasequoia glyptostroboides* 'Ogon'.** Terminal cuttings were taken from a 20-year-old stock plant between 9:35 and 11:00 am on May 26 after a rainstorm to ensure proper hydration. These cuttings were stuck at 2:30 pm on May 26 after being recut to between 10 cm (3.0 in) to 14 cm (5.5 in). They were harvested 78 days after sticking on August 12, 2022.

***Acer palmatum*.** Terminal cuttings were taken from a 30-year-old stock plant between 8:30 and 9:30 am on May 26 during a rainstorm to ensure proper hydration. These 5 node cuttings were stuck at 11:10 am on May 26 after being recut to between 8 cm (3.5 in) to 16 cm (6.3 in). They were harvested 78 days after sticking on August 12, 2022.

Zelkova serrata 'Goshiki'. Terminal cuttings were taken from a 20-year-old stock plant between 1:00 and 2:00 pm on May 27 in sunny conditions. These 6-8 node cuttings were stuck at 2:05 pm on May 27 after being recut to between 10 cm (3.9 in) and 16 cm (6.3 in). They were harvested 91 days after sticking on August 26, 2022.

RESULTS

Significant differences and ranking of rooting responses as a function of treatments varied substantially among the different taxa.

***Cornus* H2016-019-001**. The ANOVA showed that treatments were significant for all three dependent variables (percent rooting, root number and root length). The treatments with the highest rooting percentages included Midas Hydro Rooting Gel, Clonex® Rooting Gel, Rootech Cloning Gel™, Hormex powder with 16,000 ppm auxin, and the 5,000, 7,500, or 10,000 ppm auxin quick-dip liquid treatments (**Table 1**). For average root number, Rootech Cloning Gel™ significantly outperformed all other treatments with an average of 18.8 roots per cutting compared to the control with 4.2 roots. Average root length varied from 0.8 cm for Foop Clone Gel to 2.5 cm for Rootech Cloning Gel™ (that was not significantly different from Clonex® Rooting Gel or the 7,500 or 10,000 ppm auxin quick-dip treatments). The ANCOVA showed that there were significant interactions between auxin formulation and auxin concentration for all three dependent variables (α=0.05). There was a significant quadradic regression response for percent rooting as a function of auxin concentration for liquid quick-dip treatments with percent rooting = 43.16 + 0.011 auxin concentration – 0.00000073 (auxin concentration)^2, but not for powders.

Table 1. Effect of different auxin formulations and concentrations on rooting of *Cornus* H2016-019-001.

Treatment	Auxin Concentration (ppm)	Rooting Percentage (n=78)	Average Root Number (n=77)	Average Root Length (cm) (n=77)
Control	0	40.5 f	4.2 g	1.2 fg
Clonex® Rooting Gel	3,100 IBA	88.1ab	12.7 bcd	2.5 a
Midas Hydro Rooting Gel	3,500 IBA	92.9 a	13.3 bc	2.3 ab
Rootech Cloning Gel™	5,500 IBA	85.7 ab	18.8 a	2.5 a
Foop Clone Gel	0	44.4 ef	3.1g	0.80 g
Hormex (Powder)	1,000 IBA	67.9 bcd	5.9 fg	1.7 bcdef
Hormex (Powder)	3,000 IBA	59.5 def	6.0 fg	1.2 efg
Hormex (Powder)	8,000 IBA	64.3 cde	8.5 ef	1.9 bcd
Hormex (Powder)	16,000 IBA	75.3 abcd	10.6 cde	1.6 cdef
Quick-Dip	2,500 ⅔ KIBA + ⅓ KNAA	69.9 bcd	6.1fg	1.4 def
Quick-Dip	5,000 ⅔ KIBA + ⅓ KNAA	76.2 abcd	11.9 bcde	1.8 bcde
Quick-Dip	7,500 ⅔ KIBA + ⅓ KNAA	84.5 abc	9.8 de	2.3 ab
Quick-Dip	10,000 ⅔ KIBA + ⅓ KNAA	81.0 abc	14.3 b	2.1 abc

Means in the same column with the same letter are not significantly different at $\alpha=0.05$.

Visually, some cuttings treated with higher powder and liquid concentrations showed signs of auxin toxicity, defined as browning and necrosis at the dipped portion with roots forming above the dipped portion. Cuttings treated with the gels did not display this negative effect (**Fig. 1**).

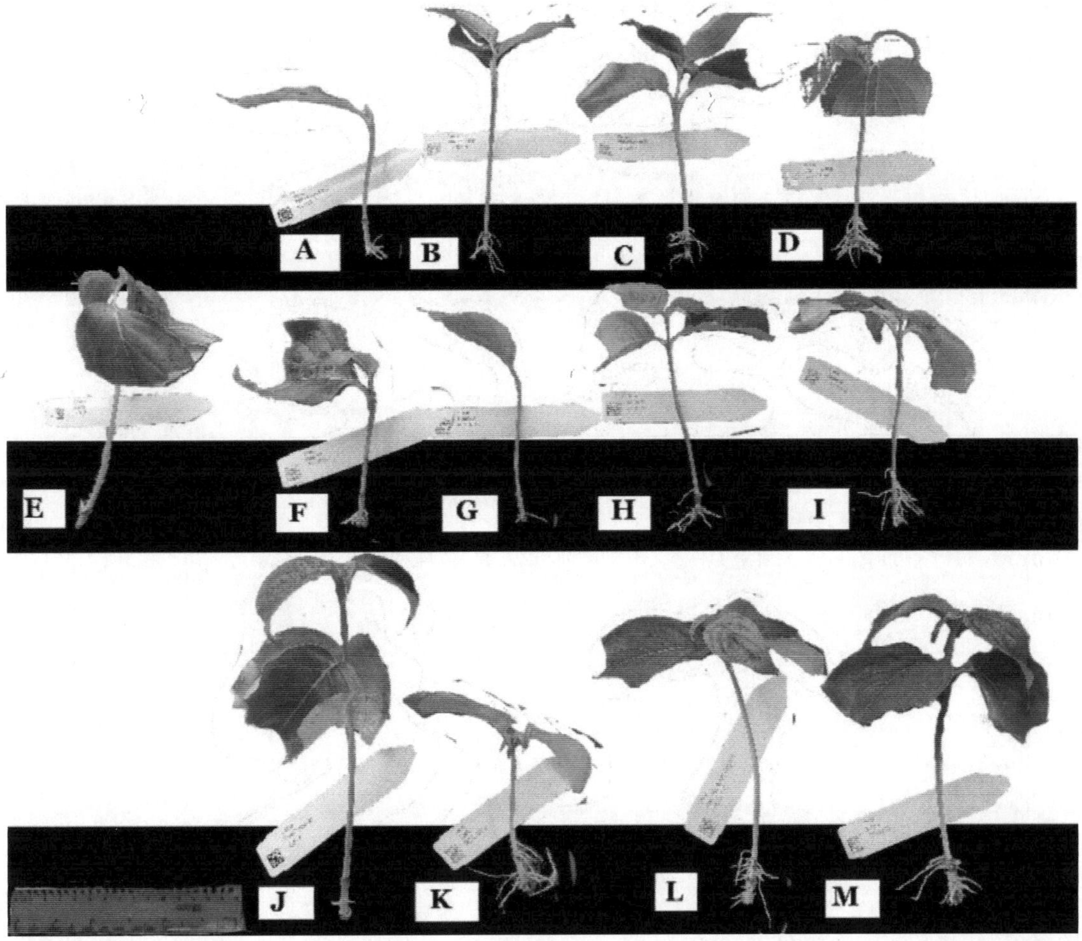

Figure 1. Representative examples of rooting of *Cornus* H2016-019-001 in response to: A-D: Liquid Quick-Dip 2,500, 5,000, 7,500, 10,000 Auxin. E: Control F-I: Hormex Rooting Powder 1,000, 3,000, 8,000, 16,000 J: FOOP Clone Gel, K: Clonex® Rooting Gel, L: Midas Hydro Rooting Gel M:Rootech Cloning Gel.

***Metasequoia glyptostroboides* 'Ogon'.** The ANOVA showed that treatments were significant for percent rooting, but not for root number and root length. The treatments with the highest rooting percentage included Clonex® Rooting Gel, Midas Hydro Rooting Gel, powders ranging from 1,000 to 16,000 ppm auxin, and liquid quick-dips between 5,000 and 10,000 ppm auxin, plus the control (**Table 2**). The ANCOVAs for rooting percentage and average root length were not significant while the ANCOVA for root number showed that there were significant interactions between auxin formulation and auxin concentration (α=0.05).

Table 2. Effect of different auxin formulations and concentrations on rooting of *Metasequoia glyptostroboides* 'Ogon'.

Treatment	Auxin Concentration (ppm)	Rooting Percentage (n=78)	Average Root Number (n=77)^{NS}	Average Root Length (cm) (n=77)^{NS}
Control	0	19.1 abc	1.1	10.8.
Clonex® Rooting Gel	3,100 IBA	35.7 a	1.4	9.1
Midas Hydro Rooting Gel	3,500 IBA	21.4 abc	1.1	11.7
Rootech Cloning Gel™	5,500 IBA	4.8 bc	4.0	1.2
Foop Clone Gel	0	2.4c	1.0	3.6
Hormex (Powder)	1,000 IBA	22.9 ab	1.6	12.2.
Hormex (Powder)	3,000 IBA	36.5 a	1.4	16.1
Hormex (Powder)	8,000 IBA	32.1 a	1.6	11.7
Hormex (Powder)	16,000 IBA	37.2 a	1.7	10.5
Quick-Dip	2,500 ⅔ KIBA + ⅓ KNAA	9.5 bc	1.5	12.8
Quick-Dip	5,000 ⅔ KIBA + ⅓ KNAA	19.1 abc	1.1	14.7
Quick-Dip	7,500 ⅔ KIBA + ⅓ KNAA	23.8 ab	1.3 .	11.3
Quick-Dip	10,000 ⅔ KIBA + ⅓ KNAA	29.5 a	1.2	6.9

Means in the same column with the same letter are nor significantly different at α=0.05. ^{NS} indicates that an LSD was not performed because the ANOVA was not significant to $\alpha = 0.05$.

There were no significant linear or quadradic regression responses for any of the dependent variables as a function of auxin concentration for liquid quick-dips or powders. Visually, cuttings treated with high concentrations of the liquid quick-dip treatment showed signs of auxin toxicity, while cuttings treated with any kind of gel showed signs of necrosis around the dipped area (**Fig. 2**).

Acer palmatum. Although percent rooting varied from 2.4 to 23.5%, there were no significant treatment effects for the ANOVA, ANCOVA, or regression analyses.

***Zelkova serrata* 'Goshiki'**. Although percent rooting varied from 0 to 9.5%, there were no significant treatment effects for the ANOVA, ANCOVA, or regression analyses.

DISCUSSION

Consistent with many propagation studies, the efficacy of different auxin formulations and concentrations on root initiation and growth varied considerably among taxa and environment (Dirr and Heuser, 2006). For *Cornus* H2016-019-001, the highest rooting percentages were achieved with rooting gels containing 3,500-5,500 ppm auxin (IBA), powder containing 16,000 ppm auxin, and liquid quick-dips containing 5,000 – 10,000 ppm auxin (mixture of 2/3 KIBA and 1/3 KNAA). It is noteworthy that cuttings treated with these gel formulations often had high rooting percentages, yet with lower auxin concentrations compared with liquid quick-dips or powders. It may be that the gels deliver a greater volume of material that adheres to the cuttings, than do liquids or powders, and thereby proved a similar dose of total auxin at a lower auxin concentration. Some of the gel formulation also include carriers/solvents like propylene glycol (e.g., Rootech Cloning Gel™) that may also enhance auxin uptake. For *Metasequoia glyptostroboides* 'Ogon', the treatments with the highest rooting percentage included two gels, powders ranging from 1,000 to 16,000 ppm auxin, and liquid quick-dips ranging from 5,000 to 10,000 ppm auxin, plus the control. That necrotic symptoms at the base of cuttings treated with gels for this conifer was a negative effect. The *Acer palmatum* and *Zelkova serrata* 'Goshiki' had low rooting percentages with no treatment effects. Consistent with previous research, these two taxa can be exceedingly difficult to root (Dirr and Heuser, 2006) and none of the treatments evaluated here proved exceptional.

ACKNOWLEDGEMENTS

Funding for this research was provided by spring Meadow Nursery, J. Frank Schmidt Family Foundation, and the North Carolina Agricultural Service. Dr. Emily Griffith provided statistical consulting. Ms. Irene Palmer and Ms. Jacquelynn Blume assisted in data collection and organization.

LITERATURE CITED

Amazon. 2022. Amazon best sellers: Best patio, lawn & garden. https://www.amazon.com/gp/bestsellers/lawn-garden/ref=pd_zg_ts_lawn-garden (Accessed: 17 July 2022).

Blythe, E.K., Sibley, J.L., Tilt, K.M., and Ruter, J.M. (2007). Methods of auxin application in cutting propagation: A review of 70 years of scientific discovery and commercial practice. J. Environ. Hort. *25*:166–185. https://doi.org/10.24266/0738-2898-25.3.166.

Bowden, A.T. *et al.* 2021. Evaluation of Surfactants for Use in One-Time Foliar Auxin Applications in the Propagation of Woody Ornamentals. Comb. Proc. Inter. Plant Prop. Soc. *71*:100–106.

Caplan, D., Stemeroff, J. Dixon, M., and Zheng, Y. (2018). Vegetative propagation of cannabis by stem cuttings: effects of leaf number, cutting position, rooting hormone, and leaf tip removal. Can. J. Plant Sci. *98*:1126–1132. https://doi.org/10.1139/cjps-2018-0038.

Dirr, M.A. and Heuser, C.W. (2006). The reference manual of woody plant propagation: From seed to tissue culture. 2nd ed. Timber Press, Portland, Ore.

Google. 2022. Google Trends, Google Trends. https://trends.google.com/trends/explore?date=all&geo=US&q=%2Fg%2F11nns0z61j (Accessed: 17 May 2022).

Hepburn, H.A. and Matthews, S. (1986). Influence of proprietary rooting compounds and basal internode length on the rooting of potato stem cuttings. Potato Res. *29*:391–393. https://doi.org/10.1007/BF02359967.

Hormex. (2020). Advantages of Hormex Rooting Powder vs. rooting gel. https://hormex.com/blogs/plant-growth-101-blog/advantages-of-rooting-powder-vs-rooting-gel. (Accessed: 25 August 2022).

Hydrodynamics International. (2022). Learn the facts about Clonex Rooting Gel. https://www.hydrodynamicsintl.com/learn-facts-clonex-rooting-gel/ (Accessed: 31 May 2022).

Ismaili, H. (2016). Study of some forms of IBA in the rooting process of the olive. Int. J. Curr. Microbiol. App. Sci. *5*:239–246. https://doi.org/10.20546/ijcmas.2016.503.029.

Mendel, K. (1992). The history of plant propagation methods during the last 70 years. *Acta Hort. 314:*19–26.

Ryals, J.B. *et al.* (2018) 'Propagation of Shumaka™ Crape Myrtle©', p. 8.

Technaflora Plant Products. 2022. *Rootech Cloning Gel™*. https://www.technaflora.com/product/rootech-cloning-gel (Accessed: 16 July 2022).

Ticknor, R.L. (1981). A comparison of several hormone formulations for rooting cuttings. Proc. Inter. Plant Prop. Soc. *31*:109–111.

Combined Proceedings IPPS 72:151-172. 2022.

Teaching in an Arboretum: Spartanburg Community College Horticulture, the First 50 years

Kevin Parris, Ph.D.

Spartanburg Community College, 107 Community College Drive, Spartanburg, SC 29303, USA

parrisk@sccsc.edu

Keywords: college enrollment, horticulture, arboretum, sustainable agriculture, hands-on experience, inspiration, persistence, Spartanburg Community College (SCC)

Summary

The green industry is experiencing a shortage of skilled labor, yet the number of institutions offering horticulture degree and certificate programs has been on the decline. In this paper the 50-year history of the Horticulture department at Spartanburg Community College (SCC) is presented as an example of a traditional horticulture program that has been able to adapt, survive, and thrive.

INTRODUCTION

During the 2021 International Plant Propagators' Society (IPPS) Southern Region in Mobile, Alabama, concerns about skilled labor shortages and enrollment decline across many of the region's horticulture programs were voiced in the Question Box Session. These concerns have been quantified in various literature. Fifty-three percent of institutions eliminated horticulture programs between 1997 and 2017 (Brown et al, 2019), and enrollment of undergraduate baccalaureate students decreased 19% between 2004 and 2012 (Reed et al, 2016).

The answer to these concerns may be the recognition of shifts in student demographics. In a study by Choloupka et al (2018), 39.7% of students were female, and 30.3% were non-traditional (25 or older). Frequency of working professionals taking part-time course loads also made it difficult to clearly capture enrollment data, especially in two-year programs (Brown et al, 2019). There is agreement among these sources that shifts in recruitment strategies and revealing details about how certain programs are thriving will be necessary to reverse the downward trend. Student enrollment at SCC has been relatively stable for most of its history. Pre-2007 numbers are incomplete, but it is known that the program began with about 14 students in 1971 and trended upward for 20 years. Student enrollment of 40 to 60 was common through the 1980s-early 2000s. A spike occurred between 2008 and 2012 with headcounts approaching 100 (**Fig. 1**).

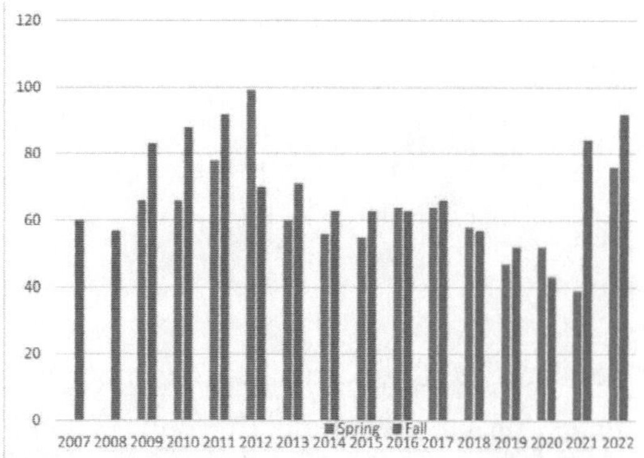

Figure 1. Student enrollment in the horticulture program at Spartanburg Community College (SCS), Spartanburg, South Carolina, 2007-2022.

A brief enrollment drop occurred just before 2020, but rebounded strongly post COVID-19 quarantines (**Fig. 1**), due in part to a tuition-free initiative made possible with COVID-19 relief funds and an institutional scholarship program made possible by community partners. The collective forces of enrollment (quantity and quality), faculty stability, administrative support, and community engagement have the program positioned for growth. I hope the history of the horticulture program at SCC will serve as an inspiration for others to stabilize their programs, grow enrollment, and provide eager minds and skilled hands for industry.

INSPIRATIONS

In 50 years of working with students in the midst of a campus arboretum (**Fig. 2**), inspiration has come from a wide array of individuals and gardens. Trends have shifted, science has advanced, and technology has evolved; however, the formula has remained the same.

1. Present the latest, relevant information inside the traditional classroom.
2. Put that information into practice in the outdoor classroom (**Fig. 3**).
3. Observe what others are doing, and assimilate (**Fig. 4**).
4. Repeat.

This loop of learning has lasted half of a century because of the interactions of people, plants, and places. The result is something that continues to exceed the sum of the parts, much like the photosynthetic process. If you ask any instructor, current student, or graduate of the program what they have been most inspired by, they are likely going to list experiences that took place outside the walls of the classroom, on a field trip, or during an event where they heard from the experiences of notable, and noble horticulturalists (**Figs. 5-10**). Exogenous inspiration has permeated the program. A universally important lesson was delivered to students during the fall 2004 field trip to the Raleigh-Durham area. In a serendipitous encounter with Dr. Michael Dirr at the J. C. Raulston Arboretum he shared he was visiting the garden while his daughter Susy was receiving treatments for her battle with cystic fibrosis. In an incredibly generous donation of personal time, Dr. Dirr taught us that a walk through a garden is good for the soul. ***Horticulture is a noble profession because gardens have the power to heal.***

Figure 2. The Giles Campus of Spartanburg Community College is nestled into 50 years of cultivated vegetation, serving as a 40-ha (100-ac) classroom.

Figure 3. Jay Moore (white beard) demonstrating tree injection to Urban Tree Care students, Fall 2016. (Right) Jason Bagwell (center) emphasizing the importance of spreader calibration to Turfgrass Management students, Fall 2019.

Figure 4. (Left) Faculty and students were treated to a tour of the J.C. Raulston Arboretum with Dr. Michael Dirr (red cap), after bumping into him on a fall day in 2004. (Right) Dr. Bruce Fraedrich (far right) demonstrates methods to test wood strength with SCC students during a spring 2016 field trip to the Bartlett Research Laboratories & Arboretum.

Figure 5. Scott McMahan (1994) and Alaina Mansueto (2019), SCC graduates, welcomed us to ABG-Gainesville on a fall 2019 field trip.

Figure 6. (Left) Mark Weathington (center) discusses the iconic specimen of *Lagerstroemia fauriei* 'Fantasy' with SCC students at the J.C. Raulston Arboretum , fall 2021. (Right) Tony Avent (white shorts) welcomes students to Juniper Level Botanical Garden and Plant Delights Nursery, fall 2016.

Figure 7. (Left) Elden LeBrun talks about the importance of codominant leader suppression at Bartlett Research Laboratories & Arboretum. (Right) Jason Bagwell and students gather under the Centennial Oak, *Quercus macrocarpa*, during a 2007 field trip to Clemson University.

Figure 8. Wayne Nicholson demonstrating the grafting and production of Japanese Maples at Pacific East Nursery, Lyman, SC is an annual field trip.

Figure 9. (Left) Retired Clemson professor Dr. David Bradshaw (holding deerskins) talked to our group about everything from sustainable agriculture, to making watercolor paint and deerskin clothing at his home in 2012 on a spring break adventure day. (Right) Andy Cabe welcomes the SCC Horticulture Department to Riverbanks Zoo and Botanical Garden in 2017.

Figure 10. (Top) Jenks Farmer (second from left), with SCC Faculty following his 2016 Arboretum Adventures presentation. Jenks has lectured to our classes on multiple occasions and provided internship opportunities. (Bottom) 2019 Arboretum Adventures guest presenter Brie Arthur (left) signs a copy of her book, *The Foodscape Revolution*, for dedicated event attendee and program supporter Dr. Larry Roël.

CAMPUS, PROGRAM, AND ARBORETUM DEVELOPMENT

The Early Years. Prior to 1930, the Spartanburg County Home and Farm were located on the property, which served as the County Poor House possibly as far back as the Civil War era. The County Home structure was last used in the mid-1980s as the Davenport Rehabilitation Center, which was demolished in the late 1980s. The site of the former structure has a landscaped perimeter and remains a potential building site.

In 1961 funding for a Technical Education Center in Spartanburg was approved with classes beginning in 1963. The TEC's 1st building is now known as the Hull Building in honor of college's first President, Dan P. Hull. A second building was added to the site in 1968 and later named in honor of former college President, Jack Powers.

This early phase of campus development preceded the existence of the Horticulture Department. Still, some very important broad strokes landscaping was initiated around the Hull and Powers Buildings. From these early plantings, venerable specimens of *Taxodium distichum*, *Quercus virginiana*, *Liquidambar styraciflua*, *Plantanus occidentalis*, *Ilex vomitoria* 'Pendula', *Ilex cornuta* 'Burfordii', and *Cornus florida* remain (**Fig. 11**). The presence of these specimens, particularly the *Taxodium* provides another universally important lesson for students, many of whom represent the first generation of their family to attend college. ***If a tree from the swamp can survive, and even thrive on a dry, west facing slope on a college campus; students can also have the power to grow and exceed expectations in a new environment.***

Figure 11. (Left) 1970 Aerial of the property. (Center) Looking into the canopies of trees planted in 1963 of a robust *Cornus florida* and (Right) a noble *Quercus virginiana*.

Horticulture Arrives. In 1971 the Horticulture Program began with a focus on pomology under the leadership of Jimmy Painter (**Fig. 12**). Jimmy ran the department single-handedly for a decade, fighting hard for its existence every step of the way. In 1980 the Gaines Building was added, and Jimmy's students were actively involved in the landscape installation. The tradition of the horticulture faculty and students being involved in campus development projects as well as projects at satellite campuses continues to this day.

Figure 12. Jimmy Painter (arrow) and the first class of horticulture students in 1971 with the old Case tractor used by the department for many years.

After graduating from Clemson in 1981, Doug McAbee, a 1977-78 student of Jimmy Painter, joined the staff. Garden development became a focus and a shift toward ornamental horticulture was in full swing after Doug's arrival (**Fig. 13**).

Figure 13. Doug McAbee (right) taught landscape design throughout his career at SCC.

The Horticulture Gardens began to have structure in the 1980s thanks to several hardscape projects implemented by students.

The 1980s also marked the timeframe of the construction of a new nursery production area. These facilities gave the department a way to grow plants for annual sales, as well as propagate plants for the campus landscape. Funds generated by these sales have helped keep learning opportunities available to decades of students by supplementing scholarships, field trip expenses, equipment acquisition and repair, and the purchase of plants and building materials for campus improvement projects (**Figs. 14-15**). Production of plants uncommon in the nursery industry, but worthy of more widespread appreciation, has been a focus for many years. This ideology was inherited from Dr. J.C. Raulston of North Carolina State University (NCSU), and we are proud to continue his mission in any small way that we can.

Figure 14. (Left) Students installing trees and shrubs in the Horticulture Gardens in 1981 and (Right) working on site preparation for the nursery production area in the early 1980s.

Figure 15. (Left) Jimmy Painter (arrow) helps customers during a 1980s plant sale. (Right) Kevin Parris (arrow) working with students Delaine Childress and Stephen Parris in the propagation house in 2016. Several thousand plants are germinated, rooted, or grafted within this structure each year.

In 1990, Kevin Parris began as an adjunct instructor and consultant for new plant additions to campus while working full time at Gilbert's Nursery as their plant propagator. Infill and diversification of the Horticulture Gardens with shrubs, ornamental grasses, and herbaceous perennials continued. Doug focused on the study and acquisition of herbaceous material while Kevin served as a pipeline of woody plant material from Gilbert's. Horticulture Garden expansion to the western property line and an Urban Forestry Grant also contributed to plantings that continued into 1999. Jimmy Painter championed these perimeter plantings and grant work. It is hard to image the arboretum today without these maturing specimens. Jimmy brought together the services of Kevin with Tipton Pitts, a former STC student and graduate of the Landscape Architecture program at Clemson University. Together they drafted the plans that helped Jimmy secure the grants. This collaboration was fitting given that Tipton's father Irvin was an architect for several of the college's first buildings.

Jason Bagwell came on board as an adjunct instructor in 1998 and became the

3rd full-time instructor in 2000. Jason's addition was vital as it coincided with a period of increased enrollment and a phase of new construction and building renovation on campus. Within eight years of Jason's arrival new projects transformed the campus landscape into what most people are familiar with today (**Fig. 16**).

Figure 16. A young Jason Bagwell in the International Peace Garden just one year after construction, 2002.

During these years Jimmy, Doug, and Jason worked steadily on the implementation of these projects in collaboration with Campus Operations Director Tommy Bulman, with support from College President Dan Terhune. This flurry of effort, following years of use of the campus landscape for education, led the Spartanburg County Commission for Technical Education to officially designate the campus as an arboretum in the spring of 2005. SCC has been a member of the American Association of Botanical Gardens and Arboreta (AABGA), now known as the American Public Garden Association (APGA) since that time.

1999. Blue Jay Way, a connector road, was built above the horticulture gardens to improve traffic through campus with access to two new parking areas. While roads and parking areas were a loss of green space, these hardscape additions created better access and opened new vistas into the Horticulture Gardens.

2000. The Health Sciences Building was completed, becoming the site's first multistory building, providing visibility and new landscaping along Business 85. The landscape partially adopted a medicinal theme.

2001. The Horticulture Pavilion was completed along with associated landscaping and additional buffers from Business 85 road noise (**Fig. 17**).

2002. The façade of the Hull Building was remodeled along with a reconfiguration of adjoining parking and green spaces (**Fig. 18**).

2003. The construction of the Terhune Student Services Building tied the core of the campus landscape to its perimeter (**Fig. 19**).

2006. The Library Learning Resources Building was built within the core of campus, which also turned some asphalt into green space. In addition to campus development driven projects, the horticulture department initiated their own projects between 2001 and 2007 to better diversify plant collections within the heart of campus. This allowed for class time to be more easily split between indoor and outdoor learning opportunities. These projects were able to come together with revenue from plants sales, private donations and the partnership of the SCC Foundation.

These planting designs were typically hammered out during brainstorming sessions while on overnight field trips to favorite garden and nursery destinations.

Figure 17. (Left) The Horticulture Pavilion under construction. (Right) Diverse plantings now grace the approach to the Horticulture Pavilion.

Figure 18. (Left) View from within the canopy of a Chinese Fringe Tree planted following the renovation of the Hull Building. (Right) An approach to the Hull Building guides you under the canopy of a Sugar Maple planted in the 1980s, toward an entrance framed by Bottlebrush Buckeye, an intersection of walkways shaded by Orange-Flowered Tea Olive, and Shawnee Brave Bald Cypress, all planted in the early 2000s, towering in the background.

Figure 19. Terhune Student Services Building. (Left) Soft pink panicles of Near East Crepe Myrtle brighten a pathway. (Right) Kay Parris Magnolias provide a screen for a utility area and loading dock.

This tradition often included students circling the round table, cheering or jeering the ideas that were being bantered back and forth. Jimmy, Doug, Jason, and Kevin would frequently use sports terminology to affirm what made it from a wild notion, to pen, and paper. Homeruns, slam dunks, and touchdowns have become a portion of the living collections within the campus arboretum.

2001. The George and Sissy Stone International Peace Garden was an extensive landscape renovation between the B and C wings of the Powers Building. Incorporation of species from many different regions of the world, in comfortable cohabitation, continues to be the theme of this niche garden, and the entire campus (**Fig. 20**).

Figure 20. (Top Left) Students adding herbaceous perennials to the International Peace Garden early in its development. (Top Right) Jimmy Painter in 2002. (Bottom Left) A 2020 rooftop view of the garden. (Bottom Right) No caption needed.

2005. The Sallie Barre James Plant Zoo, inspired by the diversity of Tony Avent's Juniper Level Botanical Garden, became the second niche garden.

A renovation of the space between the A and B wings of the Powers Buildings, the garden contains plants with animals in their names. It also includes a display of hardy palm species (**Fig. 21**).

Figure 21. (Left) Jason Bagwell (arrow) works with students on the construct of a water feature in the Sallie Barre James Plant Zoo in 2005. (Right) Rooftop view in 2020.

2007. The Garden Railroad was built on the north side of the Library Learning Resources Building in the spring semester (**Fig. 22**). A small amphitheater with a canopy of *Ulmus americana* 'Princeton' was included to provide an outdoor classroom within a secluded greenspace cloaked with *Cryptomeria japonica* 'Radicans' and *Osmanthus fragrans* 'Aurantiacus'. The railroad and amphitheater were scaled down versions of these features seen at the Morris Arboretum and Swarthmore College, respectively.

Figure 22. The train crosses the bridge in the Garden Railroad in 2007.

THE NEXT GENERATION

The year 2007 was pivotal for the horticulture program, and the arboretum. At the conclusion of the spring semester, Jimmy Painter retired after building and guiding the program for 37 years. Jimmy's contributions to the campus, his students, and the horticulture industry are immeasurable. At the time it was difficult to envision the program moving forward without him, but his vision and energy persisted in his colleagues and former students. Support for the continuation of the program was overwhelming. Jason stepped forward to assume the role of Department Chair and began to come up with a plan for the fall semester so he and Doug could manage coarse loads with the help of adjunct instructors.

Having been an adjunct for 17 years, the call to provide more help stirred Kevin's soul. Before the start of the fall semester, he sold his shares in The LandArt Design Group, Inc. and enrolled in graduate school at Clemson University to obtain the Master's degree necessary to maintain a position as a full-time instructor. From 2008-2011 he taught a full course load at SCC while doing a mix of course work at Clemson and research at The North Carolina State University Mountain Crop Improvement Lab under the direction of Tom Ranney to fulfill his degree requirements (**Fig. 23**).

Figure 23. Summer graduate school experiences, including a hike to Roan Mountain with Dr. Tom Ranney (standing, back left) and a team from the NCSU Mountain Crop Improvement Lab were important influences on Kevin Parris between 2008 and 2010.

Having become deeply involved with the study of Magnolia genetics and breeding, Kevin continued to do research and completed his Ph.D. from Clemson in Plant and Environmental Sciences with research advisement from Donglin Zhang at University of Georgia in 2018. These pursuits helped build connections with colleagues at a wide range of institutions, leading to the SCC Arboretum becoming one of the nation's most diverse collections of Magnolia taxa.

By 2008 Doug McAbee had retirement in his sights and knew exactly who he wanted to fill his shoes. He went to visit one of his favorite former students, Jay Moore, owner of Carolina Garden World. Much like Kevin had done the year prior, Jay Moore answered the call. With some graduate credit already in hand, Jay became an adjunct instructor in 2009, all while juggling the demands of teaching, owning a business, and traveling to Clemson to complete his M.S. in Plant and Environmental Sciences. For Jay and Kevin, 2008-2010 were the sleepless years, and Jason tirelessly held down the fort while they cleared all their academic hurdles. Doug's retirement in 2010 marked the beginning of Jay's full-time employment and the addition of Kelly Lewis as an adjunct to help manage course loads and take charge of greenhouse production. Despite the long days and sleepless nights, the desire to provide students with memorable experiences continued to drive projects in the arboretum, which became much more extensive thanks to Jay's construction skills. The years 2010-present, can simply be captioned "Jay Moore and students built…". These years also brought another universally important lesson to the forefront. Life will present many opportunities. ***If you can balance, prioritize, and hybridize what life brings your way, the interactions can yield products that exceed the sum of their parts.***

2008. A new parking area in front of the library created an opportunity to plant a median with *Nyssa sylvatica* 'Wildfire'. Several trees were salvaged from the planned footprint of the parking lot with a last-minute relocation. New trees were added in various locations throughout the campus. The hillside behind the Health Science Building was renovated into a *Buxus* collection following a donation by Saunder's Brothers Nursery in Virginia.

2009. "Mac's Seating Area" was built in honor of Doug's coming retirement with a new *Prunus* ×*yedoensis* planted to generate fast shade for him to sit under when he visited (**Fig. 24**). A Green Roof Gazebo was built within the Horticulture Gardens. A renovation to the lower entrance to the Ledbetter Building added *Stewartia pseudocamellia*, *Halesia tetraptera* 'Wedding Bells', and *Magnolia maudiae* to the collections. Kevin traveled to China for two weeks in May and gave a presentation to the community in October. This marked the beginning of the annual fundraising event: ***Arboretum Adventures***.

Figure 24. Horticulture faculty in "Mac's Seating Area" 2009. (Left to right) Kelly Lewis, Jason Bagwell, Doug McAbee, Kevin Parris, Jay Moore. Doug served SCC for 28 years, 21 of

those as a full-time faculty member. Doug passed away in 2018, but he will always be at the heart of the horticulture program and gardens.

2010. The pedestrian bridge and associated garden areas were renovated in the heart of the old garden area. Deciduous azaleas donated by Bob Head were planted on the creekbank flanking the pedestrian bridge. The International Peace Garden was renovated with the addition of an expanded paver entrance, and planting diversity was enhanced. Inspired by Kevin's hike to the peak of Roan Mountain, the construction of the Garden of Convergent Waters began in front of the library with the addition of a grove of *Gymnocladus dioicus*.

2011. The Convergent Waters feature was completed surrounded by a landscape reminiscent of a mountain bald (**Fig. 25**). Inspired by Jay Moore's adventures in the National Parks of Utah, construction of a Xeric Garden also began (**Fig. 26**). *Magnolia macrophylla* and *Magnolia acuminata* groves were planted near the Horticulture Garden entrance. *Magnolia officinalis* were planted behind the Health Science Building, and the first magnolia trial beds were planted along the roadway to the nursery production area.

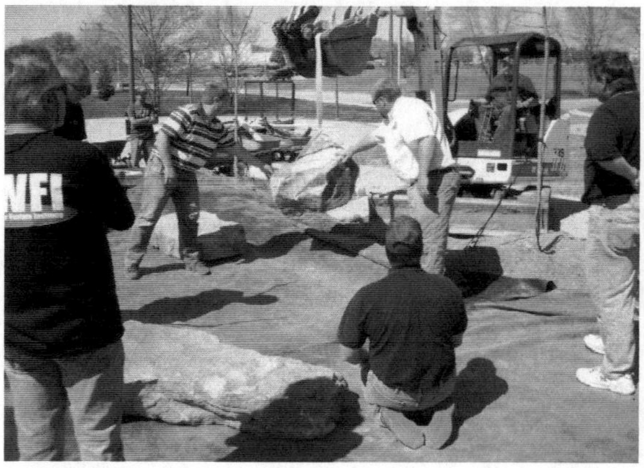

Figure 25. An industry sponsored workshop brought students, faculty, graduates, and area landscape professionals together to build the Garden of Convergent Waters in the spring of 2011.

2012. The Louie Phillips Memorial Xeric Garden was completed, and the Health Science parking lot addition provided opportunities for more diversity.

2013. The department assisted with the planning and renovation of the original Spartanburg High School, turning it into the SCC Downtown Campus/ Evans Academic Center. The landscape was completely renovated, including an interior courtyard. On the central campus a tired hillside of *Juniperus conferta* became a diverse collection known as Mt. McKinley, named for the student who did all the demolition and soil preparation by himself. The front of the Powers Building was also renovated after the removal of a monoculture of *Ilex vomitoria* 'Schelling's'. These plantings featured *Ginkgo biloba* 'Saratoga' espaliered on each section of wall.

2014. Linda's Hill was planted above the stone wall built prior to the Growing Great Gardeners event (**Fig. 27**). The Balmer Fountain and associated plantings were added to the Downtown Campus.

Figure 26. (Top Left) The courtyard between the C and D wings prior to renovations and (Top Right) Plan for the space to become a waterwise demonstration with a southwestern theme. (Bottom Left) Jay Moore (wheel barrel) and students planting the Xeric Garden wagon wheel beds in 2012. (Bottom Right) Rooftop view of the garden in 2020 while the *Agave salmiana* var. ferox was flowering. The central bed is now a crevice garden.

Figure 27. (Left) In the fall of 2013 SCC hosted The Growing Great Gardeners Symposium with Fergus Garrett (left) and Aaron Bertelsen from Great Dixter Garden in East Sussex, UK. Fergus also jumped in and helped break ground with students in an area called Linda's Hill (Right), in honor of our mutual gardening friend, Linda Cobb.

2015. Business 85/ New Cut Road Interchange project extended the arboretum into the gateway to campus. The bank of New Cut Rd, the frontage road, and the perimeter of the old stand of *Pinus taeda* were dotted with noble trees and masses of shrubs **(Fig. 28)**. The Heath Science courtyard had another renovation including the construction of swings and a stone seating wall. New parking islands in the center of campus were planted with new material. Students began referring to the largest of these as "Parris Island".

Figure 28. (Top and Bottom Left) Students planting the New Cut Road / Business 85 interchange in 2015, and (Top and Bottom Right) after establishment.

2016. After the maintenance of the train in the Garden Railroad became problematic, it was converted into an Asian Garden given there were established dwarf conifers and several *Acer palmatum* cultivars (**Fig. 29**). The new infill of evergreen groundcovers, hostas, and hardscape features transformed the space into one of the most beautiful gardens on campus. A stone staircase and new plantings were added to create an entrance to the south end of the Horticulture Gardens.

Figure 29. A peaceful view from within the Asian Garden.

2017. The Synthetic Putting Green and garden area was designed and built (**Fig. 30**). The Powers Building B-Wing Container Garden and Horticulture Television Location were developed after drainage renovation work displaced the 2014 landscaping along that wall. On October 23rd, an F2 Tornado moved across campus, severely damaging or uprooting over 60 trees within campus and over 100 *Pinus taeda* in the 85/ New Cut Rd. interchange. Miraculously, none of the new interchange landscaping from 2015 was damaged by the heavy wind and falling pines.

Figure 30. (Left) Students preparing the base for the synthetic putting green. (Right) The completed green and surrounding garden.

2018. The Sustainable Agriculture Facility was built to support a new certificate, signifying the program had come full circle (**Fig. 31**).

A new orchard, greenhouse, and garden plots were built where Jimmy's orchard was in the 1970s. A Mary Black Foundation grant supported the construction of the facility.

Figure 31. (Left) A fallow piece of ground was transformed into the Sustainable Agriculture Facility in 2018. (Right) The Red Barn contains a classroom, kitchen, and storage for harvested produce. A greenhouse is equipped for hydroponic and aeroponic production. Field plots and a small orchard complete the site. The development of this site truly brought the program full circle as this location was the original site of an orchard used to teach pomology in the 1970s.

Magnolia trial beds occupy the lower portion of this field, providing evidence of the diversity of our efforts. The Campus Green was completed, replacing a parking lot that had existed since the 1960s. At the 85/ New Cut Road Interchange a Primitive Forest was planted in the footprint of the old Pine grove eliminated during the 2017 Tornado. The Primitive Forest was funded by the Balmer Foundation and the Noble Tree Foundation. Other projects in 2018 included the Hub Garden and Xeric Garden Water Feature Renovation.

2019. The Culinary Arts Bistro Garden was a major renovation effort that started in 2018 with demolition efforts. The paver work was some of the most detailed our students have ever done (**Fig. 32**). The International Peace Garden was also renovated with a dry-stacked stone wall, new evergreen groundcovers and herbaceous perennials. New equipment sheds adjacent to the greenhouse and nursery production areas were built to better house the tools needed maintain the arboretum.

Figure 32. Seeing projects move from the conceptual stage (top left) through the construction process (top right, bottom left) and into a garden management phase (bottom right) is one of the most valuable experiences Spartanburg Community College provides to horticulture students.

2020. The parking and entrance renovations to the Terhune Student Services Building provided opportunities to install new trees. A fitting project for the spring of 2020, a semester cut short due to the COVID-19 Pandemic, was the *Parenchyma Feature*. A small number of students were allowed to return for a few hands-on instruction lab periods to complete this feature (**Fig. 33**). Parenchyma cells make up the rays and rings within a tree, providing a living impediment against the pathogens that lead to decay, defining the walls of Dr. Alex Shigo's CODIT model (Compartmentalization of Decay in Trees). Students now walk by the Shigo model daily.

Surrounding the symbolic stonework and turf is one of the most diversely planted gardens on campus.

In the fall semester project, Paisley Beds were planted with a mosaic of plants artfully bisected with stone pathways near the Terhune Student Services Building. 2020 also marked the retirement of College President Henry Giles who served the college for 50 years. To honor him, the Central Campus location is now known as the Giles Campus.

Figure 33. (Left) Students place stones symbolic of a wall of "compartmentalization of decay in trees" (CODIT) in the Parenchyma Feature. (Right) The completed garden area.

2021. The COVID-19 pandemic gave all of us a greater appreciation for our outdoor spaces. In response to a request from the Math Department, a covered patio area behind the Powers Building was expanded and seating walls were added to create a new outdoor classroom. The 50th Anniversary celebration on Sept 23rd was highlighted by completion of The Roel Bridge, the first vehicular bridge in the gardens which now provides a direct route to the Horticulture Pavilion and equipment sheds. It is a symbolic bridge to the future.

A respect for the past 50 years keeps us focused on the next 50. We prepare for the future, by caring for the past, in the present. Our campus operations staff consistently employees several graduates of the program, and their partnership is essential to the condition of our classroom. The growth of the program and arboretum have also been dependent on forward thinking administration, the support of the SCC Foundation, the SCC Marketing Department, donors from the community, and friends from arboreta and botanical gardens far and wide. We are grateful for our mentors and we are charged with the task of propagating their inspirations. We are proud of students, who in turn, carry those inspirations with them. We love what we have planted and cultivated. We are humbled when friends, graduates, colleagues, and mentors come to see what we have planted (**Fig. 34**). Every campus should be an arboretum!

Figure 34. A visit from the NCSU Mountain Crop Improvement Lab, July 2021. The Convergent Waters Garden, inspired by a 2010 hike on Roan Mountain, led by Dr. Tom Ranney (far right).

Combined Proceedings IPPS 72:173-178. 2022.

A New Way to Serve

Samantha Manning

Veteran's Farm of North Carolina, Inc., Cameron, North Carolina 28326, USA.

vetfarmofnc@gmail.com

Keywords: Career Skills Program (CSP), Farm Military Assistance Program (FARM MAP), Transition Assistance Program (TAPS), Skillbridge, leadership, nursery industry profession, superior problem-solving skills, work ethic, Veteran's Farm of North Carolina, Inc. (VFNC), Veterans Agricultural Training and Education Program (VATEP)

Summary

Agriculture, the green industry and horticulture can help returning veterans integrate back into civilian life with productive careers. Likewise, the human talent and skillsets that veterans offer can be a great employee resource for the green industry. After being first introduced to The Veteran's Farm of North Carolina (VFNC) in 2016, I became a founding board member and was inspired to pursue a career in agriculture. The Veteran's Farm of NC, Inc. (VFNC), a 501(c)(3) non-profit organization dedicated to assisting veterans and military personnel/unit commands with consultation, training and education, equipment usage, and land acquisition. Through a grant from USDA/NIFA and thanks to partnership with other industry leaders, VFNC now offers three on farm training programs: 1) Veterans Agricultural Training and Education Program (VATEP), 2) Farm Military Assistance Program (FARM MAP), and 3) Skillbridge Internships. The flagship program for VFNC is VATEP, a six month on farm training program for veterans. FARM MAP is a more concise program of on farm training lasting two to four months for transitioning service members.

Skillbridge internships offer transitioning service members on farm or on the job training in the agricultural industry https://skillbridge.osd.mil/. In addition to these on farm training programs, VFNC continues to offer mentorship, networking, and an extensive equipment loan program.

INTRODUCTION

In North Carolina, agriculture (including the nursery industry) is the number one source of income for the state, followed closely by the United State military. While historically there was much overlap between the farming and military populations - not only in NC, but in the US as a whole - these two career paths began to separate - following the passage of the GI Bill at the end of WWII. Small farms went to the wayside with the introduction of expensive massive machinery and large monoculture farms. As the average age of farmers has steadily increased, fewer people are seeking careers in agriculture and horticulture due to a number of barriers. Vet Farm seeks to remove those barriers and guide transitioning servicemembers and veterans to a fulfilling career or farm ownership.

Because of these issues, the agriculture industry (including the nursery industry) is in dire need of new farmers and professionals. Supply chain issues during COVID highlighted this need more than ever. It is vital that we have in place small local suppliers of agricultural products and skilled agricultural workers in their respective communities. Veterans are the ideal population to meet this need. They bring the strong work ethic, management experience, intuitiveness and interest required to be successful - whether they are building a small business or pursuing a career in agriculture. In return, veterans gain a new purpose and place in the community.

The barriers transitioning soldiers and veterans face when leaving the military and pursuing a career in agriculture are: 1) lack of technical agricultural skills and knowledge, 2) land acquisition, 3) lack of resources, 4) need for mentorship, 5) need for networking opportunities, and 6) obstacles to obtaining equipment. The Veteran's Farm of North Carolina (VFNC) seeks to address and remove each of these barriers. Our training programs provide a wide range of agricultural skills and knowledge in numerous production models, providing hands on practice and the practical knowledge required to develop these skills. Participants are educated in how to find land to suit their needs and how to obtain financing in creative ways to meet their goals. Curriculum includes education in navigating the wide range of resources available from government agencies such as USDA, National Resource Conservation Service (NRCS), Farm Service Agency (FSA), Cooperative Extension, as well as non-government agencies such as Farm Credit, Veteran's Business Outreach Center, and more. Participants receive ongoing mentorship throughout the program and beyond to help them achieve their goals, whether that is starting a small farm business, obtaining a college degree in the agricultural field, or seeking an agricultural career. VFNC is there to encourage and advise them every step of the way. Finally, transitioning service members and veterans have a vast networking opportunity with our partners and with each other - giving them a new military community and assistance with finding their place in the civilian community.

CURRENT TRAINING PROGRAMS

Veterans Agricultural Training Program (VATEP) is VFNC's flagship training program for transitioning service members and veterans. VATEP is a new 6-month training program at our 21.5 ha (53 ac) farm located in Cameron, North Carolina. This program is partnered with Fayetteville Technical Community College and students will complete online coursework in a variety of subjects with hands-on, practicums to be completed on the farm. The VATEP program provides participants with hands-on, on-farm training and education for numerous models of agricultural production. Main topics for the curriculum are: farm safety, soil science, livestock husbandry, crop production, horticulture, agritourism, niche-products, business plans, carpentry, mechanics, electrical and plumbing. With VATEP, VFNC seeks to bridge the gap between a beginning farmer and other educational programs, providing practical skills and knowledge required for starting a small farm business or seeking a career in agriculture (**Fig. 1**).

Figure 1. Students learning farm safety as part of Veterans Agricultural Training and Education Program (VATEP) on farm training.

The second program VFNC currently offers on farm is the Farm Military Assistance Program (FARM MAP). This is a shorter, more condensed two-to-four-month program offered as part of the Career Skills Program (CSP) in partnership with the U.S. Army at Ft. Bragg. Staff members of VFNC actively recruit transitioning service members at the Soldier for Life/Transition Assistance Program Office (SFL/TAPS) on post while they are in their last year of service. The program is also available to spouses as space allows. The FARM MAP program provides participants hands-on, on-farm training and education at VFNC's farm. This more concise program offers intensive education in the development of a working farm plan, soil science, livestock husbandry, crop production, & horticulture skills. Participants are also introduced to the variety of educational and career opportunities available in agriculture. Through a prior partnership with North Carolina State University as Soldier to Agriculture Program (STAG) and now our FARM MAP program, VFNC has been able to provide training to over 300 transitioning servicemembers and veterans so far. These classes are continuing in an ongoing basis every 6-8 weeks (**Fig. 2**).

Figure 2. Farm Military Assistance Program (FARM MAP) participants learning hands-on plant propagation in our training greenhouse.

The third program available through VFNC is our Department of Defense Skillbridge Internship opportunity. We are currently working to expand this program, offering on the job training and internships to transitioning servicemembers with agricultural employers. This program combines the on-farm training offered through VATEP and FARM MAP with career opportunities. We are actively recruiting interested employers to participate in this job placement program through VFNC. The Skillbridge Internship seeks to connect participants to the local job market and provides on-farm training for those interested in starting a small farm business. This program provides participants hands-on, on-farm training and education allowing them to work behind the scenes of all production models (**Fig. 3**).

Figure 3. Students receiving hands-on training in bee keeping. Some available farm equipment in the background.

VFNC actively recruits interested veterans both in and out of the service and exposes them to the myriad of opportunity agriculture has to offer - assessing their needs and goals, then placing them in the program best suited to meet those needs and assist in accomplishing those goals. Veterans learn that the agriculture industry has many desirable career opportunities. Recruiting also means educating a veteran before they start farming to develop the skill-sets, have a plan, minimize risks - and the foundation to develop an economically successful business. These VFNC programs help create best management practices (BMPs) through education based on past experiences and real knowledge - from experienced members of agriculture-related industries.

ADDITIONAL SERVICES

From its inception in 2016, VFNC has provided ongoing consultation and mentorship to all program participants and to any interested veterans. Frequently, this mentorship comes when a veteran has started a small farm business or is seeking further education and has not participated in our programs. Experienced farmers and those with careers in the agricultural field are available to provide advice, advocacy and encouragement. VFNC mentors accompany interested persons to assess land, enroll in programs with government agencies, or advise on educational programs to meet their needs and offer support. VFNC also offers an equipment loan program, currently maintaining an inventory of over $200,000 of donated equipment including tractors, implements, trailers, a truck, freezers, poultry processing equipment and more - all available to any farmer veteran in need for temporary use. The equipment in our program is in constant use at various veteran owned farms across eastern North Carolina - and we continue to accept financial donations, and gifting of agricultural and general construction equipment, as well as infrastructural items to help meet these needs (**Fig. 4**).

Figure 4. Veterans at Veterans Farm of North Carolina accepting a tractor donation from KIOTI Tractor.

Finally, VFNC has built a network of over 700 farmer veterans and links them to critical resources in NC. These resources include local farmers markets and outlets to sell their product along with extensive collaboration with other farmer veterans. This network provides the support, encouragement and military "family" many servicemembers find is missing from their new lives as a veteran. We believe this network, coupled with a new purpose and place in their community, is helping to reduce the alarming statistic of 22 veterans/servicemembers per day losing their lives to suicide. There has been emerging interest in the last several years in agriculture as a form of therapy, even coining the term "Agri-therapy" https://www.farmaid.org/blog/farmer-heroes/agriculture-as-therapy-for-veterans-and-all-of-us/. VFNC believes this is a vital part of our mission (**Fig. 5**).

VFNC's objective statement is, "To give veterans a new mission and America new farmers."

Figure 5. Pictured left to right: Robert Elliot, Founder and Executive Director of Veteran's Farm of NC, Inc. (VFNC); Administrative Assistant, Samantha Manning; Board President, Donny, current Farm Military Assistance Program (FARM MAP) participant, Hanna, intern for NCSU Horticulture Dept, and other volunteers at FARM AID 2022 with hydroponics and various information about VFNC. Robert also taught a workshop on mushroom production at this event.

IPPS-SRNA INVOLVEMENT

In January 2020, the IPPS Southern Region of North American (IPPS-SRNA) executive committee unanimously voted to extend free student memberships to those enrolled in the STAG program - which continues to participants in VFNC programs. This decision was an excellent first step in bringing together two groups that have much to offer each other.

We look forward to seeing this relationship grow and prosper in years to come, as IPPS members - to "Seek and Share" the collective knowledge - with veterans who wish to become colleagues. It is a win-win for all with the career opportunities the green industry offers veterans, and the human talent that veterans can bring to the nursery profession!

Combined Proceedings IPPS 72:179-190. 2022.

Alternatives to Loose-fill Media for Improved Plug Handling

Ben Sanders

Griffith Propagation Nursery, Inc., 2580 Antioch Church Road, Watkinsville, GA 30677 USA

bsanders@griffithnursery.com

Keywords: rooting, plug, liner, handling, extraction, shipping, Ellepot, Preforma, Grow Coon

Summary

We specialize in the propagation of woody ornamentals, and our primary focus is rooting cuttings quickly and uniformly. However, many factors contribute to uneven rooting and finishing, and liners always need to be graded prior to shipping. In general, rooted crops finish unevenly over the course of weeks, and it may be necessary to go through a crop several times to fill orders. Grading can also damage the plants. In many instances, extracting a plant prematurely will destroy the root system or harm it to the point that it will never be salable. This paper reports alternatives to loose-fill media in plug trays that could improve the handling process and shorten the production cycle. There were twelve plant species that were rooted as liners using five different media systems: LF: Loose-fill media, GC: Loose-fill + Growcoon, EP: Ellepot, PF: Preforma plugs, and GB: Bark fill + Growcoon. Our best liner production system was the GC: Loose-fill (30% Peat, 55% composted fines, 10% perlite, 5% vermiculite, amendments) + Growcoon liner holder.

INTRODUCTION

There are several alternatives to traditional loose-fill media in plug trays. One option is growing in individual cups. Companies such as Jiffy and Oasis offer alternative products including Jiffy Pots, Jiffy Pellets, Grow Blocks, Preforma, Fertiss, Rootcubes, and Terra Plug. Growcoon is a European-designed cutting and seed plug holder that fits into the cells of the plug tray. It is comprised of a biodegradable material and is designed to support and protect the rootball of a young plant. The Ellepot is another option that has been discussed in at least two other IPPS papers (Billingsley 2003 and Andrzejewski 2003).

This paper will compare the Preforma plugs, Ellepots, and Growcoons liner systems to our normal loose-fill media.

MATERIALS AND METHODS

Plant species and auxin treatment

Abelia x *grandiflora* 'Kaleidoscope' PP16,988 – 1000 ppm K-IBA

Aucuba japonica 'Rozannie' – 1500 ppm K-IBA

Rhododendron 'Roblez' PP28,279 Encore® Azalea Fire® - 1500 ppm K-IBA

Clethra alnifolia 'Hummingbird' – 1000 ppm K-IBA

Distylium 'PIIDIST-V' PP27,631 First Editions® Cinnamon Girl® *Distylium* – 4000 ppm K-IBA

Hydrangea macrophylla 'Nikko Blue' – 1000 ppm K-IBA

Hypericum frondosum 'Sunburst' – 1500 ppm K-IBA

Ilex verticillata 'Red Sprite' – 1500 ppm K-IBA

Illicium floridanum 'Florida Sunshine' – 1500 ppm K-IBA

Lagerstroemia x 'Sioux' – 1000 ppm K-IBA

Ligustrum sinense 'Sunshine' PP20,379 'Sunshine' Ligustrum – 1000 ppm K-IBA

Loropetalum chinense var. rubrum 'PIILC-I' PP25,534 First Editions® Crimson Fire™ – 1500 ppm K-IBA

Trial mixes

We already understood the production advantages, disadvantages, and handling benefits of Preforma, Ellepots, and Growcoon plug systems. Our goal was to better quantify the differences in time to ship for our loose-fill, the Preforma, Ellepot, and Growcoon. In order to ship, the plug needed to be fully extractable with the rootball holding together well enough to be handled without falling apart. For loose-fill plugs, this entails complete rooting. For the Preforma, Ellepots, and Grocoon, this entails that several roots should be visible on the outside of the plug.

The following liner production products were used:

LF: Loose-fill (MM 865: 30% Peat, 55% composted fines, 10% perlite, 5% vermiculite, amendments).
GC: Loose-fill (MM 865 above) + Growcoon (013-H72).
GB: Bark fill (80% fines, 20% compost, amendments) + **Growcoon** (013-H72).
EP: Ellepot (50mm x 50mm GROW Mix: 60% Peat, 30% fine perlite, 10% coir).
PF: Preforma plug (coir and peat w/ binder polymer).

The trial began on 20 May 2022 with the preparation of the trays. Cuttings were stuck on 23 May 2022. We used the TO Plastics PL-36-Star plug tray, which is our standard plug tray. Sampling of plugs began on 24 June 2022 and continued biweekly through the end of the trial. The last sample date was 3 October 2022. Rooting and extractability of the intact rooted plug were noted, and suitability for shipping was also measured.

Propagation

Most of the varieties we produce root easily, and our practices are typical of most liner propagation nurseries. Rooting percentages range from 70% to 90% for most crops. Rooting times vary by crop and season, from summer crapemyrtles rooted in 5-6 weeks and winter conifers requiring up to 5 months. About half of our production is in plug trays, which will be the focus of this paper.

Production

Once the cuttings have rooted, they are removed from the propagation houses and grown off in production areas. The production goal is to finish the crop as a fully, rooted-out liner – as quickly and uniformly as possible so orders can be shipped to customers.

Grading

A plug's readiness is largely determined by its root system. If the plug cannot be extracted from the tray and handled without falling apart, it is not shippable. Grading is a challenge for two main reasons. *First,* plants do not finish uniformly, so it can be difficult to match crops with orders. A large order may need to be delayed until enough plants are rooted out, or the customer may decide to take fewer plants than ordered. If a crop scheduled for late in the year cannot be shipped because it has not finished, it could be spring of the following year before it is ready. *Second,* processing a crop too soon will often result in the loss of many of those plants. Extracting a plug liner before it has completely rooted will result in damage that prevents further root growth into the media (**Fig. 1**).

Figure 1. Extracting a liner before it has completely rooted to the bottom of plug results in damage that prevents further root penetration into the media.

Other issues also affect grading. Even species that are typically aggressive rooters in spring and fall may not fill out the bottom of the plug if conditions are adverse due to heat, moisture, or media compaction stresses. Other plants will push new roots to the side of the tray, hit the wall, and go straight down to bottom of the plug tray instead of filling out the media evenly (**Fig. 2**). The result is a cylinder of media inside the roots that will fall away when the plug is extracted. Some plants send roots straight down, leaving a low shoulder at the point of root initiation that allows the top of the plug to fall away (**Fig. 3**).

Figure 2. Roots can grow along the plug side walls, but fail to fill out the center of the plug.

Figure 3. The plug media is not held together when roots grow to the bottom, but do not fill out the top and sides of the plug.

RESULTS

At week five, rooting evaluations were initiated and subsequently made every other week until the end of the trial. Figures 8-19 illustrate rooting of the 12 plant species, including the date of the photo, in the five different plug media systems: LF: Loose-fill media, GC: Loose-fill + Growcoon, EP: Ellepot, PF: Preforma plugs, and GB: Bark fill + Growcoon.

Rooting

In general, all of the plants rooted with greater success in the GC (Loose-fill + Growcoon) treatment (**Tables 1** and **2**).

Table 1. Summary of results by five plug media systems measured across twelve plant species.

	LF: Loose-fill	GC: Loose-fill, plus Growcoon	GB: Bark fill plus Growcoon	EP: Ellepot	PF: Preforma plug
Rooted	72%	98%	77%	71%	69%
Shippable	73%	98%	88%	77%	82%
Weeks to Extract	12.1	8.3	8.6	9.1	10.5
Weeks to Ship	15.7	10.4	12.1	13.9	14.5
Weeks < LF	---	5.3	3.6	1.8	1.2
Faster production turnover than LF	---	34%	23%	12%	8%
Salable Plants	53%	96%	68%	55%	57%

This is not surprising since the GC trial used our loose-fill media in a propagation zone that was being used to root cuttings stuck in loose-fill media. Had the EP: Ellepot, PF: Preforma plug, or GB: Bark fill + Growcoon trials been propagated in zones dedicated to those media, rooting results would likely have been similar, based on previous production experience. One surprise was how well the GC rooted. A rooting percentage of 90% would have been expected. The extra care that went into planning and observing the trial may account for increased rooting. The other surprise was the unexpectedly low rooting percentage of LF: Loose-fill media treatment. There was essentially no difference between GC and LF, and the Growcoon itself did not seem to offer any advantage beyond holding the rootball together. It may be that the process of sampling was destructive on plants in the LF group. With GC, if the roots had reached the Growcoon sleeve, the plant extracted easily. With LF, attempts at extraction prior to full rooting would have been detrimental. This reinforces our experience grading liner crops in loose-fill media.

Table 2. Weeks to ship and percent improvement of other plug systems compared to LF: loose-fill by plant species.

	LF: Loose-fill	GC: Loose- fill, plus Growcoon		GB: Bark fill, plus Growcoon		EP: Ellepot		PF: Preforma plug	
	Weeks	Weeks	%	Weeks	%	Weeks	%	Weeks	%
Abelia x *grandiflora* 'Kaleidoscope'	19	9	53%	11	42%	11	42%	11	42%
Aucuba japonica 'Rozannie'	13	9	31%	11	15%	11	15%	11	15%
Rhododendron 'Roblez'	19	11	42%	17	11%	19	0%	19	0%
Clethra alnifolia 'Hummingbird'	15	11	27%	11	27%	13	13%	11	27%
Distylium	19	15	21%	15	21%	17	11%	23	-21%
Hydrangea macrophylla 'Nikko Blue'	11	7	36%	9	18%	11	0%	11	0%
Hypericum frondosum 'Sunburst'	15	9	40%	15	0%	11	27%	11	27%
Ilex verticillata 'Red Sprite'	19	11	42%	11	42%	15	21%	19	0%
Illicium floridanum 'Florida Sunshine'	19	11	42%	11	42%	19	0%	19	0%
Lagerstroemia x 'Sioux'	11	9	18%	9	18%	9	18%	9	18%
Ligustrum sinense 'Sunshine'	9	7	22%	7	22%	7	22%	7	22%
Loropetalum chinense var. rubrum	19	15	21%	17	11%	23	-21%	23	-21%
Average	15.7	10.4	34%	12.1	23%	13.9	12%	14.5	8%

Extraction

It would have been difficult to quantify extraction without destroying plants, particularly LF plants. Therefore, weeks to extraction should be taken as estimates based on experience with the plants and the media. In general, GC and GB were the first to be extractable, followed a week later by EP, and 2-3 weeks later by PF (**Tables 1 and 2**). LF lagged behind by nearly a month. These results are reflected in the shippability of rooted liners. As with rooting, there is the caveat that these plants were not optimally grown for either GB, EP, or PF, and altering production practices in favor of those groups would likely have produced different results.

Shippability

This was the main factor on trial. On average, GC plants were shippable almost 2 weeks before GB, 3-4 weeks before EP and PF, and over 5 weeks before LF. Again, production practices that favor EP and PF over GC would likely have produced different results. However, this was what we wanted to trial – what performed best under our current growing practices compared to our current standard LF mix.

Breakdown by species

In general, the faster the plant rooted, the smaller the media effect - verses a greater media effect when the liners rooted slower. The notable exception was *Abelia* x *grandiflora* 'Kaleidoscope' PP16,988, which has a fine root system that roots out poorly into our LF media. Both *Loropetalum chinense* var. rubrum 'PIILC-I' PP25,534 First Editions® Crimson Fire™ and *Distylium* 'PIIDIST-V' PP27,631 First Editions® Cinnamon Girl® *Distylium* were surprises as well. Experience predicted that both of these slow rooters should have gained more than 4 weeks in a GC over LF, but that did not occur. A possible explanation was that while slow rooters, once they begin to root out - they do so quickly.

DISCUSSION

Performa plugs

Abelia x *grandiflora* 'Kaleidoscope' roots nearly 100% with very fine roots in loose-fill media. However, the crop takes weeks to get ready after propagation, and we still lose many liners during the grading process (**Fig. 4**).

Figure 4. *Abelia* x *grandiflora* 'Kaleidoscope' PP16,988 - in a loose-fill plug is difficult to grade and is not shippable.

A Preforma plug offered us an expensive option compared to the loose-fill plugs; but by saving two more plants per tray, we make-up the cost-differential (**Fig. 5**). 'Kaleidoscope' stuck in Preforma plugs rooted faster than loose-fill cuttings, and they were ready to ship within a month after removal from propagation. With rooting percentages close to 100%, they were all ready at the same time – and our cull rate went from 80% to less than

5%. We were also able to grade entire crops quickly.

Figure 5. *Abelia* x *grandiflora* 'Kaleidoscope' PP16,988 - in a Preforma plug grades easily and ships well.

The Preforma plug was easy to use, clean, pest free, and improved grading of rooted liners. Pre-filled trays come on pallets that can be stored for months and easily placed by a greenhouse when needed. There were also drawbacks. First, the media has a bulk density of over 900 (kg/m^3), which is twice as dense as the mixes we normally use, so watering was a challenge. The other drawback was the inability to customize. We incorporate fertilizer and bifenthrin insecticide into our loose-fill mix, which is not possible with Preforma. Dedicated irrigation zones along with additional bifenthrin and fertilizer applications corrected these issues.

Ellepots

Ellepot offered an improvement over the Preforma plugs because they were significantly cheaper, and its mix was more comparable to what we use. However, the mix we purchased is much finer than our loose-fill, in addition to being compressed, it tends to stay wetter than our typical plug - so over-watering was still an issue (**Figs. 6 and 7**).

Figure 6. Poor rooting occurs when Ellepots (1st and 3rd plugs) and Preforma plugs (2nd and 4th plugs) are over-irrigated.

Figure 7. Poor rooting of HYNB: *Hydrangea macrophylla* 'Nikko Blue' in Ellepot (Left) and Preforma plugs (Right) when over-irrigated.

Other mixes are available, including a high-porosity blend, that might be more comparable to our loose-fill. Like Preforma, it comes in clean, pre-loaded trays that are weed free and easy to handle. Although it comes with a starter charge, it cannot be customized to the degree that our loose-fill is, and it has the same drawback of not having bifenthrin in the media. Purchase of an Ellepot machine would allow the use of a custom loose-fill blend and mitigate these shortcomings.

Growcoon

The Growcoon is basically a small, biodegradable sleeve or liner that fits inside each plug. Loose-fill media is then added. These are very inexpensive. Any loose-fill media can be used. The drawback is that each sleeve must be placed in each plug; however, it is still a much cheaper option than either Preforma or Ellepot. There are also automation options to facilitate placing the sleeves in the trays.

Figure 8. *Abelia* x *grandiflora* 'Kaleidoscope' PP16,988 rooting in plugs with LF: Loose-fill media, GC: Loose-fill + Growcoon, EP: Ellepot, PF: Preforma plugs, and GB: Bark fill + Growcoon (from left to right). Rooting date was 11 July 2022.

Figure 9. *Aucuba japonica* 'Rozannie' rooting plugs with LF: Loose-fill media, GC: Loose-fill + Growcoon, EP: Ellepot, PF: Preforma plugs, and GB: Bark fill + Growcoon (from left to right). Rooting date was 25 July 2022.

Figure 10. Rhododendron 'Roblez' PP28,279 Encore® Azalea Fire® rooting plugs with LF: Loose-fill media, GC: Loose-fill + Growcoon, EP: Ellepot, PF: Preforma plugs, and GB: Bark fill + Growcoon (from left to right). Rooting date was 20 July 2022.

Figure 11. *Clethra alnifolia* 'Hummingbird' rooting plugs with LF: Loose-fill media, GC: Loose-fill + Growcoon, EP: Ellepot, PF: Preforma plugs, and GB: Bark fill + Growcoon (from left to right). Rooting date was 11 July 2022.

Figure 12. *Distylium* 'PIIDIST-V' PP27,631 First Editions® Cinnamon Girl® rooting plugs with LF: Loose-fill media, GC: Loose-fill + Growcoon, EP: Ellepot, PF: Preforma plugs, and GB: Bark fill + Growcoon (from left to right). Rooting date was 20 September 2022.

Figure 13. *Hydrangea macrophylla* 'Nikko Blue' rooting plugs with LF: Loose-fill media, GC: Loose-fill + Growcoon, EP: Ellepot, PF: Preforma plugs, and GB: Bark fill + Growcoon (from left to right). Rooting date was 11 July 2022.

Figure 14. *Hypericum frondosum* 'Sunburst' rooting plugs with LF: Loose-fill media, GC: Loose-fill + Growcoon, EP: Ellepot, PF: Preforma plugs, and GB: Bark fill + Growcoon (from left to right). Rooting date was 6 June 2022.

Figure 15. *Ilex verticillata* 'Red Sprite' rooting plugs with LF: Loose-fill media, GC: Loose-fill + Growcoon, EP: Ellepot, PF: Preforma plugs, and GB: Bark fill + Growcoon (from left to right). Rooting date was 20 September 2022.

Figure 16. *Illicium floridanum* 'Florida Sunshine' rooting plugs with LF: Loose-fill media, GC: Loose-fill + Growcoon, EP: Ellepot, PF: Preforma plugs, and GB: Bark fill + Growcoon (from left to right). Rooting date was 20 September 2022.

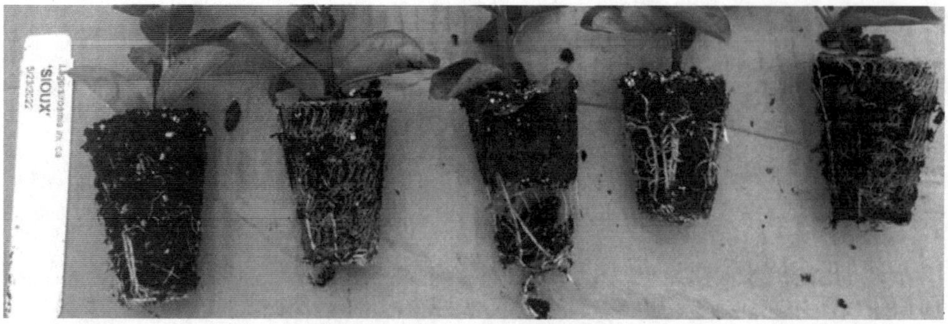

Figure 17. *Lagerstroemia* x 'Sioux' rooting plugs with LF: Loose-fill media, GC: Loose-fill + Growcoon, EP: Ellepot, PF: Preforma plugs, and GB: Bark fill + Growcoon (from left to right). Rooting date was 11 Jully 2022.

Figure 18. *–Ligustrum sinense* 'Sunshine' PP20,379 'Sunshine' Ligustrum rooting plugs with LF: Loose-fill media, GC: Loose-fill + Growcoon, EP: Ellepot, PF: Preforma plugs, and GB: Bark fill + Growcoon (from left to right). Rooting date was 11 Jully 2022.

Figure 19. *Loropetalum chinense* var. rubrum 'PIILC-I' PP25,534 First Editions® Crimson Fire™ Fringe Flower rooting plugs with LF: Loose-fill media, GC: Loose-fill + Growcoon, EP: Ellepot, PF: Preforma plugs, and GB: Bark fill + Growcoon (from left to right). The roots are difficult to see because they are red-purple and do not stand out against the media. Rooting date was 5 October 2022.

LITERATURE CITED

Billingsley, G. (2003). Increased rooting percentages and decreased transplant shock using Ellepots. Comb Proc. Inter. Plant Prop. Soc. *53*:579-581

Andrzejewski, C. (2003). Ellepots: Another tool in our propagation chest. Comb Proc. Inter. Plant Prop. Soc. *53*:487.

Combined Proceedings IPPS 72:191-195. 2022.

Softwood Cutting Propagation of Clonal Oak Trees

Dwayne Moon

Moons Tree Farm Inc., 175 Happy Hollow Road, Washington, Georgia 30763, USA

dwayne@moonstreefarm.com

Keywords: Indole butyric acid with potassium salt (K-IBA), *Quercus phellos*, *Quercus nutallii*, *Quercus lyrata*

Summary

Selected species of oak can be clonally propagated from cuttings. Clonal rooting practices at Moons Tree Farm, Inc., Washington, Georgia, for three oak species are described. Optimal rooting of *Quercus phellos* (Willow Oak) occurred with softwood cuttings propagated between May to August, quick-dipped in a solution of 3000 ppm indole butyric acid with potassium salt (K-IBA). For *Quercus nutallii* 'QNMTF', Tytlest® (Nuttall Oak), optimal rooting occurred with softwood cuttings propagated between May to June, quick-dipped with 4000 ppm K-IBA. Optimal rooting of *Quercus lyrate* (Overcup Oak) occurred between May and August with softwood cuttings quick-dipped at 2500 ppm K-IBA. The first spring flush is the best time to harvest softwood cuttings. Timing of softwood cutting collection season, and water-management of the mist irrigation system are critical and discussed in greater detail.

PROPAGATION OF *QUERCUS PHELLOS* (WILLOW OAK)

Propagation of *Quercus phellos* is done between May and August at Moons Tree Farm, Inc., Washington, Georgia (**Fig. 1**). After a new flush has set the terminal buds, 15-20 cm (4 to 6-in.) long softwood cutting are harvested from clean, healthy, vigorous growing plants which are free of disease and insects.

Figure 1. Mist propagation of *Quercus phellos* (Willow Oak) cuttings.

All cuttings are taken in early morning to keep plants well hydrated while preparing them for sticking on mist benches. Bottom leaves are removed and the cuttings are quick-dipped into a solution of indole butyric acid with potassium salt (K-IBA) at 3000 ppm (**Fig. 2**).

Figure 2. (Left) Callus development (arrow) of *Quercus phellos (*Willow Oak) cutting, after 3-weeks, and (right) rooting (arrow) of cutting.

In some selections, K-IBA treatment is as high as 6500 ppm. The propagation media is a combination of 2 parts bark fines and 1 part Pro Mix BF (**Fig. 3**).

Figure 3. (Left) Rooted liner of *Quercus phellos* 'QPMTF, Wynstar® (Willow Oak), and (right) clonal trees under field production.

PROPAGATION OF *QUERCUS NUTTALLII* 'QNMTF', TYTLEST® (NUTTALL OAK)

Propagation of *Quercus nuttallii*, is done between May and June. After a new flush has set the terminal buds, 15-20 cm (6 to 8-in.) long softwood cuttings should be harvested from clean, healthy plants (**Fig. 4**).

Figure 4. (Left) Cutting of *Quercus nuttallii* 'QNMTF', Tytlest® (Nuttall Oak), and (right) cuttings in propagation flats.

The bottom leaves should be removed and the remaining leaves reduced in size to aid in cuttings not being too congested in the propagation benches.

Cuttings are quick-dipped in a solution of 4000 ppm K-IBA, while some selections received 6000 ppm K-IBA. The propagation media should be 3 parts bark fines to 1 part Pro Mix BF (**Fig. 5**).

Figure 5. Clonal *Quercus nuttallii* 'QNMTF', Tytlest® (Nuttall Oak) propagated from cuttings and transplanted into 3-gal containers (left).Field grown *Quercus nutallii* 'QNMTF', Tytlest® (Nuttall Oak) propagated from cuttings (right).

PROPAGATION OF *QUERCUS LYRATA* (OVERCUP OAK)

Propagation of *Quercus lyrata*, is done between late May thru August. Terminal softwood cuttings should be 15-20 cim (6 to 8-in) long with the bottom leaves removed and the remaining leaves cut/reduced in size. Cuttings are quick-dipped in a solution of 2500 ppm K-IBA with some selections treated at 4000 ppm K-IBA. The propagation media is 3 parts bark to 1 part Pro Mix BF.

TIMING AND CONDITION OF CUTTING STOCK PLANTS

The timing and condition of harvesting softwood cuttings is imperative in achieving optimum results for well rooted plants that will survive throughout the winter months. *The first spring flush is absolutely the best time to harvest softwood cuttings* as these plants will have a more prolific root system and flush of growth before fall. Trees that are rooted in later months must be over wintered in the greenhouse to protect them from the cold. Frost cracks can be an issue in plants which are rooted later in the summer months.

MISTING CYCLE AND WATER REGIMENTS

Rooting response can be slow in all oaks, with some species slower than others. Water regiments are very important. Too much water can slow the development of multiple root initials. Too little water can cause cutting to only callus and never form roots.

- In Washington, Georgia, USA [latitude 33.736795, longitude -82.739309], the mist start time should be at 9:00 and off time at 19:00.
- Three seconds mist on time with 4-min delay (week one)
- Four seconds mist on time with 8-min delay (week two)
- Six seconds mist on time with 12-min delay (week three)
- Eight seconds mist on time with 16-min delay (week four). This regime is maintained until roots begin to emerge; there should be a 20% increase in delay time every 3 to 4 days until the plants are acclimated off misting.

LONG TERM CARE AND HANDLING

After rooted liners are put on a once-a-day watering schedule, a slow-release Caliber Cote 17-5-17 (N-P-K) Fertilizer is given at a low rate. Plants are then moved into a 30% shade for two weeks after which time the shade is removed. Plants will then be potted into 3-gal Redi-Root containers late in the summer, and they will be ready for field planting in the following spring.

landscapes are often slower to adopt sustainable practices than production agriculture (Doxon, 1996).

Combined Proceedings IPPS 72:196-202. 2022.

Propagation Research and Teaching for Ecologically-Friendly Landscapes and Gardens in Florida

Sandra B. Wilson

University of Florida, Department of Environmental Horticulture, PO Box 110670, Gainesville, Florida, 32611, USA

sbwilson@ufl.edu

Keywords: invasive plants, mobile application, native plants

Summary

For nearly two decades plant propagation has been central to the authors research and educational programs. Recently, a course in plant propagation was used to evaluate perceived student knowledge gains of 17 core subject areas before and after using a mobile application called PropG (htpps://propg.ifas.ufl.edu). Results revealed PropG to be a valuable tool in learning propagation concepts and terms, with an average knowledge gain of 52%. In addition to launching tools to facilitate plant propagation education, a series of propagation and production research studies have been conducted over the years to: 1) evaluate the fertility and landscape performance of cultivars and/or hybrids of ornamental invasives and 2) develop reliable propagation systems of novel or underutilized natives having ornamental and ecological value. Attractive, fruitless selections of hugely popular species such as butterfly bush (*Buddleja* sp.), heavenly bamboo (*Nandina domestica*), Mexican petunia (*Ruellia simplex*), lantana (*Lantana strigocamara*), trailing lantana (*Lantana montevidensis*), privet (*Ligustrum* sp.), maiden silvergrass (*Miscanthus* sp.) and fountain grass (*Pennisetum* sp.) have been identified as suitable non-native alternatives to the invasive or potentially invasive resident taxa. Also, as alternatives to ornamental invaders, over a dozen native species have been studied to determine their optimal propagation by seeds, cuttings, and/or micropropagation, as well as their performance in statewide landscape trials. Promising results are hoped to facilitate their increased availability and wider use in landscapes and gardens of Florida and other warm climates.

TEACHING AND RESEARCH HIGHLIGHTS

Innovative propagation teaching tool. The study of plant propagation requires a working knowledge of a significant number of terms and concepts. With this in mind, the ninth edition of *Hartmann and Kester's Plant Propagation: Principles and Practices* was updated to include a compiled glossary of nearly 500 propagation terms as a separate section following the subject matter chapters (Davies et al., 2018). The ability to readily retrieve these terms in the 1,000-page textbook at any time or place was not achievable until recently. PropG, a mobile and desktop application (htpps://propg.ifas.ufl.edu), was developed as a collaborate effort between the universities of Florida, Kentucky, and Texas A & M as a universal resource for readily accessing propagation-related glossary terms and corresponding graphics and videos. Organization of these terms began with nine categories including: 1) Biology of propagation, 2) Propagation environment, 3) Genetic selection, 4) Seed propagation, 5) Cutting propagation, 6) Budding and grafting, 7) Bulbs and other geophytes, 8) Layering and division, and 9) Tissue culture and micropropagation (**Fig. 1**).

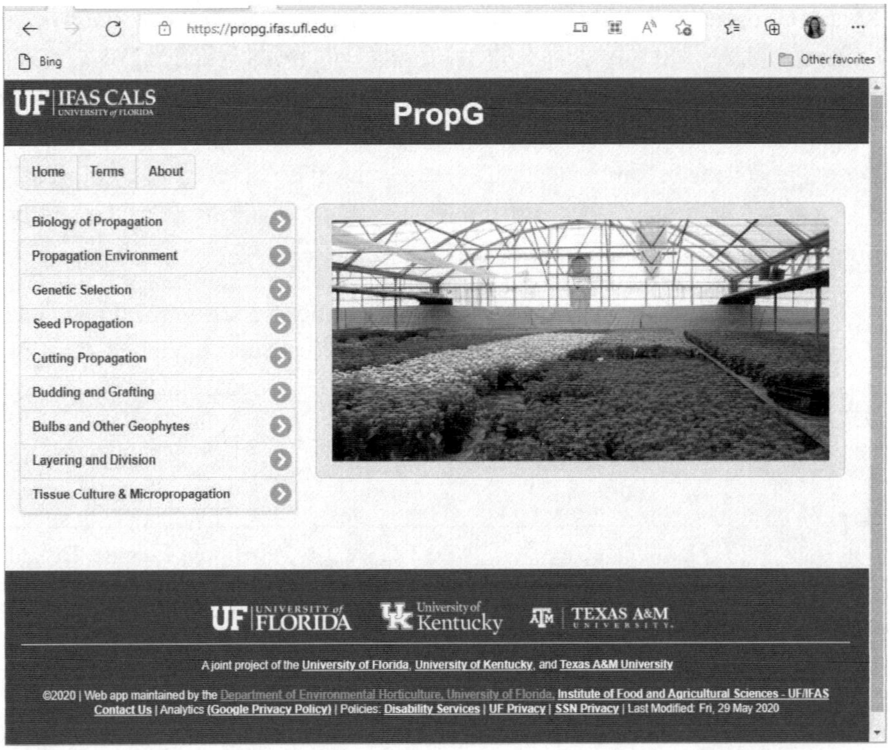

Figure 1. Screenshot of the front page of the mobile application (PropG) built using JQuery software (Mountain View, CA). Users can first select from these nine subject areas within the PropG navigational menu (htpps://propg.ifas.ufl.edu).

Each of these categories were then further divided into additional drop down menu options providing conceptual descriptors of over 1262 images. Since inception in 2021, PropG has been widely used and is gaining momentum. To date, PropG has received

136,339 original page views from over 43 countries. Further, it has been found to more than double student's perceived knowledge of each of 17 core subject areas tested before and after its use within a single semester (Wilson et al., 2023).

Evaluation of non-fruiting cultivars of popular ornamental invasive plants. Ornamental horticulture has been recognized as the fastest growing segment of U.S. Agriculture, as well as the main source of plant invasions worldwide. In the past two decades in Florida, tremendous advances have been made to identify and/or select non-invasive forms of a number of popular ornamental species such as privet (*Ligustrum* sp.), fountain grass (*Pennisetum setaceum*), heavenly bamboo (*Nandina domestica*), lantana (*Lantana camara*), maiden silvergrass (*Miscanthus sinensis*), butterfly bush (*Buddleja* sp.) Mexican petunia (*Ruellia simplex*), porterweed (*Stachytarpheta* sp.), and trailing lantana (*Lantana montevidensis*). As a result of these efforts, the invasive wild type forms are gradually being replaced with newer non-invasive, UF IFAS-approved cultivars that are superior in flowering and performance (**Table 1**).

Table 1. List of popular ornamental species in Florida along with their current ranking by the Florida Invasive Species Council (FISC) and University of Florida *Assessment of Non-native Plants* for north (N), central (C), or south (S) Florida (UF/IFAS AS, 2022). As potential non-invasive alternatives, selections with little or no fruiting are listed based on research trials. Species marked with an asterisk (*) were additionally subjected to the IFAS/AS Infraspecific Taxon Protocol plant use recommendations. Further morphological and cytological detail for each species can be obtained by downloading research publications from the authors website (https://irrecenvhort.ifas.ufl.edu/invasive_pub.html).

Species	Invasive ranking	Non-invasive selections (low to no fruiting) (https://irrecenvhort.ifas.ufl.edu/invasive_pub.html)
Buddleja lindleyana	FISC- not listed IFAS/AS- not a problem species N,C,S	*B.* × *weyeriana* × *B. lindleyana* 'Violet Eyes'; also, *B.* × *weyeriana* 'Honeycomb', 'Moonlight', and 'Sungold'
Lantana strigocamara	FISC- Category I IFAS/AS- invasive N, C, S	*T2, T3, T4, T9, Bloomify Rose, Bloomify Red, Lucious Royal Red Zone
Lantana montevidensis	FISC- not listed IFAS/AS- high invasion risk	U.S. varieties had little to no fruiting and were triploid, while the Australian form was tetraploid.
Ligustrum japonicum	FISC- not listed IFAS/AS- high invasive risk N,C,S	'Howard', 'Jack Frost', 'Lake Tresca', 'Rotundifolium', 'Texanum', 'Davidson' (all had little to no fruiting in south Florida)

Ligustrum lucidum	FISC- Category I IFAS/AS -use with caution to prevent escape, N,C,S	No fruit observed in south Florida
Ligustrum sinense	FLDACS- noxious weed FISC- Category I IFAS/AS- invasive N, C, S	'Sunshine', 'Swift Creek'
Miscanthus sinensis	FISC-not listed IFAS/AS- not a problem species N,C,S	'Morning Light' and 'Puenktchen' (south FL)
Nandina domestica	FISC- Category I IFAS/AS Invasive N,C	*'Firepower', 'Gulf Stream', 'Harbour Dwarf ', Firestorm, 'AKA' Blush Pink, 'Firehouse', 'Lemon-Lime', 'Murasaki Flirt, 'SEIKA' Obsession
Pennisetum setaceum	FISC- formerly Category II IFAS/AS- not a problem species N,C,S	'Rubrum'
Ruellia simplex	FISC- Category I IFAS/AS- invasive N,C,S	*Mayan Series (pink, white, purple, compact purple), Aztec Series (pink/white, pink, purple), 'Purple Showers' (use caution to prevent escape)
Stachytarpheta cayennensis	FISC- Category II IFAS/AS-use with caution N,C,S	'Mario Pollsa', 'Naples Lilac', and 'Violacea'

Propagation of native plants with ornamental potential and ecological value. Florida boasts abundant richness in flora with over 3,300 native plant species, yet less than a quarter of these are in cultivation. When used correctly, native plants can naturally offer desired aesthetic attributes such as color and form, while bringing biodiversity and function for ecologically friendly landscaping. In the last two decades significant progress has been made in the propagation, production, and landscape trialing of a number of native species that are either: 1) attractive in their natural areas and have potential for the ornamental industry, or 2) are already in limited cultivation, but merit widened use for landscapes and gardens. Propagation practices were explored to optimize production of natives by seed, cuttings, or tissue culture and to determine their landscape performance in multiple locations. These efforts helped to increase the native plant palette of Florida and identify ways for efficient, year-round production (**Table 2**). Opportunities remain for better consumer awareness, marketing and promotion of environmentally friendly plants that can offer similar form, flowering, fruiting, and growing requirements (sun, soil, moisture) as popular, non-invasive exotics.

Table 2. List of ornamental species native to Florida that merit wider use in landscapes and gardens based on landscape evaluations. Propagation systems were evaluated using seed, cuttings, or micropropagation with key findings briefly described for each species. Greater detail can be found in the associated publications downloadable from the authors website (https://irrecenvhort.ifas.ufl.edu/nativeplant_pub.html).

Common name	Species	Propagation technique https://irrecenvhort.ifas.ufl.edu/nativeplant_pub.html
Coastal plain honeycombhead	*Balduina angustifolia*	Seeds germinated under light or dark conditions and germination was influenced by temperature and population. Seeds are orthodox and retained high viability after a year of storage. Gibberellic acid improved germination of some populations. This species can be propagated by cuttings. Use of substrates with sand improved container quality of plants.
Florida scrub roseling	*Callisia ornata*	Propagation by seed is possible but vegetative propagation results in a fuller plant that performed well in the landscape trials. Plants grown in container media with a high proportion of vermiculite (low air-filled porosity) did not perform as well as other substrates tested.
Woody goldenrod	*Chrysoma pauciflosculosa*	Seeds prefer cooler alternating temperatures to germinate best. Cutting propagation is possible from softwood or hardwood cuttings. Auxin is not necessary but will improve rooting quality. Plants can grow in a variety of substrates.
Godfrey's goldenaster	*Chrysopsis godfreyi*	Optimal seed germination was in fall or winter with light. Substrates with low peat improved container production.
Feay's prairie clover	*Dalea feayi*	Seed scarification was necessary to alleviate physical dormancy. This species had very good visual quality ratings when container-grown both peat and bark-based media.
Gopher apple	*Geobalanus oblongifolius*	Seeds are nondormant preferring warm alternating temperatures for best germination. Cutting propagation is possible using softwood cuttings with auxin for best rooting.
Squareflower	*Paronychia erecta*	Seeds germinate readily to high percentages without pretreatments. Germination is promoted by exposure to light although some germination occurs in the dark. Seeds prefer moderate to cooler temperatures compared to summer. This species has been successfully propagated by cuttings and also by micropropagation.

October flower	*Polygonum polyga-mum*	Seeds have non-deep physiological dormancy that can be overcome by after ripening, warm stratification or application of GA. This species can be easily propagated by softwood cuttings stuck in late May.
Largeflower jointweed	*Polygonum nesomii*	Seeds have non-deep physiological dormancy. The population from where cuttings are collected may affect rooting percent and quality, with a combination of different NAA and IBA concentrations useful.
Wild coffee	*Psychotria nervosa*	In controlled studies, spring and summer temperatures were ideal for seed germination, but seeds had sporadic emergence over time. Cutting propagation is a reliable and efficient method of production, with auxin producing the high-quality root systems. A cultivar of this species is in commercial micropropagation production.
Softleaf wild coffee	*Psychotria tenuifolia*	A high proportion of cuttings can root fairly quickly with or without auxin, but auxin increases rooting response.
Bahama wild coffee	*Psychotria ligustrifolia*	Cuttings can likely be taken year-round with minimal concentrations of talc auxin.
Sweet acacia	*Vachellia farnesiana*	Seed scarification is needed prior to germination to alleviate physical dormancy. Cutting propagation is possible but not ideal. This species can be easily micropropagated using multiplication medium with BA and rooting media with IBA and NAA.
Wild lime	*Zanthoxylum fagara*	A portion of the seeds have physiological dormancy that must be overcome before germination. With proper stock management, semi-hardwood/softwood cuttings root when using moderate levels of auxin. Micropropagation has been a challenge.

CONCLUSIONS

In summary Prop-G is an effective mobile application for learning or reviewing plant propagation terms and concepts. Along with technological advances such as PropG as a novel teaching tool, significant research progress has also been made in identifying safer native and non-native alternatives to ornamental invasives and understanding their reproductive potential. Education remains key. It is hopeful that the newly released 'Plant This not That' guidebook of 22 invasive plant entries paired with research-based non-invasive alternatives (McIntyre et al., 2021) will help research, teaching and extension personnel, homeowners, and industry alike to make informed decisions of future plant selection and use.

ACKNOWLEDGEMENTS

The summary of research presented herein was a collaborate effort with colleagues Robert Geneve and Fred Davies (PropG), Zhanao Deng, Gary Knox and Rosanna Freyre (Invasive ornamentals), and Mack Thetford and Gabriel Campbell (Native propagation) along with many graduate students is gratefully acknowledged.

LITERATURE CITED

Davies, F.T., Geneve, R.L., and Wilson, S.B. (2018). Hartmann and Kester's Plant Propagation-Principles and Practices. Pearson Education Inc., NY.

McIntyre, T., Gutner, R., Pinkerton, M., and Wilson, S. (2021). Plant This Not That: A Guide to Avoiding Invasive Plant Species in Florida. University of Florida Institute of Food and Agricultural Sciences Extension (Accessed 17 Nov 2022). //ifasbooks.ifas.ufl.edu/p-1658-plant-this-not-that.aspx

University of Florida, Institute of Food and Agricultural Sciences. (2022). Assessment of Non-native Plants in Florida's Natural Areas (Accessed 17 Nov. 2022). https://assessment.ifas.ufl.edu

Wilson, S.B., Geneve, R.L, Davies, F.T., and Sotala, A. (2023). PropG-A new mobile application for learning plant propagation glossary terms. HortTech. 33:181-185. https://doi.org/10.21273/HORTTECH05128-22

Combined Proceedings IPPS 72:203-206. 2022.

Zoo Horticulture: Growing Plants with Wild Appeal

Eric Shealy

Riverbanks Zoo and Garden, 500 Wildlife Pkwy. Columbia, SC 29210, USA

eshealy@riverbanks.org

Keywords: Animal/plant interactions, conservation, plant toxicity, aesthetic use, functional use, plant database, *Hymenocallis coronaria*

Summary

This is a general overview of zoo horticulture and the challenges that horticulturists and plant production workers face dealing with this side of public horticulture. Riverbanks Zoo and Garden provides a unique confluence of public horticulture having both the botanical garden and a zoo, both of which are heavily planted. Challenges arise in each area of design and production to account for plant toxicity, aesthetic use, functional use, and conservation. Plant toxicity is chief among the concerns with the zoo and aesthetics, of course, are the main concerns for the botanical garden. We have a database that keeps all information about plant placement and installation, as well as other helpful cultural and conservation information. Animal interactions with plants can be complex and exhibit design needs to be able to meet these challenges while providing the animal with a wonderful habitat and the visitor with a great guest experience. Plants that we particularly find helpful in these endeavors will be discussed, as well as our conservation efforts towards two native perennials.

INTRODUCTION

Zoo horticulture is gardening and landscaping activities that occur within zoos and aquarium settings. It is a complex area of horticulture that focuses immensely on creating habitat for zoo animals while engaging guests with excellent displays and framing the animal in a setting that, as closely as possible, mimics the animal's native habitat (**Fig. 1**).

Figure 1. (Left) African lowland gorilla with *Arundo donax* and (right) kangaroo with *Acca sellowiana*

Plant toxicity to animals is always at the forefront of plant selection, followed closely by plant/animal interaction. Toxicity is very dependent on the animal as well as the plant. Since we have mammals, invertebrates, reptiles, fish, and birds on display, all these exhibits can be very diverse, plant-wise. What is toxic to one category of animal, is not to another. Reptiles are known to eat many things that would be deadly to mammals, so you will often find some commonly known toxic plants in those exhibits.

Horticulture in zoos is also public gardening, in that there are a wide variety of public spaces that need to be planted or screened or direct the guest's view toward exhibits. Guest experience is key with this side of public horticulture, and although guests might not realize it, horticulture plays a key role in making the zoo feel like "habitat."

Plant selection

Toxicity is always key in plant selection, but having animals interact with plants is one of the keys to successful exhibit spaces. Choosing plants that animals have direct access to is determined by how the animal will utilize the plant. Plants in exhibits can also provide a route for an animal escape, and that with toxicity are the ultimate concerns. Hotwire and electrical fencing are commonly used to keep animals away from key plants in an exhibit, mostly so those plants have a chance to reach their full aesthetic potential. Some plants are made to be sacrifices for the good of the design. An animal on exhibit for extended periods of time will ultimately become bored and will destroy plants. Planting in substantial numbers is a strategy to help plants establish and recover, along with hotwire. Horticulturists in exhibit spaces are also responsible exhibit "furniture," which means anything non-living that is brought in for animals to perch on or interact with in a permanent capacity. Examples of this include boulders, tree trunks, wooden structures, etc.

Database

Riverbanks has a plant registrar that utilizes Filemaker® Pro to organize and catalogue all plants that are installed in exhibits, public spaces, and the botanical garden. The database also generates all production information and facilitates the growing of all plants for the zoo, garden, and our annual plant sale. Riverbanks stores conservation information for any plants that are in our collection and are significant to state or federal agencies.

Growing challenges

Growing for a zoo and botanical garden is a wide varying job and requires the grower to be very versatile. Individual gardener requests and exhibit needs drive plant production at the growing center, and monoculture is not seen in our production facility. Timing plants to be installed in public displays and in exhibits is hampered by exhibit maintenance or renovation and staff time to plant. Riverbanks uses a river water irrigation system that delivers water from the Saluda River to both the botanical garden and the zoo. This system has its own challenges ranging from debris encroachment from the river on high flow days to pump failures that can take the system down for hours to weeks. Staffing shortages have also affected our ability to maintain and install plants and exhibits.

Challenges in exhibits

With new exhibit design, horticulturists are usually the last crew to get onto the project. But while horticulturists are the last ones to be on site, they usually must lay out clear guidelines for other crews to follow. Chief among these are access points for exhibit maintenance to allow for people and machinery to move easily into these spaces. Irrigation access points are also a major concern. In established exhibits, the time when a horticulturist can maintain plants on exhibit is dictated by animal welfare and keeper allowance. If a bird is nesting, or keepers are unable to pull animals off exhibit, then horticulturists cannot maintain plants.

Common plants used at riverbanks

- *Sabal minor*
- *Taxodium* sp.
- *Eucalyptus nicholii*
- *Muhlenbergia dumosa*
- *Daphniphyllum macropodum*
- *Vaccinium darrowii* 'Rosa's Blush'
- *Leucophyllum frutescens*
- *Rudbeckia maximillianii*
- *Hemerocallis* 'Autumn Minaret'
- *Acca sellowiana*
- *Carex divulsa*
- *Olea* 'Arbequina'
- *Hibiscus syriacus*
- *Chimonanthus praecox*
- *Camellias/Sasanqua*

Conservation at riverbanks

Riverbanks, with the help of state entities, conserves two different species: *Helianthus schweinitzii* and *Hymenocallis coronaria* (**Fig. 2**). Schweinitz's Sunflower grows mostly on disturbed roadsides and forest edges and is usually endangered from construction of roads and bridges in the state. It is a valuable plant to the Catawba Native Americans, and we have worked to propagate and multiply plants for reintroduction to Catawba lands near Charlotte, North Carolina. *Hymenocallis coronaria* is a threatened species that exists in rocky outcrops in rivers above or on the fall line in South Carolina, Georgia, and Alabama. A group of horticulturists collect seed every year and grow them out for reintroduction in the spring of the following year.

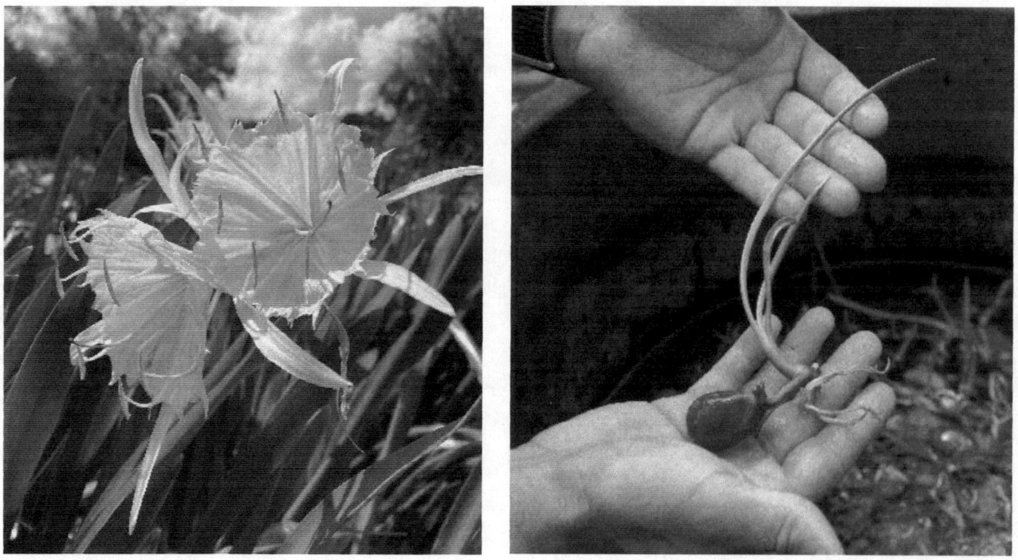

Figure 2. (Left) *Hymenocallis coronaria* observed in the field and (right) growing on from seed in the nursery.

Combined Proceedings IPPS 72:207-214. 2022.

Windows of Opportunity for Rooting Woody Stem Cuttings

Yiping Zou[1], Donglin Zhang[2], and Zach Hutzell[2]

[1]College of Forestry, Nanjing Forestry University, Nanjing 210037, China.
[2]Department of Horticulture, University of Georgia, Athens, GA 30602, USA

donglin@uga.edu

Keywords: Phytohormones, ornamental plants, plant exploration, propagation, regeneration, seasonal timing

Summary

To share newly developed woody ornamental plants with the public, it is absolutely essential to be able to regenerate them. In the past 10 years, we have propagated more than 60,000 plants in our woody ornamental plant breeding programs. Our successes and failures indicated that our first consideration should be optimal seasonal timing in taking stem cuttings. Timing and types of woody stem cuttings significantly impacted the rooting ability of different species. Depending on the taxa, the higher rooting percentages were largely from semi-hardwood cuttings, collected during late spring and summer. In addition to timing, stock plant source, hormone application, plant materials, hormone cofactors, and other environmental factors should also be considered.

INTRODUCTION

The most crucial step to market newly developed woody ornamental plants is developing realistic regeneration strategies for commercial production. We have collected specimens from the wild, conducted controlled crosses, and germinated seeds from open pollination. In the process, we developed many new clones, with cultivars of great market potential. Unfortunately, recalcitrant propagation of these plants can be a difficult challenge that significantly delays or prevents the commercial introduction of promising new cultivars.

The challenges associated with propagation increase when plants are collected from wild populations. After multiple collection trips to various Asian countries - thousands of plant taxa were brought back to the University of Georgia (UGA) after USDA inspection. Each year, approximately 6,000 potential new woody plants are propagated and evaluated at the UGA Woody Plant Research Lab. This paper presents data from the Lab, including a review of the window of opportunity for rooting woody stem cuttings.

MATERIALS AND METHODS

Sources of New Ornamental Plants. For the past 10 years, we have explored and collected ornamental plants from local, state, national, and international locations. Before the pandemic, many plant collection trips were completed (at least twice a year) to Asian countries, especially China, Japan, Korea, Nepal, and Vietnam. With USDA import permission and immense help from local hosts, around 150-250 accessions of potential ornamental plants were brought back to the United States each year. To trial and make available sufficient new plant materials, regeneration of these species is a primary concern.

Controlled crosses yielded many new clones, and open pollinated seedlings were a great resource to select elite ornamental plants. Again, some of these newly developed plants are exceedingly difficult to propagate, limiting further evaluation.

General Propagation Approaches. To be able to regenerate new plants and share them with gardeners and the Green Industry, it is important that we approach plant propagation using appropriate technologies (Davies et al., 2018). Since our collection areas had similar climates and other natural environmental conditions, we managed to germinate seeds either with a one to four-month cold stratification, or directly sowing outdoors with protection of wired mesh (to prevent animal damage) during late fall or winter. For other propagules, such as bulbs or rhizomes, we placed them in a cooler with moist peat moss.

For clonal propagation of new woody plants, attempting to root stem cuttings has provided the best results. The ability to successfully root woody stem cuttings typically guarantees the feasibility of commercial nursery production. When taking woody stem cuttings, a primary consideration is the seasonal collection time and cutting types. Semi-hardwood stem cuttings during late spring and summer months are preferred. For certain woody plants, softwood stem cuttings root more vigorously. A lot of hardwood cuttings (both evergreen and deciduous plants) were also collected; it took a significantly longer period to root the cuttings but sometimes proved a better approach for regenerating new plants. Rooting hormone composition and concentration along with rooting co-factors should also be considered for the propagation of new woody plants.

RESULTS AND DISCUSSION

***Helwingia chinensis*.** This new potential woody ornamental shrub [0.9-1.2 m (3-4 ft) tall] is from north China with unique flowers and red berry-like drupes on the middle of upper leaves (Fig. 1). The plant is performing well in partly shade areas in piedmont Georgia. To propagate this plant, the most vigorously rooting cuttings are from the softwood stem, which rooted in 3-4 weeks with or without rooting hormones.

Figure 1. *Helwingia chinensis* (red pearl on palm) has unique ornamental traits, including red berry-like drupe on the middle of leaves.

***Nandina* 'Coolglow Peach'.** A patented new cultivar with outstanding foliage color and compact habit (Fig. 2). Only a few seeds have been observed. The plant can be rooted year-round via stem cuttings with or without a rooting hormone of 1,000 ppm indole-3-butyric acid (IBA). Unfortunately, very few stem cuttings are available per plant; hence, tissue culture may be a feasible method for commercial production.

Figure 2. Foliage color and habit of *Nandina* 'Coolglow'.

***Zelkova schneideriana* 'Gold Goblin'.** A dwarf and compact seedling mutation was selected from thousands of seedling populations (Zhang et al., 2022). The grafted plant is about 1.8 m (6 ft) tall and 1.8 m (6 ft) wide after 10 years (Fig. 3). To control the height, this plant should be grafted on the regular *Zelkova* seedlings at the height you desire. We conducted softwood, semi-hardwood, and hardwood cuttings, and the rooting percentages were respectively, 100%, 93%, and 62%. The rooted liners grew as ground-cover plants if no stake is provided.

Figure 3. Fall foliage color and habit of *Zelkova schneideriana* 'Gold Goblin'.

***Ilex rotunda* 'Peace Time'.** A new, yellow fruited cultivar of evergreen Lord's holly (**Fig. 4**). The plant was selected from wild seedling populations and evaluated for 9 years. The optimum window for rooting this plant is via hardwood stem cuttings, collected during winter. With aid of 3,000 ppm IBA, rooting percentage of 40-70% can be reached after about 6 weeks (Zou et al., 2022).

Figure 4. Fruit branches of *Ilex rotunda* 'Peace Time'.

***Myrica (Morella) rubra* clones.** Yummy berry, or yang berry, is a small evergreen landscape tree originated from China (**Fig. 5**). The plant is dioecious, so we can select male plants or female plants for landscapes and edible fruit production. For the past nine years, we have evaluated 75 seedling plants (35 females, 21 males so far). Spring's new growth is gorgeous, and the fruits are delicious. In an attempt to propagate this plant, both Atlanta Botanical Garden and UGA Woody Plant Research Lab took stem cuttings from 2016 to 2018 - without success.

Figure 5. Fruits, new foliage color, and habit of *Myrica rubra*.

To find the best time to root stem cuttings, we took woody stem cuttings monthly from May 2019 to February 2020 and treated them with 8,000 ppm-IBA. The results indicated that the semi-hardwood stem cuttings from the first flush of the year could be rooted at 33% (**Fig. 6**). Obviously, the window of rooting this newly introduced plant should be May to July, when the foliage of the first sprouted branches of the year was fully extended. During 2021, we collected stem cuttings of *M. rubra* in May, June, and July with the additional step of soaking them in Superthrive® solution (one teaspoon per gallon) for 30 minutes https://superthrive.com/ . We then applied rooting hormones at 8,000 ppm-IBA with a rooting percentage of 78%.

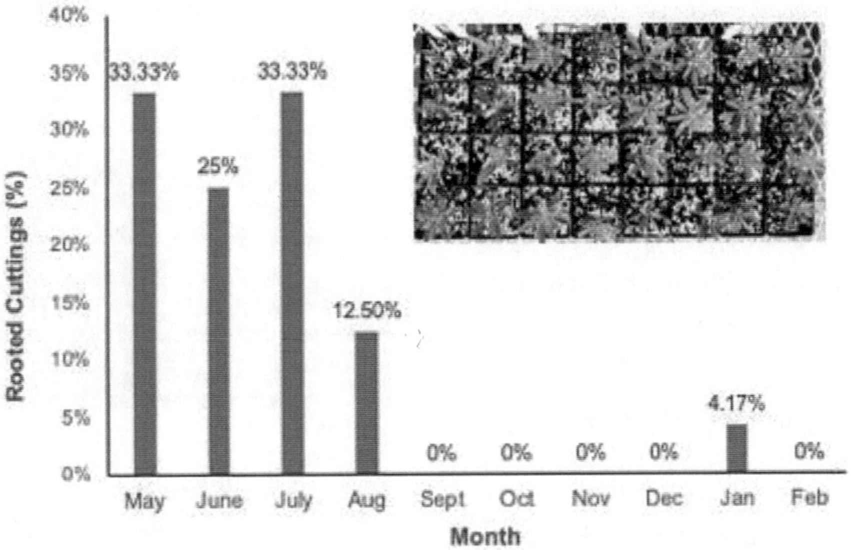

Figure 6. Effect of timing on the rooting of *Myrica rubra* stem cuttings under the treatment Hormodin 3 during 2019-2020.

We approach the propagation of new woody ornamentals systematically by investigating the best timing for taking stem cuttings, rooting hormone type, formulation, and concentration, co-factors - and finally sources of plant materials (roots, juvenility, etc.) and other factors (Ruchala et al., 2002). For *Myrica rubra*, we recommend taking stem cuttings in May, June, or July, soaking them in a Superthrive® solution (as a co-factor surce), and applying rooting hormones at 8,000 ppm-IBA or higher.

***Loropetalum chinense* var. *rubrum* 'Ever Red'.** The stem cuttings should easily root, however the stock plants for stem cutting collection should be replenished every 5 years. If stem cuttings continue to be collected after 5 years, the rooting percentage could be 33% or lower.

***Magnolia* 'Kevin1609'.** This plant is a controlled cross seedling from *Magnolia sieboldii* 'Colossus' x *M. insignis* 'Anita Figlar' with beautiful red flowers (**Fig. 7**). As soon as new branches are long enough [more than 13 cm (5-in) long] - stem cuttings should be taken and treated with 3,000 or 8,000 ppm-IBA. Rooting percentages were between 27% - 60% after 2-4 months. The rooted cuttings should be placed on extended light and warm conditions to push their bud break and grow new shoots before the winter months. Rooted cuttings without new flush growth had 100% mortality next spring!

Figure 7. Habit and flowers of *Magnolia* 'Kevin1609'.

***Camellia sinensis* (tea).** Tea ranks as the most consumed beverage in the world excluding water. Although we have grown tea in the U.S. for more than 200 years, there was only one commercial tea farm in the continental U.S. before 2010. To increase the popularity of tea and select better adapted cultivars for the southeastern U.S., we collected 147 clones and propagated them for further evaluation (**Fig. 8**).

Figure 8. Propagation of *Camellia sinensis* using stem cuttings by Dr. Ming Cai, Dr. Junhuo Cai, Dr. Junjun Fan, Dr. Xiaohong Zhou, Dr. He Li, Ms. Qian Song, Ms. Jinying Dong, Dr. Zhilong Hao, Dr. Donglin Zhang, and Dr. Jieming Wang. More than 10,000 rooted cutting were transplanted into one-gallon pots.

The rooting percentages were 25% - 100% and highly variable among the clones (Hao et al., 2020). 'Rosea' is a named cultivar with red new growth and pink flowers. The leaves are small and the mother plant was 40 years old. 'McArthur' was brought back from Asia and planted in the late 1800s. The age of the mother plants played a significant role in the rooting percentage of stem cuttings. The lowest rooting percentages were from older, mature stock plants (**Table 1**).

Table 1: Rooting percentage of *Camellia sinensis* clones. Significant different at $p < 0.05$ level with different letters.

Clone	Rooting %	Clone	Rooting %	Clone	Rooting %	Clone	Rooting %
SZ28	100.0a	SZ21	93.0ab	SH26	87.5abc	SZ11	68.8cdef
SH29	100.0a	SH41	92.2abc	SZ07	84.4abc	'Assamic'	63.3def
CC17	99.2a	'Kunming'	91.4abc	SH47	81.3abcd	SH33	59.4rfg
SH36	97.7a	CC15	90.6abc	SH08	78.1bcde	'McArthur'	39.8fg
CC06	96.9a	SZ16	89.9abc	SZ09	75.0bcde	SZ03	37.5fg
SZ27	95.3ab	SH51	88.3abc	CC04	75.0bcde	'Carswell'	33.6fg
CC11	93.8ab	SZ22	87.5abc	SH13	71.9cdef	'Rosea'	25.0g

When cuttings were taken, the number of nodes per cutting (length) had significant effects on rooting percentage and root quality. For tea plants, 2-3 nodes per cutting yielded significantly higher rooting percentage at 90% or higher. Both one-node or 5-nodes reduced the rooting percentage to 75% or lower. Rooting quality was significantly better as the number of nodes increased. Generally, rooted cuttings rated 3 or higher were acceptable root quality for transplanting (**Table 2**). Plant growth increased as the number of nodes increased (data not presented; Zhou et al., 2020). We recommend that tea stem cuttings should be taken from October to November, rooted in warm greenhouses and transplanted the next April or May. In July or August, tea liners could be sold in one-gallon containers at 41-51 cm (16-20 in.) tall.

Table 2. Effect of cutting length/nodes on rooting of *Camellia sinensis* 'Kunming'. Root rating was from 1 (no roots) to 5 (excellent rooting). Rooted cuttings rated 3 or higher were acceptable root quality for transplanting.

Node	Rooting %	Root Quality
1	75.0c	1.8c
2	96.9a	2.5b
3	90.6ab	3.2ab
4	84.4bc	3.6a
5	75.0c	3.9a

Some plants, such as *Anneslea fragrans*, *Cercis chuniana*, *Ilex buxoides*, *Kalmia latifolia*, and *Osmanthus fragrans*, have not been rooted yet or have an extremely low rooting percentage from stem cuttings. However, not all newly introduced plants are difficult to propagate by stem cuttings. We had higher than 75% rooting success for *Camellia rosthorniana* 'Xulin166' (prolific flowering during winter), *Chimonanthus* 'ZackK07-23' (evergreen and very compact), *Ilex decidua* 'GuihongC04-36' (persistent bright red berry-like drupes to July), *Lagerstroemia* cultivars, *Populus deltoides* 'Hongye1501', *Viburnum* 'YujieK02-33' (very compact), 'Pinkie51' (pink flowers), 'Jinying1572' (evergreen and cold hardy), 'LongqingC04-31' (prolific flowering with rounded habit).

It is our immense pleasure to share the newly bred woody ornamental plants and our experiences of propagating them. Together, we are seeking your help to better propagate these plants and quickly bring them to the ornamental market.

ACKNOWLEGEMENTS

Support from The University of Georgia Research Foundation, Georgia Seed Development Commission, Georgia Green Industry Association, The University of Georgia Horticulture Department, and Center for Applied Nursery Research are forever appreciated.

LITERATURE CITED

Davies, F. T., Geneve, R. L., and Wilson, S. B. (2018). Hartmann and Kester's Plant Propagation: Principles and Practices. Ninth edition. Pearson, NY, NY.

Hao, Z., Zhou,T., and Zhang, D. (2020). Clonal impact of rooting on *Camellia sinensis* (L.) Kuntze stem cuttings. HortScience 55:S34-S35.

Ruchala, S., Zhang,D., Mitchell,W., and Li, J.. (2002). Improving vegetative propagation techniques of sweet fern (*Comptonia peregrina*). Comb. Proc. Intern. Plant Prop. Soc. 52:381-387.

Zhang, D., Dirr, M.A., and Dong, J. (2020). US PP32397 P2, *Nandina* plant named 'Zhnan28'. Application date: December 18, 2019. Awarded date: October 27, 2020.

Zhang, W., Jin, X., Zhang, D., Zhang, M., and Xing, W. (2022). New *Zelkova schneideriana* Hand.-Mazz. cultivar 'Gold Goblin'. HortScience 57:372-373.

Zhou, T., Hao,Z. Bi, G.,and Zhang,D. (2020). Rooting and production of *Camellia sinensis* (L.) Kuntze. HortScience 55:S54.

Yiping,Z., Li,Q., Zhang, D., Liang, Y., Hao,M., and Yang, Y. (2022). Peace Time: a new yellow-fruit holly cultivar. HortScience 57:328-329.

Combined Proceedings IPPS 72:215-229. 2022.

New Introductions from the Mountain Crop Improvement Lab

Irene Palmer

The Mountain Crop Improvement Laboratory, 455 Research Drive, Mills River, NC 28759, USA

iepalmer@ncsu.edu

Keywords: Plant breeding, new plant introductions, non-invasive alternatives

Summary

The North Carolina State University Mountain Crop Improvement Laboratory (MCIL) is an integrated research and development driven program where faculty, staff, students, and interns work together to develop a greater understanding of plant breeding and genetics in a wide array of nursery, bioenergy and emerging crops. In this paper, new plant introductions are discussed including native hydrangeas, new non-invasive introductions in *Spiraea japonica*, *Berberis sp.*, and *Miscanthus*, tough evergreen rhododendrons, new cold-tolerant evergreen azaleas, and deer-resistant *Illicium*.

INTRODUCTION

The Mountain Crop Improvement Laboratory (MCIL) led by Dr. Tom Ranney, is located in Mills River, NC. Housed within the Mountain Horticultural Crops Research and Extension Center and research farm, the MCIL shares in the larger mission of NC State to prepare students, create and apply unbiased knowledge through advances in science and technology, drive economic development, and improve the quality of life in North Carolina and the wider world.

The MCIL works in diverse plant groups with a focus on plant breeding and genetics in nursery and bioenergy crops. This includes the exploration of relationships within plant groups: genome sizing, ploidy surveys, cytology, plant phylogenetics, nomenclature etc., plant fertility and reproductive biology (including non-invasive potential), and the development of tissue culture protocols. As we look ahead to the future, the MCIL is part of complex interdisciplinary teams and working groups hoping to apply gene-editing technologies to nursery and bioenergy crops.

Informed by our research, we utilize traditional plant breeding to improve native plants, develop non-invasive alternatives to weedy nursery crops, and expand the way popular woody shrubs and trees can be used within the landscape. We also train future plant scientists, breeders, enthusiasts, and educators.

A few introductions are highlighted below. See **Table 1** for a complete list of MCIL introductions.

Native Hydrangeas. *Hydrangea arborescens* is a popular, adaptable native hydrangea that blooms on new wood. The MCIL has five pink mophead introductions and two dwarf white mophead introductions (**Figs. 1 and 2**).

Figure 1. List of native *Hydrangea arborescens* pink mophead introductions from the MCIL: Invincibelle Mini Mauvette® pictured.

Figure 2. Compact white *Hydrangea arborescens* introductions from the MCIL.

Newer non-invasive introductions. *Spiraea japonica*, while not yet on invasive plant ban lists, is considered to be an emerging invasive. Double Play® Candy Corn® and Doozie® are non-invasive introductions with other exciting ornamental benefits, including bright foliage and continuous blooming (Figs. 3 and 4).

Figure 3. *Spiraea japonica* Double Play® Candy Corn®.

Figure 4. *Spiraea japonica* Double Play Doozie®.

Invasive *Berberis thunbergii* and *Miscanthus sinensis* have been banned in multiple states. Replicated, multi-year field trials of Sunjoy Todo™ and Mini Maroon® have shown these non-invasive barberries exhibit a 99% reduction in female fertility compared to traditional *B. thunbergii* introductions (**Fig. 5**). Non-invasive *Miscanthus* introductions Bandwidth (compact introduction) and High Frequency exhibit attractive golden banding (**Fig. 6**).

Figure 5. *Berberis* Sunjoy Mini Maroon®.

Figure 6. *Miscanthus sinensis* High Frequency.

Rhododendrons. The MCIL spends a lot of time evaluating evergreen rhododendrons and azaleas! MCIL introductions have been selected for durability in the landscape and excellent container presentation (**Figs. 7 and 8**).

Figure 7. Rhododendron Dandyman Color Wheel®.

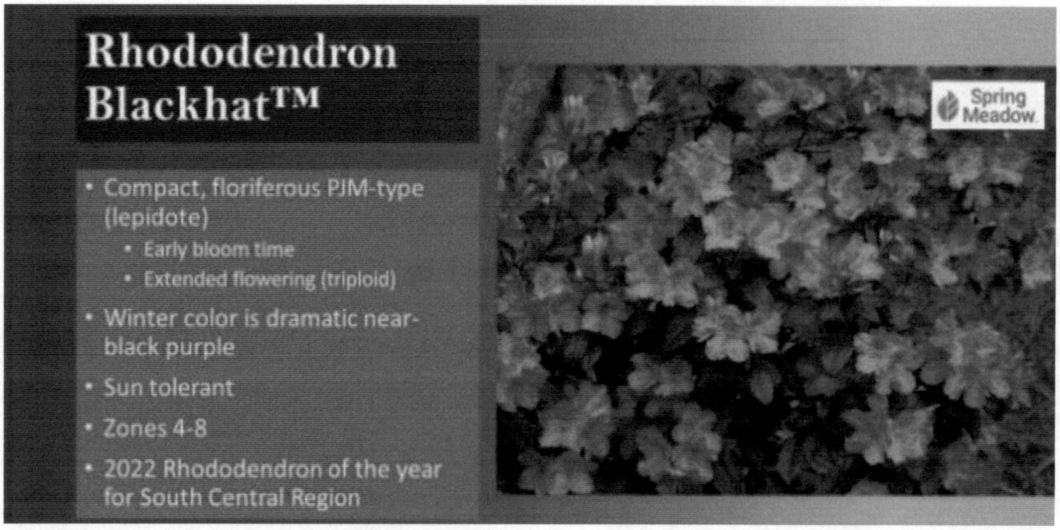

Figure 8. PJM rhododendron Blackhat™

Perfecto Mundo® Evergreen azaleas. The MCIL has been working on azaleas for over 15 years with a focus on cold-hardiness, longevity in bloom, floral coverage and strong field and container performance. See **Table 1** for the complete list of Perfecto Mundo® azaleas (**Fig. 9**).

Figure 9. Perfecto Mundo Double Pink evergreen azalea.

Miscellaneous newer introductions. Some miscellaneous newer introductions include barberries with neon orange new growth Sunjoy® Neo® and Orange Pillar (Fig. 10).

Figure 10. *Berberis* Sunjoy Neo®.

Illicium. Illicium is an exciting underrepresented genus with star-shaped flowers. Interspecific hybrid introductions Scorpio and Orion are adaptable, reblooming, broadleaf evergreens with a dense habit (Fig. 11).

Figure 11. *Illicium* Scorpio and Orion.

Latest Introduction. Our newest introduction is El Nino™ Desert Orchard a unique intergeneric hybrid (*xChitalpa*) die-back shrub to multi-stemmed tree with long racemes of showy purple flowers (Fig. 12).

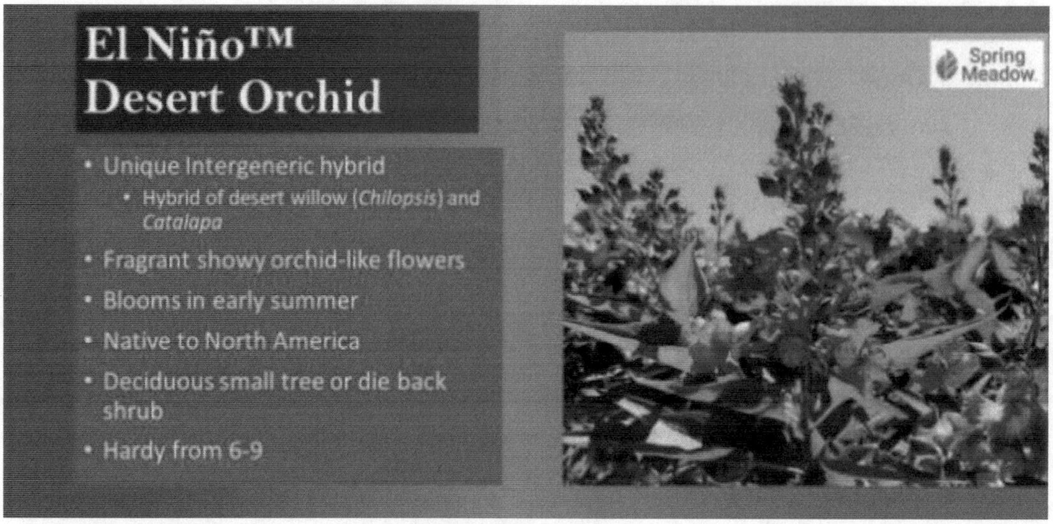

Figure 12. Intergeneric hybrid x*Chitalpa* El Niño™.

Table 1. Mountain Crop Improvement Lab Current Plant Introductions.

Plant Name	Species	Licensee	Cultivar	Patent	Zone	Height and Spread	Characteristics
Chocolate Fountain® silk tree	*Albizia julibrissin*	Star Roses and Plants	'NCAJ1'	USPP25813	6-10	15ft-20ft x 15ft	Purple weeping selection with pink flowers. Drought resistant once established.
Sunjoy Todo® barberry	*Berberis* hybrid	Proven Winners: Spring Meadow Nursery	'NCBX1'	USPP29504	4-8	2ft x 2ft	Compact dark purple foliage, reduced fertility 0% relative female fertility(2021), durable color, deer resistant
Sunjoy Mini Maroon® barberry	*Berberis thunbergii*	Proven Winners: Spring Meadow Nursery	'NCBT1'	USPP30330	4-8	3ft x3ft	Compact purple red foliage, reduced fertility 99% reduction in relative female fertility; durable color, deer resistant
Sunjoy Neo™ barberry	*Berberis thunbergii*	Proven Winners: Spring	'NCBT2'	USPP33272	5-8	3ft x 3ft	Electric orange new growth fades to a deep orange, deer

		Meadow Nursery					resistant, low maintenance.
Sunjoy® Orange Pillar barberry	*Berberis thunbergii*	Proven Winners: Spring Meadow Nursery	'NCBT3'	PPAF	4-8	4ft x 3ft	Same electric new growth as Neo™ but with an upright habit. Deer resistant.
Pearl Glam® beautyberry	*Callicarpa* hybrid	Proven Winners: Spring Meadow Nursery	'NCCX2'	USPP28312	5-8	4ft-5ft x 3ft-4ft	Unique purple foliage, upright form, fall and winter interest with white flowers and purple fruit.
'Aphrodite' sweetshrub	*Calycanthus* hybrid	Proven Winners: Spring Meadow Nursery	'Aphrodite'	USPP24014	5-9	5ft-8ft x 6ft-8ft	Large red-pink fragrant flowers. Vigorous and easy to grow.
'Venus' sweetshrub	*Calycanthus* hybrid	Proven Winners: Spring Meadow Nursery	'Venus'	USPP15925	5-9	5ft-6ft x 6ft-8ft	Fragrant white flowers in the summer, vigorous selection
'Chastity'	*Campsis ×tagliabuana*	Various	'Chastity'	Not patented	5-9	15ft-30ft (spreads)	Non-invasive triploid selection with 99% reduction in fertility. Large orange/peach flowers bloom throughout the summer
Carolina Sweetheart™ redbud	*Cercis canadensis*	Star Roses and Plants	'NCCC1'	USPP27712	4-9	15ft x15ft	Tricolor foliage that emerges electric pink with white foamy variegations. Fades to green and white. Incredibly unique. Native.
Double Take® Orange flowering quince	*Chaenomeles speciosa*	Proven Winners: Spring Meadow Nursery	'Orange Storm'	USPP20950	5-9	4ft-5ft x 3ft-4ft	Orange double flowers, thornless, extended flowering, fruitless, largest flowers
Double Take® Peach flowering quince	*Chaenomeles speciosa*	Proven Winners: Spring Meadow Nursery	'NCCS4'	USPP30231	5-9	4ft-5ft x 3ft-4ft	Peach double flowers, thornless, extended flowering, fruitless
Double Take® Pink flowering quince	*Chaenomeles speciosa*	Proven Winners: Spring Meadow Nursery	'Pink Storm'	USPP20920	5-9	4ft-5ft x 3ft-4ft	Pink double flowers, thornless, extended flowering, fruitless

Double Take® Scarlet flowering quince	*Chaenomeles speciosa*	Proven Winners: Spring Meadow Nursery	'Scarlet Storm'	USPP20951	5-9	4ft-5ft x 3ft-4ft	Crimson double flowers, thornless, extended flowering, fruitless
El NIÑO™ Desert Orchid	*xChitalpa*	Proven Winners: Spring Meadow Nursery	'NCXC1'	PPAF	6-9	5ft-8ft x 4ft-6ft	Large, orchid like, purple flowers.
Sugartina® 'Crystalina' summersweet	*Clethra alfnifolia*	Proven Winners: Spring Meadow Nursery	'Crystalina'	USPP21561, Can4160	4-9	2ft-3ft x 2ft-3ft	Fragrant white flowers in midsummer. Yellow fall color, compact. Attracts butterflies
Winecraft Black® smokebush	*Cotinus coggygria*	Proven Winners: Spring Meadow Nursery	'NCCO1'	USPP30216	4-8	4ft-6ft x 6ft-10ft	Semi-dwarf smokebush with deep purple foliage that darkens to a near black. Bright orange fall color. Rounded habit.
Lucky Leprechaun® dogwood	*Cornus elliptica*	Star Roses and Plants	'NCCE1'	USPP28998	7-9	13ft x 5ft	Disease resistant, evergreen dogwood, strong bloomer, compact upright form
Yuki Cherry Blossom® deutzia	*Deutzia* hybrid	Proven Winners: Spring Meadow Nursery	'NCDX1'	USPP25916	5-8	1ft-2ft x 1ft-2ft	Extra heavy bloomer in pink, attractive fall color, tight form, compact
Yuki Snowflake® deutzia	*Deutzia* hybrid	Proven Winners: Spring Meadow Nursery	'NCDX2'	USPP28347	5-8	1ft-2ft x 2ft-3ft	Extra heavy blooming variety with a profusion of white flowers in spring, attractive fall color, tight form
Snow Day® 'Blizzard' pearlbush	*Exochorda*	Proven Winners: Spring Meadow Nursery	'Blizzard'	USPP23329	3-8	6ft x7ft	Large flowers, compact plant. Can be trained as a small tree or a shrub.
Legend of the Small®	*Fothergilla ×intermedia*	Proven Winners: Spring Meadow Nursery	'NCFI1'	USPP34139	5-9	2.5ft x 2-4ft	Compact, small-leafed selection that produces a profusion of fragrant white flowers in spring and bright fall color in yellows, oranges and reds; native

'Sweet Tea'	*xGordlinia grandiflora*	Various	'Sweet Tea'	Not patented	7-10	20ft-30ft x 8ft-10ft	Large showy white flowers visible July-September with delicate fragrance. Often works best as a multi-stem tree. Unique intergeneric hybrid.
Incrediball® Blush smooth leaf hydrangea	*Hydrangea arborescens*	Proven Winners: Spring Meadow Nursery	'NCHA4'	USPP28280	3-8	4ft-5ft x 5ft-6ft	Strong stems with exceptionally large light pink blooms; long bloom season. Blooms on new wood; native.
Invincibelle Garnetta® smooth leaf hydrangea	*Hydrangea arborescens*	Proven Winners: Spring Meadow Nursery	'NCHA6'	USPP33142	3-8	2.5ft x 2.5ft	Compact, season-extending deep garnet buds that open to pink, blooms a bit later than other members of the series; native
Invincibelle Limetta® smooth leaf hydrangea	*Hydrangea arborescens*	Proven Winners: Spring Meadow Nursery	'NCHA8'	USPP30431	3-8	3ft-4ft x 3ft-4ft	Cheery chartreuse flowers fade to light green. Plant has a compact form and long bloom season; native.
Invincibelle Mini Mauvette® smooth leaf hydrangea	*Hydrangea arborescens*	Proven Winners: Spring Meadow Nursery	'NCHA7'	USPP30358	3-8	3ft-4ft x 4ft-6ft	Strong stems with purple pink blooms and a long bloom season; native
Invincibelle Wee White® smooth leaf hydrangea	*Hydrangea arborescens*	Proven Winners: Spring Meadow Nursery	'NCHA5'	USPP30296	3-8	1ft-2.5ft x 3ft	Dwarf Annabelle type with strong stems and a compact habit; native.
Invincibelle® Ruby smooth leaf hydrangea	*Hydrangea arborescens*	Proven Winners: Spring Meadow Nursery	'NCHA3'	USPP28317	3-8	3ft-4ft x 4ft-6ft	Strong stems and a compact form. Brightest of the Invincibelle series with bright pink flowers; native
Invincibelle® Spirit II smooth leaf hydrangea	*Hydrangea arborescens*	Proven Winners: Spring Meadow Nursery	'NCHA2'	USPP28316	3-8	4ft-4.5ft x 4ft-6ft	Strong stems, pink blooms, long bloom season, benefits Breast Cancer Research Foundation; native
Orion Star Flower	*Illicium* hybrid	Star Roses and Plants	'NCIH2'	USPP29938	6-9	5ft x 5ft	Broad leaf evergreen rebloomer

							with white star-anise like flowers
Scorpio Star Flower	*Illicium* hybrid	Star Roses and Plants	'NCIH1'	USPP29939	6-9	5ft x 5ft	Broad leaf evergreen rebloomer with red star-anise like flowers
Golden Ticket® privet	*Ligustrum* (vicary privet)	Proven Winners: Spring Meadow Nursery	'NCLX1'	USPP27301	5-9	4ft-6ft x 4ft-6ft	Gold-leafed, fragrant reduced fertility privet that is durable and compact.
Bandwidth maiden grass	*Miscanthus sinensis*	Darwin Perennials	'NCMS2B'	USPP29460	5-9	3ft x 2ft	Highly reduced fertility, attractive yellow banding on a compact plant; good in the ground or in containers
My Fair Maiden™ maiden grass	*Miscanthus sinensis*	Darwin Perennials	'NCMS1'	USPP26387	6-9	3ft-4ft x 7ft-9ft	Reduced fertility selection with an attractive vase-shaped form and showy flowers.
High Frequency™ maiden grass	*Miscanthus sinensis*	Darwin Perennials	'NCMS3'	PPAF	5-9	6ft-7ft x 3ft	Tall, columnar variegated miscanthus with bright yellow banding; non-invasive introduction.
Mercury™ Magnolia	*Magnolia*	J Frank Schmidt	'NCMX1'	USPP29218	5-8	25ft x15ft	Deep pink buds open lavendar. Flowers up to a month later than traditional M. soulangeana, greatly reducing flower loss to frost. Deciduous.
Good Vibes Mahonia	*Mahonia* hybrid	Star Roses and Plants	'NCMH1'	USPP34442	7b-9	4ft x 2ft	Compact habit with dense branching; soft, dark evergreen foliage with a refined look; and long racemes of showy yellow flowers produced sporadically in fall, winter, and spring.
Groovy Glow Mahonia	*Mahonia* hybrid	Star Roses and Plants	'NCMH2'	USPP34443	7b-9	4ft x 2ft	Compact habit; narrow, evergreen foliage; showy racemes with multicolored flowers that bloom sporadically

							throughout fall, winter, and spring; and showy red new foliage
Pink Cascade® weeping cherry	*Prunus* hybrid	J Frank Schmidt	'NCPH1'	USPP27579	5-8	12ft x12ft	Easy to grow weeping cherry with a prolific spring show of light pink flowers; orange fall color.
Javelin® Pear	*Pyrus*	J Frank Schmidt	'NCPX1'	USPP26539	5-8	35ft x10ft	Fastigiate pear with purple new growth in the spring. Fireblight resistant. Excellent urban tree.
Chastity® Pear	*Pyrus ×triploida*	J Frank Schmidt	'NCPX2'	USPP30788	5-8	35ft x 25ft	A non-invasive flowering pear with dark green flossy foliage, orange/red fall color and profuse white flowers in spring.
Perfecto Mundo® Double Pink evergreen azalea	*Reblooming Azalea*	Proven Winners: Spring Meadow Nursery	'NCRX3'	USPP32819	6-9	3ft x 3ft	Double shell pink, strong rebloomer, lacebug resistant, sun tolerant, compact form; generally adorable
Perfecto Mundo® Double Purple evergreen azalea	*Reblooming Azalea*	Proven Winners: Spring Meadow Nursery	'NCRX4'	USPP33205	6-9	3.5ft-3.5ft	Reblooming, evergreen double purple azalea with large flowers and an extended bloom time.
Perfecto Mundo® Double White evergreen azalea	*Reblooming Azalea*	Proven Winners: Spring Meadow Nursery	'NCRX5'	USPP33900	6-9	3ft x ft	Reblooming azalea with large clean white double flowers and extended bloom time. Dense habit.
Perfecto Mundo® Double Dark Pink evergreen azalea	*Reblooming Azalea*	Proven Winners: Spring Meadow Nursery	'NCRX7'	USPP33923	6-9	2.5ft-3ft x 3ft-4ft	Reblooming azalea with large dark pink double flowers and dark green foliage. Exceptionally long in flower.
Perfecto Mundo® Orange evergreen azalea	*Reblooming Azalea*	Proven Winners: Spring Meadow Nursery	'NCRX6'	USPP33945	6-9	3ft-4ft	Reblooming azalea with bright orange-red single flowers with a dense, rounded habit.

Perfecto Mundo® Epic Coral evergreen azalea	*Reblooming Azalea*	Proven Winners: Spring Meadow Nursery	'NCRX10'	PPAF	6-9	3ft-4ft x 4ft-5ft	Reblooming azalea with exceptionally large flowers in a soft pink. Long in flower, strong re-bloomer; dense habit.
Perfecto Mundo® Red ever-green azalea	*Reblooming Azalea*	Proven Winners: Spring Meadow Nursery	'NCRX8'	USPP33922	6-9	2.5 ft x 3ft-4ft	Reblooming azalea with a dense, com-pact habit and an explosion of single red flowers in the spring and fall.
Perfecto Mundo® Epic Pink evergreen azalea	*Reblooming Azalea*	Proven Winners: Spring Meadow Nursery	'NCRX9'	PPAF	6-9	3ft-4ft x 4ft-5ft	Reblooming azalea with large deep pink single flowers and dark green foli-age. Long in flower, dense habit.
Black Hat™ PJM Rhodo-dendron	*Rhododen-dron* hybrid	Proven Winners: Spring Meadow Nursery	'NCRX2'	USPP31898	4-8	3ft x 3ft	Compact PJM type with dark winter fo-liage. Consistent flowering year to year with extended bloom time.
Dandyman Color Wheel™ Rhododen-dron	*Rhododen-dron* hybrid	Proven Winners: Spring Meadow Nursery	'NCRX1'	USPP31150	5-9	8ft x 8ft	Sun-tolerant ever-green Rhododen-dron with unique color-changing pink to white flowers; thrives in tough conditions.
Mountain Schimlinia	×*Schimlinia*	Various	'Schima Lina Ding Dong'	None	5-9	12ft x 6ft	Intergeneric hybrid between *Schima re-mote-serrata* and *Franklinia alatamaha*
Double-play® Candy Corn® spi-raea	*Spiraea japonica*	Proven Winners: Spring Meadow Nursery	'NCSX1'	USPP28313 P2	4-7	2ft-2.5ft x 3ft	Interesting multi-tone foliage transi-tioning from bright orange red to char-treuse gold. Bright magenta flowers, low maintenance; non-invasive.
Double-play® Doozie® spiraea	*Spiraea japonica*	Proven Winners: Spring Meadow Nursery	'NCSX2'	USPP30953 P3	3-8	3ft x 3ft	A revolutionary spiraea: a dense, continuous bloomer with red new growth; seedless.

Sweet Talker™ viburnum	*V. bodnantense x suspensum* hybrid	Proven Winners: Spring Meadow Nursery	'NCVX4'	USPP33229	7-8	8ft-10ft [upright]	Fragrant pink flowers in early spring, handsome evergreen foliage. Easy to grow, tolerant of heat and salt.
Yang™ viburnum	*V. davidii x V. propinqumum* hybrid	Proven Winners: Spring Meadow Nursery	'NCVX3'	USPP33203	7b-9	2ft x 4ft	Shade tolerant, deer tolerant evergreen foliage plant with white flowers and blue fruit (with Yin). Foliage is blue-green.
Yin™ viburnum	*V. davidii x V. propinqumum hybrid*	Proven Winners: Spring Meadow Nursery	'NCVX2'	USPP33479	7b-9	2ft x 2ft-4ft	Shade tolerant, deer tolerant evergreen foliage plant with white flowers and blue fruit (with Yang). Foliage is blue-green.
Shiny Dancer® viburnum	*Viburnum*	Proven Winners: Spring Meadow Nursery	'NCVX1'	USPP28095	6-8	3ft-5ft x 3ft-5ft	Compact selection with red flush in the spring. Shiny, heavily-textured foliage turns to deep burgundy in the fall. Needs 'Huron' or 'Chippewa' to fruit
Steady Eddy®	*Viburnum plicatum var. tomentosum*	Proven Winners: Spring Meadow Nursery	'NCVP1'	USPP33899	5-8	5ft x 5ft'	An improvement on 'Summer Snowflake' with dense flowering and good display in containers.
My Monet® 'Sunset' weigela	*Weigela florida*	Proven Winners: Spring Meadow Nursery	'Sunset'	USPP23212	5-7	1ft-1.5ft x 1ft-1.5ft	Dwarf weigela with variegated foliage, rosy pink new growth. Bright reds and oranges in the fall.
Juke box®	*xPyracomeles*	Proven Winners: Spring Meadow Nursery	'NCXP1'	USPP31409	7-9	2ft-3ft x 2ft-3ft	Boxwood alternative with unique foliage, durable, low maintenance

Combined Proceedings IPPS 72:230-234. 2022.

Translating the European Approach to Domestic Plant Production

Laura Kline

The Plant Company, 3038 Stuarts Draft Hwy., Stuarts Draft, VA 24477, USA

lkline@theplantcompany.com

Keywords: Automation, mechanization, European production, environmental controls, beneficial insects

Summary

This paper details how European practices can be applied to plant production in North America. After multiple trips to the Netherlands and investigating how plants are produced overseas, only the most fitting automation and production ideas were brought to our greenhouse and implemented. From our state-of-the-art greenhouse to our choice of environmental controls, we strive for top quality plants and ease-of-work and greater efficiency of our team. Our team is vital in our success, and if our automation and equipment can make their life easier and our plant quality better - we have succeeded.

INTRODUCTION

The Plant Company was founded in 2020 by Frank Paul (COO) and brothers Jason and Wes VanWingerden. The 2-ha (5-ac) state of the art greenhouse is for the production of indoor foliage and tropical plants. We currently produce multiple varieties of Philodendron, Alocasia, Ficus, Calathea, Syngonium, Aglaonema, pothos, Ctenanthe, Dieffenbachia, Tradescantia, and Monstera. Some 85% of our plant liners come from tissue culture. We purchase our plantlets from multiple labs throughout the world. We do this to ensure virus free plants for the end consumer. The 15% of the propagules that we propagate as unrooted cuttings are tested with Agida® virus test strips https://www.agdia.com/ upon arrival to our greenhouse - and watched closely in a quarantined area.

PRODUCTION AUTOMATION

We receive plant material the first half of the week and inspect each variety. All unrooted cuttings are treated before they are stuck. Tissue culture liners go straight to the sticking line (**Fig. 1**).

Figure 1. (Left) Ergonomic sticking tables, (center) potting line, and (right) a Gantry robot machine https://www.sagerobot.com/gantry-robots/

We use 3 sizes of trays: 72, 40 and 28. We run our trays through the tray filling machine, after filling, the media is irrigated once before sticking. Each seated team member will take one tray from the conveyor above and place it directly in front of them on their table for ease of sticking. After the tray is filled, it will be placed back on the conveyor to head to the next irrigation point before being placed on a grow rack. Once the trailer is full, the grow rack heads out to the liner area. Our trailers are pulled by battery powered tuggers (**Fig. 2**) to eliminate any possible fumes inside the facility. All equipment inside the greenhouse is battery powered or electric (**Fig. 2**). Once the trailer reaches its assigned location our "flying forks" lift the grow rack from the trailer and place it on the floor (**Fig. 3**). This equipment requires only one person to operate. Once the racks are placed on the floor, we enclose them in poly to create a microclimate. This eliminates overhead irrigation and algae growth. After 10 days the poly will be vented. Once the liners are rooted, the poly comes off the grow racks.

Figure 2. (Left) Battery powered tugger, (center) forklift, and (right) and electric boom.

Figure 3. (Left) Flying forks (arrows) moving grow racks. (Right) Operator controlling grow racks.

This same equipment will be used to pick up grow racks of fully rooted plants for planting into 12-cm or 17-cm recycled plastic pots. We load our trailers with finished liners and head to the potting line (**Fig. 1**). Our machine has a pot dispenser that releases the pots onto the conveyor, fills the pot with media and dibbles a hole. Once planted the pot will go through a series of irrigation nozzles to lightly water the plant for its trip to the greenhouse. The potted plants are then gathered on the rail for the robot to place onto the trailer (Fig. 1). When a line on the trailer is full the trailer moves forward the required space for the next row of plants. When two trailers are full of plants, we drive them to the greenhouse for placement in the proper environment using the flying forks (**Fig. 4**).

Greenhouse

The climate-controlled Venlo greenhouse https://www.venloinc.com/ has six separate compartments with a glass roof. We keep each compartment at different light levels and humidity for optimal plant growth. We have photosynthetically active radiation (PAR) light sensors above and below the shade curtain. There is an energy curtain and a shade curtain in each compartment along with varying degrees of whitewash in the warmer months to achieve this. Our "par perfect" energy curtain diffuses the light evenly over the crop - even if the shade is partially open. We use a high-pressure fog machine to keep the humidity within the correct parameters in the greenhouse (**Fig. 5**).

Figure 5. (Left) Fog machine, (center) fresh water tank, and fertilizer recipe tanks for flood-floor, and (right) fertilizer injectors.

Our irrigation can be done with our mobile boom, overhead or with our flood floors. The electric powered boom is used when small areas need irrigation (**Figs. 2 and 4**). This piece of equipment can also be used to apply chemicals if needed for a precise even pattern. When using the overhead irrigation, we can program the pH and EC needed for each individual bay. When we use the flood floors, we have two fertility recipes to choose from: a nitrogen rich formula and a potassium rich formula (**Fig. 6**).

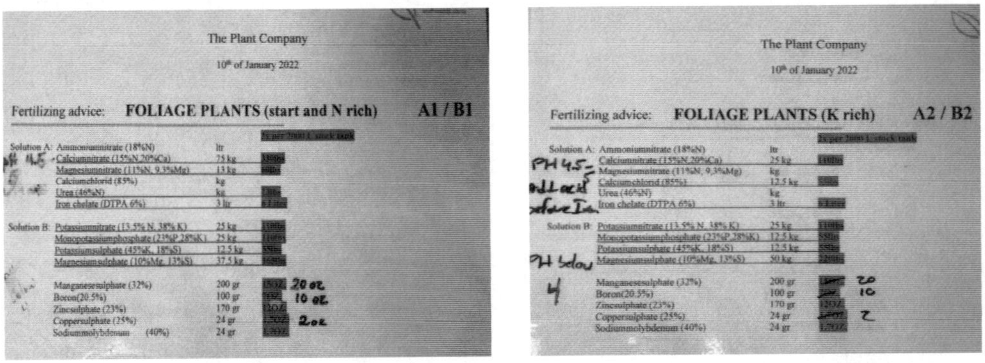

The Plant Company

10th of January 2022

Fertilizing advice: **FOLIAGE PLANTS (start and N rich)** **A1 / B1**

			[redacted]	
Solution A:	Ammoniumnitrate (18%N)	ltr		
	Calciumnitrate (15%N 20%Ca)	75 kg	[redacted]	
	Magnesiumnitrate (11%N, 9.3%Mg)	13 kg	[redacted]	
	Calciumchlorid (85%)	kg		
	Urea (46%N)	kg	[redacted]	
	Iron chelate (DTPA 6%)	3 ltr	[redacted]	
Solution B:	Potassiumnitrate (13.5% N, 38% K)	25 kg	[redacted]	
	Monopotassiumphosphate (23%P 28%K)	25 kg	[redacted]	
	Potassiumsulphate (45%K, 18%S)	12.5 kg	[redacted]	
	Magnesiumsulphate (10%Mg 13%S)	37.5 kg	[redacted]	
	Manganesesulphate (32%)	200 gr	[redacted]	20 oz
	Boron(20.5%)	100 gr	[redacted]	10 oz
	Zincsulphate (23%)	170 gr	[redacted]	
	Coppersulphate (25%)	24 gr	[redacted]	2 oz
	Sodiummolybdenum (40%)	24 gr	[redacted]	

The Plant Company

10th of January 2022

Fertilizing advice: **FOLIAGE PLANTS (K rich)** **A2 / B2**

			[redacted]	
Solution A:	Ammoniumnitrate (18%N)	ltr		
	Calciumnitrate (15%N 20%Ca)	25 kg	[redacted]	
	Magnesiumnitrate (11%N, 9,3%Mg)	kg		
	Calciumchlorid (85%)	12.5 kg	[redacted]	
	Urea (46%N)	kg		
	Iron chelate (DTPA 6%)	3 ltr	[redacted]	
Solution B:	Potassiumnitrate (13.5% N, 38% K)	25 kg	[redacted]	
	Monopotassiumphosphate (23%P 28%K)	12.5 kg	[redacted]	
	Potassiumsulphate (45%K, 18%S)	12.5 kg	[redacted]	
	Magnesiumsulphate (10%Mg, 13%S)	50 kg	[redacted]	
	Manganesesulphate (32%)	200 gr	[redacted]	20
	Boron(20.5%)	100 gr	[redacted]	10
	Zincsulphate (23%)	170 gr	[redacted]	
	Coppersulphate (25%)	24 gr	[redacted]	2
	Sodiummolybdenum (40%)	24 gr	[redacted]	

Figure 6. Recipes for (left) N-nitrogen-rich and (right) K-potassium-rich fertilization.

We mix the recipe ingredients individually so we can adjust if needed based on nutrient testing. Our flood floors are not made of concrete. They have a plastic layer at the bottom with crushed limestone in the middle and a fiber membrane on top. We program the volume of water by adjusting the time the flood runs in each bay. We can use this floor to cool the greenhouse if needed by running a short program that will not irrigate our plants. All irrigation water is captured, whether overhead or flood, and returned to its appropriate silo for reuse (**Fig. 5**). To minimize our chance of disease we inject Kleengrow™ https://www.greenbook.net/pace-chemicals-ltd/kleengrow directly into our water lines that are connected to the greenhouse (**Fig. 5**). Our heating is done with boilers that is a closed loop system with a tank outside that houses the water (**Fig. 7**). The heated water warms the warehouse, shipping area and greenhouse. In the greenhouse we have heat pipes in the floor, the gables, and the top of the greenhouse. Our heat is moved downward by vertical fans. Our plants benefit from higher CO_2 levels - so we are able to scrub the CO_2 from our natural gas boilers and send it back into the greenhouse. This is done by running our boilers during the day while storing the heat in our hot water storage tank.

Figure 7. (Left) Hot water holding tank, and (right) boilers for heating the facility.

Our entire greenhouse environment is controlled by IIVO Hoogendoorn https://readysetgrow.nl/ . Just a few of the benefits of IIVO are ease of use, weather forecasting that will close vents in case of a storm, and my favorite - it can be controlled remotely with your phone. Our growers are able to access real time information from the greenhouse environment. We are also able to look at a year's worth of data, in an easy-to-read graph.

Since our greenhouse range is very open, we utilize beneficial insects to keep the pest populations in check. This program is modified with the seasons and pest pressure. We have used *Aphidius colemani*, lacewings, *Orius, Atheta, Swirski*, beneficial nematodes and *Persimilis*. If we need to apply chemicals, we use ones that work with our beneficial program. Some of those chemicals we can use, and then apply beneficials four hours later! Our employees greatly appreciate the low use of any chemicals and the elimination of all harsh chemistries.

The mindset at The Plant Company is intense quality control - focused on plants and our staff. Our owners and directors are out in the facility observing the tasks to see if there is an easier way to get the job done. We believe and follow through with creating an environment where people have quality time outside of work. This European mindset has been communicated to our community and we are not at a loss for job applicants. With the back breaking tasks automated, we have people that enjoy coming to work every day and can go home at a reasonable time. As we say: "Together we grow stronger"!

Combined Proceedings IPPS 72:235-248. 2022.

How to Help Your Plants Hold Their "P" in Container-Based Nursery Production

Jacob H. Shreckhise[1] and James E. Altland[2]

[1]U.S. Department of Agriculture, Agricultural Research Service, U.S. National Arboretum, Floral and Nursery Plants Research Unit, Otis L. Floyd Nursery Research Center, 472 Cadillac Lane, McMinnville, TN 37110, USA. [2]U.S. Department of Agriculture, Agricultural Research Service, Application Technology Research Unit, 1680 Madison Ave., Wooster, Ohio 44691, USA

jacob.shreckhise@usda.gov

Keywords: phosphorus (P), ferrous sulfate, leaching, soilless substrate, cationization, best management practices (BMPs), controlled release fertilizer (CRF)

Summary

Pine bark substrate used for container-based nursery crop production poorly retains phosphorus (P), resulting in much of the applied P leaching from containers. Research was conducted to evaluate the effect of $FeSO_4 \cdot 7H_2O$ (ferrous sulfate heptahydrate)-amended pine bark in nursery containers, added as a bottom layer (50% volume) or sole substrate, on P leaching and plant growth of economically important nursery crops. Freeman maple (*Acer ×freemanii* 'Jeffersred' Autumn Blaze®), panicle hydrangea (*Hydrangea paniculata* 'SMNHPRZEP' Zinfin Doll®), shrub rose (*Rosa* ×'HORCOGJIL' At Last®), nandina (*Nandina domestica*), and arborvitae (*Thuja* ×'Green Giant') were grown for 13 weeks in 6.1-L (#2) containers with surface-applied controlled-release fertilizer [(CRF); 16N−2.6P−9.1K + micronutrients] and received daily overhead irrigation that was periodically adjusted to achieve a 0.35 leaching fraction. Plants were grown in one of four substrate treatments comprised of dolomite-amended pine bark with: 1) no $FeSO_4 \cdot 7H_2O$ (control); 2) 1.5 kg/m^3 (2.5 lbs/yd^3) $FeSO_4 \cdot 7H_2O$ (FS-1.5); 3) 3 kg/m^3

(5 lbs/yd^3) $FeSO_4 \cdot 7H_2O$ (FS-3); or 4) stratified substrate (FS-3St) in which containers had a 2.5-L layer of FS-3 in the bottom and a 2.5-L layer of the control substrate on top. All leachate from Freeman maple was collected from each container weekly and analyzed for P. Relative to the control, the FS-1.5, FS-3, and FS-3St treatments reduced P leaching by 32%, 57%, and 54%, respectively. Shoot and root dry weight of panicle hydrangea, nandina, shrub rose, and arborvitae were unaffected by substrate treatments. Freeman maple had highest dry weight when grown in the control, but there were no differences in visual quality among treatments. Pine bark amended with 3 kg/m^3 $FeSO_4 \cdot 7H_2O$ either layered in the bottoms of nursery containers or used as the sole substrate can substantially reduce P leaching without affecting growth of four economically important shrub taxa; however, additional fast-growing taxa with high nutrient requirements (like Freeman maple) should be evaluated.

INTRODUCTION

Approximately 80% of U.S. nursery operations produce crops in above-ground containers (USDA, 2019). Substrates commonly used for container-based production, predominantly pine bark in the eastern U.S., are inherently low in plant-essential nutrients and have poor nutrient-holding capacities (Majsztrik et al., 2011). Frequent replenishment of the substrate with nutrients, whether by liquid feeding or applying a controlled-release fertilizer (CRF), is therefore essential to producing a salable crop. However, the constant presence of a soluble or solubilizing fertilizer in a substrate that poorly retains nutrients, paired with frequent (often daily) irrigation and periodic rainfall, results in excess nutrients leaching from containers. Phosphorus (P) is particularly prone to leaching from pine bark-based substrates (Cole and Dole, 1997; Godoy and Cole, 2000; Yeager and Barrett, 1984, 1985a, 1985b). For example, Yeager and Barrett (1984) showed that when 3 kg/m^3 superphosphate was incorporated into a substrate composed of 2 pine bark: 1 peatmoss: 1 sand, 76% of the applied P leached from the substrate in just three weeks of once-daily irrigation.

Nutrients that drain from nursery containers can subsequently runoff to surface waters. Phosphorus contamination of surface waters has been linked to eutrophication and harmful algal blooms that are responsible for annual "dead zones" that plague the Gulf of Mexico, Chesapeake Bay, Lake Erie, Florida Everglades, Lake Okeechobee, and other economically and ecologically important water bodies (Wurtsbaugh et al., 2019). The impact of agricultural P runoff on surface water quality has resulted in increased environmental regulation, a trend that will likely continue in an effort to remediate and preserve impaired waterways. The nursery industry is not immune to state-mandated nutrient management laws. For example, Maryland's Water Quality Improvement Act of 1998 requires all agricultural operations (including ornamental plant nurseries) grossing ≥$2,500 to submit nitrogen (N) and P management plans and file annual reports on N and P applications (Majsztrik and Lea-Cox, 2013). More recently, Florida enacted Senate Bill 712 (the "Clean Waterways Act") in 2020 which requires all agricultural landowners and growers to submit N and P application records to the Florida Department of Agriculture and Consumer Services; individuals who fail to do so may be reported to the Florida Department of Environmental Protection for "regulatory action."

Best Management Practices (BMPs), i.e., voluntary activities, prohibitions, and cultural practices designed and implemented to preserve and/or remediate water resources, have been widely adopted by containerized nurseries in the U.S. (Bilderback et al., 2013). Fertilizing with a CRF instead of soluble forms is among the most widely implemented BMPs for fertilizer management according to survey studies in Virginia and Alabama (Fain et al., 2000; Mack et al., 2017). However, P leaching from CRF-fertilized containerized crops can be substantial. Broschat (1995), Million et al. (2007a, 2007b), Tyler et al. (1996a, 1996b), and Million and Yeager (2021) reported that 7% to 47% of P applied in controlled-release fertilizer was found in the leachate of containerized crops grown in pine bark-based substrates. Closely monitoring and managing irrigation to avoid excessive leaching (e.g., maintaining a leaching fraction of <0.15) can reduce P leaching (Owen et al., 2008; Tyler et al., 1996b). However, precisely managing irrigation to minimize water and nutrient leaching from container-grown crops is challenging for even the most experienced growers. Furthermore, when using micro-irrigation for outdoor containerized nursery production, rainfall may void the nutrient retention benefits of managing irrigation to maintain a low leaching fraction (Million and Yeager, 2021).

Modifying the charge properties of conventional substrate components (e.g., pine bark) through a process called cationization is a novel but simple approach to reducing P leaching losses from containerized crops, even during heavy rainfall events or when irrigation is overapplied. Cationization can be accomplished by amending an organic material with a metal salt. As the metal salt dissolves, the metal cations rapidly adsorb to the surface of the organic material resulting in an increase in anion binding sites. Metal-loaded agricultural byproducts (e.g., sugarcane bagasse, coir pith, wood particles, okara) have been studied extensively for their capacity to sequester phosphate (PO_4^{3-}) from wastewater (Nguyen et al., 2014; Pokhrel et al., 2019). Relative to other metal compounds that have been used to cationize organic materials [e.g., $ZnCl_2$, ZrO_2Cl, $La(NO_3)_3$], iron salts are less expensive, non-toxic, and more readily available for purchase (Pokhrel et al., 2019). Ferrous sulfate heptahydrate ($FeSO_4 \cdot 7H_2O$) is one such soluble iron compound that is often used in containerized nursery production as a pre-plant Fe fertilizer (usually a component of a complete micronutrient fertilizer). When mixed into pine bark substrate along with superphosphate, $FeSO_4 \cdot 7H_2O$ has been shown to reduce the amount of water-extractable P from the substrate (Handreck, 1992). However, the effects of amending substrate with relatively high rates of $FeSO_4 \cdot 7H_2O$ (i.e., 1.5 to 3 kg/m^3) on P leaching and growth of containerized nursery crops has not yet been investigated.

Depending on the nature of the bark-Fe-P complexes, the adsorbed P may or may not be available for plant uptake. Adding a layer of $FeSO_4 \cdot 7H_2O$-amended pine bark to the bottoms of the plant containers with the remainder of the container filled with a standard pine bark substrate may be a way to avoid possible reductions in P availability to plants while still mitigating P leaching from the container. In this stratified substrate system, P that is released from the fertilizer and leaches past the root zone would be adsorbed to the $FeSO_4 \cdot 7H_2O$-amended bark before draining from the container. Growing nursery and greenhouse crops in containers with strategic layers of substrate has recently been proposed as a means to decrease fertilizer requirements (Fields et al., 2021), reduce weed growth and weed seed germination (Khamare et al., 2022a and 2022b), and mitigate crop water stress during drought (Criscione et al., 2022). However, stratifying substrate to reduce nutrient leaching from containers has not yet been investigated. The objectives of this research were to investigate the effect of a $FeSO_4 \cdot 7H_2O$-amended pine bark as a bottom layer or throughout container on P leaching and plant growth of five economically important nursery crops.

MATERIALS AND METHODS

Three pine bark batches containing varying rates of agricultural-grade $FeSO_4 \cdot 7H_2O$ and pulverized dolomite were produced using a ribbon mixer [Twister I (single phase); Bouldin & Lawson, McMinnville, TN] with a 5-minute mixing time after all amendments had been added. Batch 1 contained 0.79 kg/m^3 (1.34 lbs/yd^3) pulverized dolomite and no $FeSO_4 \cdot 7H_2O$; Batch 2 contained 2.37 kg/m^3 (3.99 lbs/yd^3) pulverized dolomite and 1.5 kg/m^3 (2.5 lbs/yd^3) $FeSO4 \cdot 7H_2O$; and Batch 3 contained 3.27 kg/m^3 (5.51 lbs/yd^3) pulverized dolomite and 3 kg/m^3 (5 lb/yd^3) $FeSO_4 \cdot 7H_2O$. The dolomite rate varied between batches to achieve a starting pH of ~5.5 in all batches while compensating for the acidifying effect of the $FeSO_4 \cdot 7H_2O$ as it dissolves and hydrolyzes. Black plastic, 6.1-L (#2) containers were each filled to 5 L with the substrate batches according to the treatments shown in **Fig. 1**.

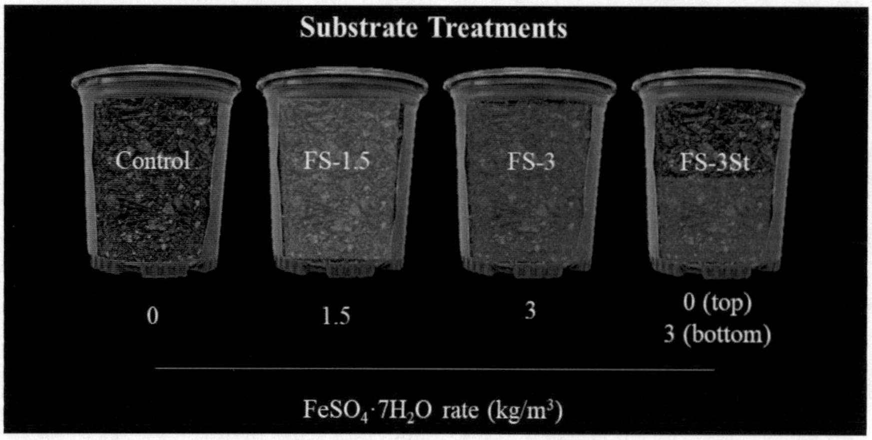

Figure 1. Illustration of substrate treatments, including the location and rate of $FeSO_4 \cdot 7H_2O$ (ferrous sulfate heptahydrate) incorporated into the pine bark. In addition to $FeSO_4 \cdot 7H_2O$, the control, FS-1.5, and FS-3 substrates were amended with 0.79, 2.4, and 3.3 kg/m^3 pulverized dolomite, respectively.

Containers that represented the control, FS-1.5, and FS-3 treatments contained batches 1, 2, and 3, respectively, throughout the container profile, whereas FS-3St containers were stratified with a 2.5-L layer of Batch 3 in the bottom and a 2.5-L layer of Batch 1 on top.

Ten replicate rooted cuttings per treatment from each of five woody plant taxa (200 plants total) were planted into the substrate-filled containers. Taxa in this experiment included a Freeman maple (*Acer ×freemanii* 'Jeffersred' Autumn Blaze®), panicle hydrangea (*Hydrangea paniculata* 'SMNHPRZEP' Zinfin Doll®), shrub rose (*Rosa* ×'HORCOGJIL' At Last®), nandina (*Nandina domestica*), and arborvitae (*Thuja* ×'Green Giant'). All newly potted plants were top-dressed with 33.8 g (medium rate) of a 5- to 6-month CRF [16N–2.6P–9.1K + micronutrients (16N–6P_2O_5–11K_2O); Harrell's, Lakeland, FL].

On 15 June 2021, plants were placed on an open-air gravel pad separated according to taxa, and plants within each taxon were completely randomized. Due to differences in water requirements, Freeman maple, panicle hydrangea, and shrub rose were placed in an irrigation zone separate from arborvitae and nandina. All plants were irrigated daily at 5:00 via overhead sprinklers (High-angle Xcel-Wobbler, #7 nozzle, Senninger Irrigation, Clermont, FL). Leaching fraction (water volume leached ÷ water volume applied) was measured on five randomly selected plants per taxa every two weeks, and irrigation duration was adjusted to maintain a leaching fraction of 0.35. While this target leaching fraction (0.35) is higher than BMPs recommendation of 0.1 to 0.15 (Bilderback et al., 2013) – a leaching fraction of 0.35 more closely mimics commercial nursery settings (personal observation).

Leachate from each Freeman maple was collected continuously for 13 weeks. To collect the leachate resulting from all irrigation and rainfall, Freeman maples were nested in black, 18.9-L (5-gallon) buckets with 30-cm-deep basket lids (HG10MESHPOT; Hydrofarm, Philadelphia, PA) such that the bottom of the plant container was suspended ~6.5 cm above the bottom of the bucket (**Fig. 2**).

Figure 2. Illustration of the leachate collection system.

Plastic capes were taped around the plant containers at ~2 cm below the container lip and draped over the bucket to prevent evaporation of the leachate, minimize sunlight reaching the leachate, and deflect rain and irrigation water from running directly into the leachate buckets. The capes were 63 cm × 63 cm squares of 6 mil black plastic sheeting (Poly-America, Grand Prairie, TX). Reflective bubble insulation (BP48025, Reflectix, Markleville, IN) was wrapped around each leachate bucket to prevent high leachate temperatures. Every 7 days, leachate was weighed to approximate volume (1 g ≈ 1 mL), sampled for later P analysis, and the remaining leachate was discarded. Leachate samples were stored at -20 °C until the end of the study and then thawed, digested to solubilize any particulate P, and analyzed for total P.

At 13 weeks after potting, all plants were measured to calculate growth index [(height + widest width + perpendicular width) ÷ 3], and foliar samples were harvested according to species-specific protocols described by Bryson et al. (2014). Foliar samples were oven-dried at 65 °C, ground to a <0.5 mm particle size using a Cyclone Sample Mill (model 3010-030; UDY, Fort Collins, CO), and sent to a commercial laboratory to be analyzed for P concentration. Plant shoots were severed level with the substrate and roots were separated from the substrate via compressed air. Shoots and roots were oven-dried at 65 °C and separately weighed to obtain shoot dry weight (SDW) and root dry weight (RDW).

Data were subjected to analysis of variance (ANOVA), and post-hoc means separation was accomplished using Tukey's Honestly Significant Difference test (α = 0.05). Statistical analyses were performed using JMP Pro 17 software.

RESULTS AND DISCUSSION

P leaching. During the first four weeks after potting, leachate P from Freeman maples potted in FS-1.5, FS-3, or FS-3St was 71% to 94% lower than that from Freeman maples planted in the control substrate (**Fig. 3**).

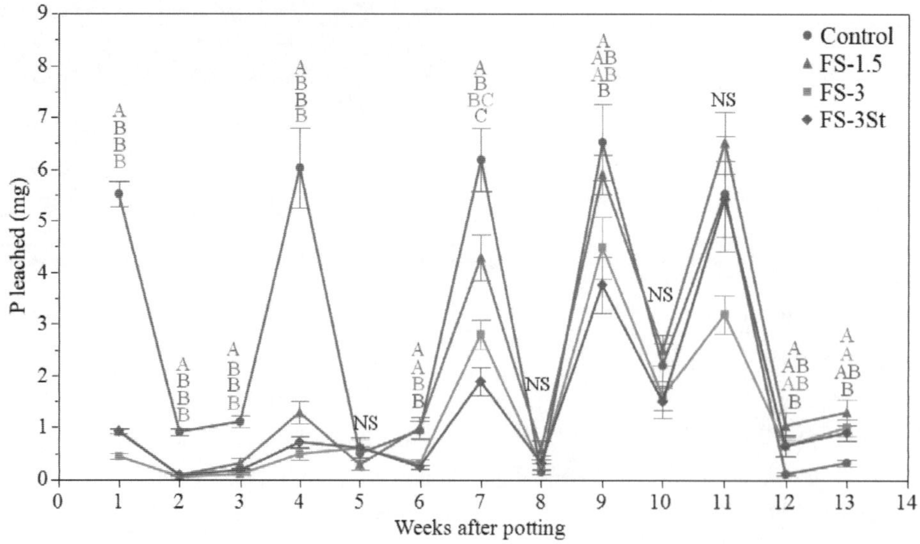

Figure 3. Phosphorus (P) content (± SE) in 7-day cumulative leachate collected once weekly for 13 weeks from containerized (#2) Freeman maples (*Acer ×freemanii* 'Jeffersred' Autumn Blaze) grown in in dolomite-amended pine bark with 1) no $FeSO_4·7H_2O$ (control), 2) 1.5 kg/m^3 $FeSO_4·7H_2O$ (FS-1.5), 3) 3 kg/m^3 $FeSO_4·7H_2O$ (FS-3), or 4) stratified substrate (FS-3St) in which containers had a 2.5-L layer of FS-3 in the bottom and a 2.5-L layer of the control on top. Different letters stacked at a given week (colored to match the treatment they represent) indicate means are significantly different according to Tukey's Honestly Significant Difference Test (α = 0.05). NS = not significantly different.

Thereafter, the FS-1.5 treatment did not significantly reduce P leaching relative to the control, whereas FS-3 and FS-3St were generally effective through weeks 7 and 9, respectively. Diminishing efficacy of the $FeSO_4 \cdot 7H_2O$ treatments over time suggests the adsorption sites for P were becoming saturated. Interestingly, at weeks 12 and 13, the non-stratified $FeSO_4 \cdot 7H_2O$ treatments tended to leach more P than the control. This may have been a result of higher root and shoot biomass (**Table 1**) and thus greater P uptake of the Freeman maples growing in the control substrate relative to those in the FS-1.5 and FS-3 treatments. Resolubilization of P from bark-Fe-P complexes is another possible explanation for more P leaching from FS-1.5 and FS-3 versus the control at weeks 12 and 13. However, evidence of delayed P release from bark-Fe complexes has not been observed in previous, longer-term (19-week) experiments during which leachate was collected from fallow containers with substrate treatments similar to those in the current study (unpublished data).

Freeman maples potted in the control substrate leached a total of 37 mg P over the course of the experiment (**Fig. 4**). By contrast, Freeman maples in the FS-1.5 substrate leached 25.2 mg P (32% reduction) and those in the FS-3 and FS-3St substrates leached between 16 and 17 mg P (57% to 54% reduction, respectively). Similar P retention by the FS-3 and FS-3St treatments, despite the FS-3 containers having twice the amount of $FeSO_4 \cdot 7H_2O$, suggests that $FeSO_4 \cdot 7H_2O$-amended pine bark in the upper half of the container had a nominal effect on P retention. This is further supported by our finding that less P leached from FS-3St compared to FS-1.5 even though these two treatments had the same amount of $FeSO_4 \cdot 7H_2O$ per container. Poor P adsorption by Fe-charged pine bark in the upper portion of the container may be related to its lower moisture content relative to substrate near the bottom of the container.

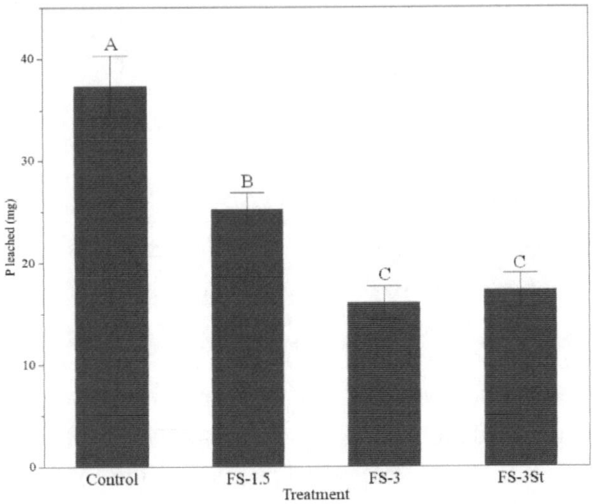

Figure 4. Cumulative phosphorus (P) leached (± SE), on average, from containerized (#2) Freeman maple (*Acer ×freemanii* 'Jeffersred' Autumn Blaze) grown for 13 weeks in dolomite-amended pine bark with 1) no $FeSO_4 \cdot 7H_2O$ (control), 2) 1.5 kg/m^3 $FeSO_4 \cdot 7H_2O$ (FS-1.5), 3) 3 kg/m^3 $FeSO_4 \cdot 7H_2O$ (FS-3), or 4) stratified substrate (FS-3St) in which containers had a 2.5-L layer of FS-3 in the bottom and a 2.5-L layer of the control on top. Different letters above bars indicate mean values are significantly different according to Tukey's Honest Significant Difference Test (α = 0.05).

Table 1. Growth index [(height + widest width + perpendicular width) ÷ 3], shoot dry weight (SDW), root dry weight (RDW), and foliar phosphorus (P) concentrations of containerized Freeman maple (*Acer ×freemanii* 'Jeffersred' Autumn Blaze®), panicle hydrangea (*Hydrangea paniculata* 'SMNHPRZEP' Zinfin Doll®), shrub rose (*Rosa* ×'HORCOGJIL' At Last®), nandina (*Nandina domestica*), and arborvitae (*Thuja* ×'Green Giant') after being grown for 13 weeks in dolomite-amended pine bark with 1) no $FeSO_4 \cdot 7H_2O$ (control), 2) 1.5 kg/m^3 $FeSO_4 \cdot 7H_2O$ (FS-1.5), 3) 3 kg/m^3 $FeSO_4 \cdot 7H_2O$ (FS-3), or 4) stratified substrate (FS-3St) in which containers had a 2.5-L layer of FS-3 in the bottom and a 2.5-L layer of the control on top.

Taxa	Treatment	Growth index (cm)	SDW (g)	RDW (g)	Foliar P (%)
Freeman maple	Control	53.1 a[z]	64.8 a	48.3 a	0.22 a
	FS-1.5	42.6 b	50.9 b	41.8 ab	0.21 ab
	FS-3	43.0 b	50.6 b	40.1 b	0.19 b
	FS-3St	47.0 ab	53.6 b	37.7 b	0.20 b
	P-value	0.0051	0.0001	0.0079	0.0013
panicle hydrangea	Control	31.9	35.1	10.20	0.24
	FS-1.5	31.3	28.7	7.01	0.23
	FS-3	30.8	31.6	8.33	0.19
	FS-3St	32.2	34.2	8.49	0.24
	P-value	0.9010	0.2400	0.1542	0.0563
nandina	Control	45.9	31.6	7.21	0.19
	FS-1.5	43.0	29.4	6.85	0.19
	FS-3	44.1	33.0	7.02	0.18
	FS-3St	43.0	25.4	6.25	0.18
	P-value	0.4690	0.4380	0.7809	0.7666
shrub rose	Control	30.7	28.0	13.8	0.20 a
	FS-1.5	29.8	26.9	13.6	0.17 ab
	FS-3	28.7	27.0	14.0	0.16 b
	FS-3St	27.9	22.4	13.9	0.19 a
	P-value	0.2730	0.1928	0.8542	0.0038
arborvitae	Control	33.1	18.1	3.28	0.26
	FS-1.5	32.3	19.8	3.15	0.25
	FS-3	32.8	23.7	3.93	0.24
	FS-3St	31.4	17.7	3.32	0.27
	P-value	0.5460	0.0998	0.1216	0.1084

[z]Mean values with different letters within the same column and taxon are significantly different according to Tukey's Honestly Significant Difference test ($P < 0.05$).

Rainfall effect on P leaching. The P leaching pattern over time was strongly influenced by rainfall. Weeks during which plants received >3 cm of rainfall (weeks 1, 4, 7, 9, 10, and 11; **Fig. 5**) corresponded with spikes in P leaching (**Fig. 3**). These "rainy weeks", which represented less than half the total number of samplings, accounted for 89% of the cumulative P leached from the control plants over the 13-week period. Greatest reductions in P leaching by the $FeSO_4 \cdot 7H_2O$ treatments relative to the control also occurred during rainy weeks. For example, on weeks 1, 4, and 7 (weeks with 3.5 to 6 cm rainfall), Freeman maples in FS-3 leached, respectively, 5.1, 5.5, and 3.4 mg less P than the control; during weeks 2, 3, 5, and 6 (weeks with 0.2 to 1.3 cm rainfall), the FS-3 treatment reduced P leaching, relative to the control, by 0.1 to 1 mg.

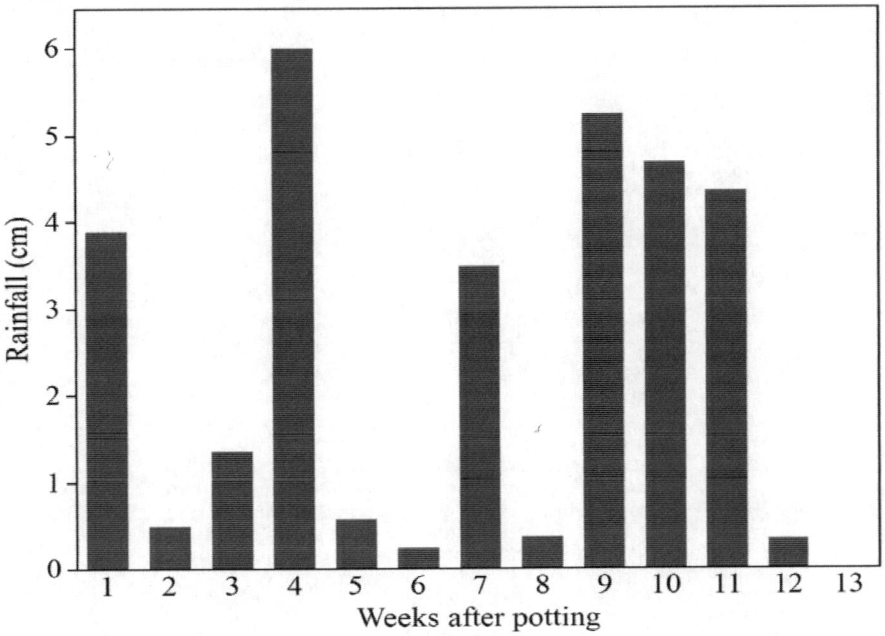

Figure 5. Cumulative rainfall measured once weekly over the course of the 13-week experiment conducted from 15 June 2021 to 14 Sept. 2021 at the Tennessee State Nursery Research Center in McMinnville, TN.

Plant growth, biomass, and foliar P. Growth index, SDW, and RDW of panicle hydrangea, nandina, shrub rose, and arborvitae were unaffected by substrate treatments (**Table 1; Fig. 6**). In contrast, Freeman maples grown in the control substrate had a higher GI than those grown in FS-1.5 or FS-3, higher SDW than those grown in all other treatments, and higher RDW than those grown in FS-3 or FS-3St. Despite these differences in growth among Freeman maples produced in the various substrate treatments, differences in visual quality were not apparent (**Fig. 7**).

One possible explanation for reduced growth and biomass of Freeman maples in the FS treatments is that they were mildly deficient in P. Foliar P concentrations of Freeman maples grown in the control were equal to the lower limit of the survey range (0.22% to 0.29% P) for healthy 'Jeffersred' (Autumn Blaze) Freeman maples (Bryson and Mills, 2014), and Freeman maples grown in FS-3 or FS-3St had lower foliar P concentrations than those in the control. If the survey range is representative of the true sufficiency range, even a slight reduction in P availability to plants that are already near the critical deficiency concentration could result in growth limitation.

Figure 6. Panicle hydrangea (*Hydrangea paniculata* 'SMNHPRZEP' Zinfin Doll®), nandina (*Nandina domestica*), shrub rose (*Rosa* ×'HORCOGJIL' At Last®), and arborvitae (*Thuja* ×'Green Giant') after being grown in 6.1-L containers for 13 weeks in dolomite-amended pine bark with 1) no $FeSO_4 \cdot 7H_2O$ (control), 2) 1.5 kg/m^3 $FeSO_4 \cdot 7H_2O$ (FS-1.5), 3) 3 kg/m^3 $FeSO_4 \cdot 7H_2O$ (FS-3), or 4) stratified substrate (FS-3St) in which containers had a 2.5-L layer of FS-3 in the bottom and a 2.5-L layer of the control on top.

Figure 7. Freeman maple (*Acer ×freemanii* 'Jeffersred' Autumn Blaze) after being grown in 6.1-L containers for 13 weeks in dolomite-amended pine bark with 1) no $FeSO_4 \cdot 7H_2O$ (control), 2) 1.5 kg/m^3 $FeSO_4 \cdot 7H_2O$ (FS-1.5), 3) 3 kg/m^3 $FeSO_4 \cdot 7H_2O$ (FS-3), or 4) stratified substrate (FS-3St) in which containers had a 2.5-L layer of FS-3 in the bottom and a 2.5-L layer of the control on top.

Shrub rose also had lower foliar P concentrations when grown in FS-3 versus the control. However, since lower foliar P concentrations did not correspond with a reduction in plant growth or visual deficiency symptoms, the reduction in foliar P concentration was likely inconsequential to the plant. These corroborated results reported by Johansson (1978) who reported that *Rosa* 'Parel van Aalsmeer' showed no signs of P deficiency until foliar P concentrations fell to $\leq$ 0.14%. Panicle hydrangea, nandina, and arborvitae foliar P concentrations were not affected by substrate treatments, suggesting that there was enough soluble P remaining after complexation with bark-Fe that P uptake was unrestricted.

CONCLUSION

Amending pine bark with $FeSO_4 \cdot 7H_2O$ can substantially reduce P leaching from container-grown nursery crops, and the magnitude of this effect increases with increasing $FeSO_4 \cdot 7H_2O$ rate (i.e., from 1.5 kg/m^3 to 3 kg/m^3). Reductions in P leaching provided by the $FeSO_4 \cdot 7H_2O$-amended pine bark were especially important when rainfall was a major contributor to the total leachate volume and excessive leaching was unavoidable. By placing a layer of $FeSO_4 \cdot 7H_2O$-amended pine bark in the lower portion of the container instead of amending the entire substrate volume with $FeSO_4 \cdot 7H_2O$ - the total amount of $FeSO_4 \cdot 7H_2O$ applied can be reduced by half while achieving the same reductions in P leaching. Although the plant availability of P associated with bark-Fe complexes was not measured directly, foliar P analyses indicate that 0.6 kg/m^3 $FeSO_4 \cdot 7H_2O$ may reduce P uptake for some taxa (e.g., Freeman maple and shrub rose) but not others (e.g., arborvitae, panicle hydrangea, and nandina). This reduction in P uptake may not necessarily lead to a reduction in plant growth or quality, as was observed in shrub rose.

Moreover, layering the $FeSO_4 \cdot 7H_2O$-amended bark in the bottom of the container effectively avoided foliar P reductions that were present when the same $FeSO_4 \cdot 7H_2O$ rate was incorporated throughout the substrate. In contrast, reduced biomass and foliar P concentrations of Freeman maples grown in $FeSO_4 \cdot 7H_2O$-amended pine bark (stratified or non-stratified) were likely a consequence of their relatively high nutrient demand (Fulcher et al., 2004), particularly when the plants become rootbound as was observed at harvest.

ACKNOWLEDGEMENTS

This research could not have been possible without the technical support provided by Benji Moore and Joseph Davis.

LITERATURE CITED

Bilderback, T., Boyer, C., Chappell, M., Fain, G., Fare, D., Gilliam, C., Jackson, B.E., Lea-Cox, J., LeBude, A.V., and Niemiera, A.X. (2013). Best management practices: Guide for producing nursery crops. Southern Nursery Assn., Acworth, GA.

Broschat, T.K. (1995). Nitrate, phosphate, and potassium leaching from container-grown plants fertilized by several methods. HortScience *30*: 74–77. doi: 10.21273/HORTSCI.30.1.74.

Bryson, G.M., Mills, H.A., Sasseville, D.N., Jones, J.B., and Barker, A.V. (2014). Plant analysis handbook IV: A guide to plant nutrition and interpretation of plant analysis for agronomic and horticultural crops. Micro-Macro Publishing: Athens, GA.

Bugbee, G.J., and Elliott, G.C. (1998). Leaching of nitrogen and phosphorus from potting media containing biosolids compost as affected by organic and clay amendments. Bull. Environ. Contam. Toxicol. *60*: 716–723. doi: 10.1007/s001289900685.

Cole, J.C. and Dole, J.M. (1997). Temperature and phosphorus source affect phosphorus retention by a pine bark-based container medium. HortScience *32*:236-240.

Criscione, K.S., Fields, J.S. and Owen, J.S. (2022). Root exploration, initial moisture conditions, and irrigation scheduling influence hydration of stratified and non-stratified substrates. Horticult. *8*: 826. doi: 10.3390/horticulturae8090826.

Fain, G.B., Gilliam, C.H., Tilt, K.M., Olive, J.W., and Wallace, B. (2000). Survey of best management practices in container production nurseries. J. Environ. Hort. *18:* 142–144.
doi: 10.24266/0738-2898-18.3.142.

Fields, J.S., Owen, J.S. and Altland, J.E. (2021). Substrate stratification: layering unique substrates within a container increases resource efficiency without impacting growth of shrub rose. Agrono. *11*: 1454. doi: 10.3390/agronomy11081454.

Fulcher, A., Dunwell, W., McNiel, R., Wolfe, D., and Murdock, L. (2004). Effect of fertilizer rate on growth of seven tree species in pot-in-pot production. Proc. Sout. Nur. Assoc. Res. Conf. *49*:114-116.

Godoy, A., and Cole, J.C. (2000). Phosphorus source affects phosphorus leaching and growth of containerized Spirea. HortScience *35*: 1249–1252. doi: 10.21273/HORTSCI.35.7.1249.

Handreck, K. (1992). Iron-phosphorus interactions in the nutrition of seedling macadamia in organic potting media. Aust. J. Exp. Agric. 32(6): 773. doi: 10.1071/EA9920773.

Johansson, J. (1978). Effects of nutrient levels on growth, flowering and leaf nutrient content of greenhouse roses. Acta Agricul. Scandin. *28*:363-386.

Khamare, Y., Marble, S.C., Altland, J.E., Pearson, B.J., Chen, J., et al. (2022a). Effect of substrate stratification on growth of common nursery weed species and container-grown ornamental species. HortTech. *32*: 74–83.
doi: 10.21273/HORTTECH04965-21.

Khamare, Y., Marble, S.C., Altland, J.E., Pearson, B.J., Chen, J., et al. (2022b). Effect of substrate stratification without fine pne bark particles on growth of common nursery weed species and container-grown ornamental species. HortTech. *32:* 491–498. doi: 10.21273/HORTTECH05113-22.

Mack, R., Owen, J.S., Niemiera, A.X., and Latimer,J. (2017). Virginia nursery and greenhouse grower survey of best management practices. HortTech. 27: 386–392. doi: 10.21273/HORTTECH03664-17.

Majsztrik, J.C and Lea-Cox, J. (2013). Water quality regulations in the Chesapeake Bay: Working to more precisely estimate nutrient loading rates and incentivize best management practices in the nursery and greenhouse industry. HortScience *48*:1097-1102.

Majsztrik, J.C., Ristvey, A.G., and Lea-Cox, J.D. (2011). Water and nutrient management in the production of container-grown ornamentals. In: Janick, J., editor, Horticultural Reviews. John Wiley & Sons, Inc., Hoboken, NJ, USA. p. 253–297

Million, J., Yeager, T., and Albano, J. (2007a). Effects of container spacing practice and fertilizer placement on runoff from overhead-irrigated sweet viburnum. J. Environ. Hort. *25:* 61–72.
doi: 10.24266/0738-2898-25.2.61.

Million, J., Yeager, T., and Albano, J. (2007b). Consequences of excessive overhead irrigation on runoff during container production of sweet viburnum. J. Environ. Hort. *25*: 117–125. doi: 10.24266/0738-2898-25.3.117.

Million, J. and Yeager,T. (2021). Leaching fraction-based microirrigation schedule reduced water use but not N and P loss during production of a container-grown shrub. HortScience *56*:147-153.
doi: 10.21273/HORTSCI15503-20

Nguyen, T.A.H., Ngo, H.H., Guo, W.S., Zhang, J., Liang, S., Lee, D.J., Nguyen, P.D., and Bui, X.T. (2014). Modification of agricultural waste/by-products for enhanced phosphate removal and recovery: potential and obstacles. Biores. Tech. *169*: 750–762. doi: 10.1016/j.biortech.2014.07.047.

Ogutu, R.A., Williams, K.A., and Pierzynski, G.M. (2009). Phosphate sorption of calcined materials used as components of soilless root media characterized in laboratory studies. HortScience *44*: 431–437. doi: 10.21273/HORTSCI.44.2.431.

Owen, J.S., Warren, S.L., Bilderback, T.E., and Albano, J.P. (2007). Industrial mineral aggregate amendment affects physical and chemical properties of pine bark substrates. HortScience *42*: 1287–1294. doi: 10.21273/HORTSCI.42.5.1287.

Owen, J.S., Warren, S.L., Bilderback, T.E., and Albano, J.P. (2008). Phosphorus rate, leaching fraction, and substrate influence on influent quantity, effluent nutrient content, and response of a containerized woody ornamental crop. HortScience *43*: 906–912. doi: 10.21273/HORTSCI.43.3.906.

Pokhrel, M.R., Poudel, B.R., Aryal, R.L., Paudyal, H., and Ghimire, K.N. (2019). Removal and recovery of phosphate from water and wastewater using metal-loaded agricultural waste-based adsorbents: A. Rev. J. Inst. Sci. Tech. *24*: 77–89. doi: 10.3126/jist.v24i1.24640.

Tyler, H.H., Warren, S.L.,and Bilderback, T.E. (1996a). Cyclic irrigation increases irrigation application efficiency and decreases ammonium losses. J. Environ. Hort. *14:* 194–198. doi: 10.24266/0738-2898-14.4.194.

Tyler, H.H., Warren, S.L., and Bilderback. T.E. (1996b). Reduced leaching fractions improve irrigation use efficiency and nutrient efficacy. J. Environ. Hort. *14*: 199–204. doi: 10.24266/0738-2898-14.4.199.

U.S. Department of Agriculture, National Agricultural Statistics Service. (2020). 2019 Census of Horticultural Specialties. https://www.nass.usda.gov/Publications/AgCensus/2017/Online_Resources/Census_of_Horticulture_Specialties/HORTIC.pdf (accessed 11 November 2022).

Williams, K.A., and Nelson, P.V. (1996). Modifying a soilless root medium with aluminum influences phosphorus retention and chrysanthemum growth. HortScience *31*: 381–384. doi: 10.21273/HORTSCI.31.3.381.

Wright, R.D. (1986). The pour-through nutrient extraction procedure. HortScience 21:227–229.

Wurtsbaugh, W.A., Paerl, H.W., and Dodds, W.K. (2019). Nutrients, eutrophication and harmful algal blooms along the freshwater to marine continuum. WIREs Water 6(5). doi: 10.1002/wat2.1373.

Yeager, T. and Barrett, J. (1984). Phosphorus leaching from ^{32}P-superphosphate-amended soilless container media. HortScience *19*:216–217.

Yeager, T.H. and Barrett, J.E. (1985a). Phosphorus and sulfur leaching from an incubated superphosphate-amended soilless container medium. HortScience *20*:671-672.

Yeager, T.H. and Barrett, J.E. (1985). Influence of incubation time on phosphorus leaching from a container medium. J. Environ. Hort. *3*:186–187. doi: 10.24266/0738-2898-3.4.186.

Combined Proceedings IPPS 72:249-253. 2022.

Getting Started with the H2A Visa Program

Dan Bremer

AgWorks H2, 2110 Paterson Street, Suite C, Valdosta, GA 31602, USA

dbremer@agworksh2.com

Keywords: Foreign - seasonal agricultural workers, Immigration Reform and Control Act of 1986 (IRCA), form I-9, Federal Employer Identification Number (FEIN), three-quarter work guarantee, farm labor contractor

Summary

The Immigration Reform and Control Act (IRCA) enables agricultural producers to employ foreign, agricultural workers – which is becoming an increasing important source of employees for the Green/Nursery industries. It is important to develop a plan that includes: housing, transportation, workers compensation, record keeping, appropriate supervisors, the application process, recruitment of workers, and expenses of participating in the program. A specialized attorney or consultant can be helpful in planning and staying in compliance of regulations.

INTRODUCTION

The Immigration Reform and Control Act of 1986 (IRCA) created several reforms that employers in the USA use extensively today. The most used item in the law is the requirement that all new hires complete a form called the I-9. These forms are used to determine the identity of all new employees and to determine if they are eligible to work in the USA.

The IRCA also created an opportunity for employers to hire seasonal agricultural workers from foreign countries under the H2A Visa Program. If an agricultural employer is unable to find enough local and domestic workers to do the work, then the employer can apply to the U.S. Department of Labor, the U.S. Department of Homeland Security and the U.S. Department of State to bring workers from other countries. All of these agencies have various rules to follow to actually get the visas needed to come to the USA to do agricultural work.

The H2A Visa Program annually brings over 300,000 workers from foreign countries to do agricultural work. Without this program many family horticultural operations would not be able to continue - because they could not find workers to plant, tend and harvest the seasonal crops grown.

Planning for the h2a program

In order to be ready for workers from other countries to work on your agricultural operation you must have a plan. Usually, this plan starts at least a year in advance and covers many areas and aspects of the operation.

Housing. Housing is one of the most important aspects of this plan. Housing, free of charge, must be provided to workers who are unable to return to their home every night. This housing can be provided in many different ways, but it must meet federal and state guidelines prior to the workers arrival. Most of these guidelines can be found in the OSHA 1910.142 "Temporary Labor Camps" publication https://www.ecfr.gov/current/title-29/subtitle-B/chapter-XVII/part-1910/subpart-J/section-1910.142 . Most agricultural employers build barracks type housing that will safely house multiple workers in one building. Some employers use mobile homes or they rent houses for the workers. The use of public accommodations such as hotels and motels will work if the need for the workers is of a short duration.

The main point is that you must have a plan for the kind of housing that anticipates the cost effectiveness of each kind - and the problems with stringent rules that apply to each. AgWorks has many plans available for building housing for H2A visa workers.

Transportation. The H2A rules require the employer to have safe transportation for workers to get from housing to field and to take care of other business the workers may need.

All drivers of vehicles used to transport migrant agricultural workers must have a proper license. If the vehicle is just a normal car or truck with a capacity of 15 or less, a simple driver's license is all that is needed. If the vehicle will carry more than 15 passengers, a commercial drivers license with a passenger endorsement must be possessed by the driver.

Most vehicles used to transport workers must have a safety inspection done by the state, or, if no state safety inspection is required, then by a certified mechanic. The driver also must usually have a medical examination done by a doctor, either under a form from the U.S. Department of Labor or the U.S. Department of Transportation.

Workers Compensation. All employers who use the H2A Visa Program must have a worker's compensation policy that covers all workers engaged in the duties that are listed in the application. Many states do not require workers comp for agricultural employers, however, if you are certified under the H2A Visa Program, you *must* have workers comp.

Record Keeping. H2A employers must have specific payroll records for all their employees. There are certain items that must be present in the records such as total hours worked, piece rates, deductions and net pay, etc. Listing of accurate start and stop times are crucial to workers being paid correctly. There are also requirements that workers be given check stubs. The check stubs must include the employer's name, address and the employer identification number, also known as the Federal Employer Identification Number or the Federal Tax Identification Number, (FEIN number).

Supervisors. Most workers who arrive from other countries do not speak English. Therefore, there must be a plan for giving directions and information in a language that can be understood by the workers. Usually a trusted employee already employed by the grower can be used to translate from the grower to the workers. Sometimes a worker from the foreign country speaks enough English to translate the instructions.

Can I Afford This? There are many costs associated with the H2A Visa Program. One of these is the Adverse Effect Wage Rate that is set by the U.S. Department of Agriculture. This rate is set so that the use of foreign workers does not suppress local wages. For example, in Georgia, the rate is $11.99 per hour. This rate becomes the minimum wage that must be paid to any worker who does any of the duties listed on the H2A application. It usually goes up every year by a percentage of what other workers in the area are being paid.

The wages are only some of the costs of the program. The employer must pay the transportation for the worker to come from their hometown in the foreign country to the work location here in the USA. The cost of the visa purchase from the U.S. Department of State is now $190 per person. There are costs that are charged by other government agencies like the U.S. Department of Labor and the U.S. Department of Homeland Security. The average cost is around $1,000 per worker - to travel from their home country to the USA and back, and for U.S. Government fees.

Application Process. There are many decisions to be made before the application is sent to the government. For example, the number of workers needed is a critical decision. If you request too few, the work may not get done. If you request too many, there may not be enough work for each worker to comply with the three-quarter work guarantee. That guarantee says that during the duration of the contract period, each worker must average at least three quarters of the hours you told the government you were going to work. If you put down 40 hours per week in your application, then the workers, all workers, must average at least 30 hours per week.

The time of year and the crops being worked must also be in the application. H2A visa workers can only be employed for

10 months or less in the USA, in your operation. These workers may not work year around on your locations. Many employers need the workers on a year around basis but the H2A visa workers can only be at your location for 10 months or less. The question is what 10 months are the most optimal for you?

Recruitment of Workers. The employer who has a H2A certification must hire all qualified local and domestic workers that apply. There are certain requirements that can be integrated into the application to eliminate workers who could not do the work. These rules must apply to both domestic workers and H2A visa holding workers. Your job must be listed on the government's recruitment site. The hiring of all U.S. workers who may be qualified is a critical part of the H2A visa program. If you do not hire qualified U.S. workers, the government has severe penalties - and they could remove you from eligibility to participate in the H2A Visa Program.

Since you are bringing workers from a foreign country, a decision must be made on how to find these workers. Some employers may have a current worker or supervisor that may know some workers in that foreign country. There are also recruitment companies that find workers for you. Whoever does the recruiting, the recruiter must sign a document that says they will not charge workers any fees for anything. The employer must pay all fees and costs of the program. The workers are not allowed to pay fees to be recruited.

Alternative Ways to Use the H2A Visa Program. Individual employers working solely for their own operations are eligible to use the program. However, there are other ways to have H2A workers do your work. A Farm Labor Contractor could have multiple growers and work locations on a single application. Different growers could be on one master application and could share the workers. If only one grower asks for the H2A workers, then that grower may not share the workers with any other employer. The use of a Farm Labor Contractor or a group application is the only way to share workers with other employers. In an application for an association, many growers would form a legal entity that is the actual employer of the workers. This must be for agricultural work. The definition of agricultural work is sometimes very complicated.

Potential scrutiny from the U.S. Dept of Labor - Wage and Hour Division. As a participant in the H2A Visa Program, you become a bigger target for enforcement efforts by the U.S. Department of Labor, Wage and Hour Division. The Wage and Hour Division is the agency that makes sure you are complying with the things you promised in your application.

It is important that you know what rules apply and how to make sure you are doing everything in compliance with the law. To do that you need someone or someway to ask questions and evaluate your compliance posture. An employment attorney may be able to help you, but most are not familiar with the H2A Visa Program. A specialized attorney or consultant is usually necessary to help you.

The Future. The H2A Visa Program has not really changed from its beginning in 1986. The wages have risen dramatically and the use of the program has increased exponentially. But there are some changes being proposed now that could change things.

The U.S. Department of Labor is trying to change the way growers can participate in group applications by limiting the fulltime employment in these applications. They are also trying to force the various states to do their own surveys to determine wages and practices applicable to agricultural workers. The Occupational Safety and Health Administration (OSHA) has a new heat safety program that growers need to be aware of - especially in hot climates.

There are even proposed programs that give amnesty to undocumented workers who have been working in agriculture for several years. So far, not much has changed in the H2A Visa Program. It is a viable program that works very well for thousands of agricultural employers.

Combined Proceedings IPPS 72:254-259. 2022.

From Curios to Champions: Delayed Value in Plant Collections

Richard T. Olsen

USDA Agricultural Research Service, U.S. National Arboretum, Washington, DC 20002, USA

richard.olsen@usda.gov

Keywords: germplasm, plant genetic resources, plant breeding, Frank Meyer, *Metasequoia glyptostroboides, Acer truncatum, Viburnum macrocephalum, Ilex cornuta, Loropetalum chinense, Syringa oblata, Syringa meyeri, Hydrangea arborescens*

Summary

Modern plant breeding relies not only on new technologies and new germplasm, but new thoughts on existing, overlooked, or underappreciated plants. Often, these passé plants are the domain of arboreta, who collect, catalog, and curate these plant genetic resources for use by enterprising nursery professionals, public and private breeders, and academics. A given plant's perceived value fluctuates over its cultivated lifetime, whether the initial introduction was an immediate success or not. A vast majority of introductions and selections are relegated to the historical record, persisting as curios for collectors, but in many cases, their traits are awaiting a new trend, threat, or technology to rerelease their importance to American landscapes.

INTRODUCTION

Breakthrough breeding achievements, or "aha" moments, are often made possible by traits residing in less celebrated, forgotten, or even discarded plant selections, species, or even entire genera found in cultivation. Frequently, these plants are maintained through the long-term efforts in collecting, characterizing, documenting, preserving, and distributing by botanical collections. Support and advocacy for these institutions, like the U.S. National Arboretum (USNA), Washington, DC. and our colleagues in the American Public Garden Association (APGA) Plant Collections Network (PCN) program (www.publicgardens.org/programs/about-plant-collections-network) is critical for sustained growth and success in the nursery industry. But let us not forget, that much diversity available is in some state of cultivation or preservation outside of public gardens in the hands of nursery professionals, amateur collectors, and private gardens and arboreta.

The path from novel to established, from curio to champion is not a given. The norm for most plants' popularity settles into a middling ground: some descending from on high after a bit of excitement or hype, others ascending from more meager beginnings with slow recognition of utility and worth. And that popularity distribution shifts on many tides, influenced by fads and fashions, as well as changing social values that carry over to our gardens and landscapes. More consistently perhaps, is the ever-present danger of pests, diseases, and environmental stressors that knock plant champions from their perch. So herein, we step backward in time to provide a longer view and perspective at plant introductions and the delayed value or contributions common to many. Some stories will be known, others new, some insightful, some speculative, but on the whole worthy of telling.

DISCUSSION

Perhaps no plant went from unknown to commercial success as fast as dawn redwood (*Metasequoia glyptostroboides*). Dawn redwood's scientific discovery and introduction is well-known (Hu, 1998; Merrill, 1948; Nelson, 1998) with recent updates and corrections (Ma, 2003; Ma and Guofan, 2003). However, when viewed through the lens of commercial acceptance, its speedy path from an unknown to commercial champion is unheralded. Fossils were first discovered in 1938 and scientifically described in 1941. Also in 1941, unrelated researchers discover an unknown living conifer in Sichuan, however, herbarium specimens were not collected until 1943. In 1946, researchers match the herbarium and fossil specimens. In 1947, seeds and plants are collected and distributed to botanical institutes around the globe, including the Missouri Botanical Garden, St. Louis, MO and the Arnold Arboretum, Boston, MA in the U.S. The Arnold Arboretum distributes the bulk of the material, including to the USNA. In 1948 the species is officially described and announced to the world (Merrill, 1948). By 1958, USNA scientists select a clone from the original seedlings with a uniform, upright-pyramidal growth, which is officially released in 1963 as *M. glyptostroboides* 'National' (conifersociety.org/conifers/metasequoia-glyptostroboides-national/). Thus, within a span of 25 years, a plant went from unknown Mesozoic conifer, to living fossil, to commercial success and clonal selection. National dawn redwood found little traction in the nursery trade and is itself a curio today, along with other cultivars like Waasland, a compact upright plant introduced through Bömer Nursery, Zundert, Netherlands. New selections likely to become commercial champions include 'Raven' (SHAW'S LEGACY ®), like National selected from the original seed distribution in 1948, and 'JFS-PN3Legacy' (Jade Prince ®) selected from commercially available seed.

The USNA was established by an act of Congress in 1927, the culmination of decades of lobbying by USDA officials, academia, industry, and garden clubs. David Fairchild, USDA scientist and plant explorer created and managed the Office of Foreign Seed and Plant Introduction and established experimental gardens across the United States. This effort included the National Arboretum, with a focus on ornamental plant introductions. Fairchild understood that cultivated varieties were the backbone of modern agriculture, and their discovery, genetic improvement, and distribution are fundamental aspects of a public garden devoted to economic work. The USNA has introduced over 650 new plant varieties, including floricultural crops, herbaceous annual and perennials, and all manner of woody shrubs, vines, and trees (www.usna.usda.gov/science/plant-introductions-and-releases/). Unofficially, the list grows to 850 with the inclusion of germplasm distributions that resulted in cultivar introductions by receiving nurseries and public gardens. We have certainly had our share relegated to curios and historical records (e.g., *Magnolia* 'Maryland' and *Ulmus parvifolia* 'Ohio;), however, many are still commercially relevant (e.g., *Ilex crenata* 'Sky Pencil' and *Deutzia gracilis* 'Nikko'), and still others contribute to ongoing developments within their respective genera (e.g., *Lagerstroemia* and *Viburnum*).

If the USNA's experience is reflective of plant introductions in general, then most selections are relegated to the historical record, persisting as curios for collectors, but whose traits are awaiting a new trend, threat, or technology to rerelease their importance to American landscapes. Take the Shantung maple (*Acer truncatum*). Introduced at the end of the 19th century and made available through USDA distributions of Frank N. Meyer's 1905 collections, the species remained obscure and underutilized for most of the 20th century. Horticulturists and breeders looking for heat-tolerant, and non-invasive substitutes for Norway maple (*Acer platanoides*) rediscovered Shantung maple. The first commercially viable selections, 'Fire Dragon' from Keith Johannson, MetroMaples, TX and 'WTF-AT1' (Main Street ®) from Worthington Farms, NC were introduced in 2006, and 2011, respectively. They were presaged by two *A. truncatum* × *A. platanoides* cultivars, Warrenred (Pacific Sunset^{TM}) and Keithsform (Norwegian Sunset^{TM}), of spontaneous hybrid origin, selected and introduced by Keith Warren of J. Frank Schmidt & Son Co., Boring, OR in 1990. Finally, after a century, Shantung maple went from curiosity in botanical collections to commercial champion.

Frank N. Meyer's success as a plant explorer for David Fairchild does not receive the attention it deserves by horticulturists. Meyer's tragic death in China cut short his career, and unlike other famed explorers (E.H. "Chinese" Wilson), Meyer covered all agronomic and horticultural crops, not just ornamental ones. However, Meyer's acumen and impact on American landscapes is illustrated in just one shipment in one year. In 1908, Meyer hand delivered a large shipment of plants he collected from various sources and locations in or near Suzhou, Jiangsu, China to the USDA Plant Introduction Station, Chico, CA (Galloway, 1909). Within this shipment are introductions that saw immediate success, slow success, or a much-delayed success in contributing to American landscapes:

***Viburnum macrocephalum* 'Sterile'** PI22978. Meyer's collection is a reintroduction of Robert Fortune's original 1844 English import of the Chinese snowball viburnum. A classic Southern landscape plant, it has been in near continual commercial production since Meyer's introduction. The straight species (*V. macrocephalum* f. *keteleeri*) was not known in cultivation until collected and introduced by USNA botanist Ted Dudley in conjunction with the 1980 Sino-American Botanical Expedition from Zhejiang, China.

Ilex cornuta. PI 22979. Here, an unknown cultivated variety of Chinese holly. The species was another introduction by Robert Fortune in 1846. The USDA received *I. cornuta* seeds on several occasions from other sources or collectors (PI24638 in 1909; PI65860 in 1923; PI70979 and PI70980 in 1927) that by the 1950s yielded several introductions just beginning to receive commercial interest (Hume, 1953). Not only did the species contribute important cultivars (e.g., Burfordii, Carissa, Dwarf Burford, and Rotunda) but also genetics in holly breeding programs (Kosar, 1957). Today, older Chinese holly cultivars are curios, with some exceptions, but largely superseded by advanced hybrids carrying on their bloodlines.

Loropetalum chinense. PI22982. Chinese fringe-flower. Here, Meyer comments on the ornamental traits of *Loropetalum,* observes it is rarely cultivated and notes its transplanting difficulty. First introduced by Veitch collector Charles Maries to England in 1880, the species relishes heat and languished in cool English landscapes. Even in the U.S. the species rarely makes an appearance although old specimens are noted in South Carolina (Dirr, 1998) and Alabama (Martin Van Der Giessen, person. comm.). The 1980 rediscovery of the pink flowered, red-foliaged variety (*L. chinense* var. *rubrum*) in Hunan, China elevated its mass appeal (Wu et al., 2021). A Chinese industry developed around the variety focused on its introduction, propagation, and cultivation (Wu et al., 2021). From here it, quickly went global through the international nursery scene, with Chinese, Japanese, Canadian, and American nurseries, and horticulturists distributing clones, often the same ones under different cultivar names (Gawel et al., 1996). In 1986, there were no cultivars of any form of *Loropetalum chinense* in the U.S. As of 2022, there are 71 recognized cultivars and counting (Hatch, 2021).

Syringa oblata. PI23030 and 23031. Purple and white early blooming lilacs. Technically, another reintroduction of Fortune who first sent plants of both color forms to England in 1859 (Fiala, 1988). Although hardy, the species and its varieties were not fully appreciated before the demand for heat-tolerant lilacs. The early bloom and thus lower chill requirements, makes *S. oblata* valuable in heat-tolerant breeding programs. The white flowered clone (PI23031) was used in at least 17 crosses in the early USNA lilac breeding program (Lura et al., 2013), before Fiala had named the clone, *S. oblata* 'Frank Meyer' after its collector in 1988.

Syringa sp. PI23032 and 23033. These PIs were later determined to be a new species, *S. meyeri,* although its origins, taxonomy, and native range is debated (Fiala, 1988). Despite Meyer's observations, the "species" is quite hardy while being heat tolerant. Additional introductions from other collectors did little to popularize, perhaps due to a muddled taxonomy. The "species" remained a collector's item in lilac circles for much of the 20th century, taking a back seat to the common lilac (*S. vulgaris*) and its closer relatives. *Syringa meyeri* 'Paliban' was registered in 1980 and received awards from the Royal Horticultural Society in 1984 and 1993 (DeBard, 2022). Taxonomy aside, the "species" as the cultivar Paliban is an important component of modern breeding programs, contributing powdery mildew resistance, compact habit, and early to repeat blooming as in the popular BLOOMERANG ® cultivars Dark Purple ('SMSJBP7'), Dwarf Pink ('SMNJRPI') and new PURPINK™ ('SMNSPTP).

CONCLUSION

As we have seen, plant introduction is a cycle of rediscovery and recreation depended on curated botanical collections, horticulturists and nursery professionals seeking novel and underutilized plant genetic resources. Introductions like the forgotten 1907 Frank Meyer collection of *Buxus harlandii* (PI23013, Hangzhou, Zhejiang, China) is, over a century later, a critical source of tolerance to boxblight (*Calonectria pseudonaviculata* and *C. henricotiae*) in current breeding programs. Thankfully, it was preserved in the USNA's National Boxwood Collection, part of the APGA Plant Collections Network. Properly curated collections found at public gardens connect plant records with herbarium specimens and associated data. These, too, can be mined by plant breeders for discovering lost curios and relics, some of which are just under our noses. Although a pink form of smooth hydrangea was described by C.A. Rafinesque in 1838 (as *H. vulgaris* var. *carnea*), subsequent authors ignored his work and contributions, including Rehder (1949) who listed it as synonym of *H. arborescens,* even as he recognized the mopheads *H. arborescens* f. *grandiflora* and f. *sterilis* in the same publication. Uttal (1986) validated the proper combination, *H. arborescens* f. *carnea,* after finding pink forms in natural populations in Tennessee. The first pink cultivar was Eco Pink Puff, introduced by the late Don Jacobs, Eco Gardens, Decatur, GA. Essentially unknown, it was cultivated in at least one garden (Juniper Level Botanic Gardens, Raleigh, NC) by 1998 (Olsen, personal observation). A second cultivar, Wesser Falls, was introduced by the author in 2000 after its discovery in native populations near the Nantahala River in western NC (Ranney and Olsen, 2009). So, 162 years after its discovery and documentation, followed by taxonomic lumping relegating it to botanical obscurity did *H. arborescens* f. *carnea* leave the curio cabinet and contribute to American landscapes through plant breeding. By 2009, the world had its first pink, mophead smooth hydrangea, with a whole range of cultivars available in the INVINCIBELLE ® series from Spring Meadow Nursery, Grand Haven, Michigan in 2022. Imagine what other forgotten curios reside in the botanical collections of the United States and abroad awaiting rediscovery and reintroduction, as is or through plant breeding!

LITERATURE CITED

Debard, M.L. (2022). International register and checklist of cultivar names in the genus *Syringa* L. (Oleaceae). International Lilac Society (www.internationallilacsociety.org/public-register/).

Dirr, M.A. (1998). Manual of woody landscape plants: their identification, ornamental characteristics, culture, propagation, and uses. Stipes, Champaign, IL.

Fiala, J.L. (1988). The lilacs. The genus *Syringa.* Timber Press, Portland, OR.

Galloway, B.T. (1909). Seeds and plants imported during the period from April 1 to June 30, 1908: Inventory No. 15; Nos. 22511 to 23322. U.S. Department of Agriculture, Bureau of Plant Industry. Bulletin 142. Washington, DC. Government Printing Office.

Gawel, N.J., Johnson, G.R., Suave, R. (1996). Identification of genetic diversity among *Loropetalum chinense* var. *rubrum* introductions. J. Environ. Hort. *14*:38-41.

Hatch, L.C. (2022). Hatch's cultivars of woody plants. *Leiphyllum* to *Lyonia*. 2021-2022 edition. (www.cultivar.org).

Hu, H.H. (1998). How *Metasequoia*, the "living fossil", was discovered in China. Arnoldia

58:4-7.

Hume, H.H. (1953). Hollies. Macmillian, NY.

Kosar, W.F. (1957). Hybridizing hollies. Nat.l Horticul. Mag. *36*:121-129.

Lura, S.B., Pooler, M.R., Kirkbride, J.H. (2013). Description and history of *Syringa oblata* subsp. *oblata* 'Frank Meyer'. Hanburyana. *7*:37-40.

Ma, J. (2003). The chronology of the "living fossil" *Metasequoia glyptostroboides* (Taxodiaceae): a review (1943-2003). Harvard Papers Bot. *8*:9-18.

Ma, J., Shao, G. (2003). Rediscovery of the 'first collection' of the 'living fossil', *Metasequoia glyptostroboides.* Taxon *52*:585-588.

Merrill, E.D. (1948). A living *Metasequoia* in China. Science *107*:140.

Nelson, E. C. (1998). *Metasequoia glyptostroboides*, the dawn redwood: some Irish glosses on its discovery and introduction into cultivation. Curtis's Botan. Mag. *15*:77-80.

Ranney, T.G. and Olsen, R.T. (2009). Breaking the color barrier: the classic Annabelle hydrangea now comes in pink. Nurser. Notes *43*:44-45.

Rehder, A. (1949). Bibliography of cultivated trees and shrubs hardy in the cooler temperate regions of the Northern Hemisphere. Arnold Arboretum, Harvard University, Jamaica Plain, MA.

Uttal, L.J. (1986). A recombination in *Hydrangea* L. (Saxifragaceae)—*Hydrangea arborescens* L. f. *carnea* (Raf.) L.J. Uttal comb. et stat. SIDA 11:352.

Wu, W., S. Wen, T. Feng, G. Chen, and B. Yang. (2021). Review of *Loropetalum chinense* as an industrial, aesthetic, and genetic resource in China. HortScience *56*:1148–1153.

PROCEEDING'S PAPERS

EASTERN REGION OF NORTH AMERICA

Dr. Charles Heuser, Jr., Regional Editor

Seventy-first Annual Meeting - 2022

Long Island, New York U.S.A.

Combined Proceedings IPPS 72:260-262. 2022.

New, Exciting and Superior Flowering Trees and Shrubs for the Forward-Thinking Horticulturist

Vincent A. Simeone

Planting Fields Arboretum, Oyster Bay, New York, 11771 U.S.A.

vasimeone@aol.com

Keywords: Historic park, newer promising plants, old standby plants, underutilized and exceptional mainstays.

Summary

This presentation puts into focus some of the newer and most promising selections of trees and shrubs available on the market as well as some old favorites that offer exceptional garden merit.

INTRODUCTION

Each year hundreds of new species, varieties and cultivars of plants are introduced into commerce. These choice plants offer superior flowers, foliage, fruit, improved vigor and cultural adaptability. At the same time, there are many underutilized and exceptional mainstays that should be used more in the landscape. This presentation puts into focus some of the newer and most promising selections of trees and shrubs available on the market as well as some old favorites that offer exceptional garden merit. These plants no doubt offer great versatility to the professional horticulturist and designer as well as the home gardener with a true appreciation for the best and brightest plants available. With so much emphasis these days on climate change, the need to support pollinators and birds, deer pressure and the demand for exceptional plant qualities and performance, plant selelction and breeding has become more important than ever before. Buckle up and enjoy the journey!

PLANTS

Deciduous Trees and Shrubs

Abelia chinensis; *A.* × *grandiflora*, 'Rose Creek', 'Hopley', 'Ruby Anniversary', 'Golden Fleece'; *A. mosanensis*

Acer buergerianum; *A. triflorum*

Aesculus parviflora var. *serotina*; *A. pavia*

Amelanchier arborea; *A. canadensis*; *A.* × *grandiflora*, 'Robin Hill', 'Princess Diana'

Aralia 'Sun King'

Calycanthus floridus 'Athens', 'Michael Lindsey' C. 'Hartlage Wine', 'Aphrodite', 'Venus'

Cercis canadensis, 'Appalachian Red', 'Covey', Flame Thrower® red bud, 'Merlot', 'Royal White', The Rising Sun™ red bud, 'Ruby Falls'

Cercidiphyllum japonicum 'Heronswood Globe', 'Red Fox'

Clethra alnifolia 'Compacta', Hummingbird', 'Ruby Spice', 'Sixteen Candles', *C. barbinervis*

Cornus alternifolia 'Argentea', Golden Shadows™ dogwood

Cornus florida 'Appalachian Spring', 'Cherokee Brave', 'Cherokee Princess', 'Red Pygmy'

Cornus kousa, 'Gold Heart'. 'Summer Stars', 'Satomi', Scarlet Fire® dogwood, 'Heart Throb', 'Wolf Eyes'.

Cornus mas, 'Variegata', 'Aurea'

Cornus Rutgers hybrids, Venus® dogwood, Saturn® dogwood

Forsythia viridissima 'Bronxensis', 'Kumson', 'Fiesta'

Fothergilla gardenii 'Mount Airy', 'Blue Mist', 'Blue Shadow', 'Red Licorice'

Heptacodium miconioides

Hydrangea arborescens Incrediball™ hydrangea, Invincibelle Spirit II® hydrangea, Bella Anna™ hydrangea, Ruby® hydrangea

Hydrangea macrophylla Bloomstruck®, Endless Summer® hydrangea, 'Dooley', 'Lady in Red', Twist & Shout® hydrangea, Let's Dance™ hydrangea

Hydrangea paniculata 'Phantom', Baby Lace™, Bobo™ hydrangea, Little Lime™ hydrangea, 'Limelight', Little Hottie® hydrangea, 'Pinky Winky', Vanilla Strawberry™ hydrangea

Hydrangea quercifolia 'Alice', 'Little Honey', 'Snowflake', 'Ruby Slippers', Jet Stream™ hydrangea, 'Sike's Dwarf'

Indigofera amblyantha; *I. kirilowii*

Kolkwitzia amabilis Dream Catcher™ beautybush

Lagerstroemia 'Pocomoke', Razzle Dazzle® Series, 'Coral Magic', 'Plum Magic'

Liquidambar styraciflua 'Slender Silhouette'

Magnolia Little Girl Series, 'Ann', 'Betty', 'Jane', 'Judy', Pinkie', 'Randy',

'Ricki', 'Susan'; *M.* 'Elizabeth', 'Judy Zuk'

Malus 'Red Jewel', 'Sugar Tyme', 'Centurion', 'Pink Spires'

Prunus 'Hally Jolivette', 'Shirofugen', 'Snow Fountains', 'Snow Goose'; *P. subhirtella* 'Autumnalis'; *P.* × *blireana*; *P.* × 'Okame'

Quercus palustris 'Green Pillar', Kindred Spirit® oak, 'Regal Prince'

Rosa Knockout™, Carefree Series®, Carpet Series™, Drift® rose, 'Purple Rain', 'Lady Elsie May', 'Mandarin Ice'

Stewartia pseudocamellia; *S. koreana*; *S. ovata*

Syringa microphylla 'Superba', Bloomerang® lilac, Tinkerbelle™ lilac, Prince Charming™ lilac, Sugar Plum Fairy™ lilac; *S. meyeri* 'Palibin'; *S. laciniata*

Viburnum × *bodnantense* 'Dawn'

Viburnum × *burkwoodii* 'Conoy', 'Mohawk'

Viburnum carlesii

Viburnum dentatum Blue Muffin™ viburnum, Chicago Lustre® viburnum, 'Emerald Luster'

Viburnum dilatatum Autumn Jazz® viburnum, 'Erie', Cardinal Candy™ viburnum

Viburnum plicatum var. *tomentosum* 'Summer Snowflake', 'Molly Schroeder', 'Mary Milton'

Viburnum plicatum 'Opening Day', 'Spellbound', Spring Lace™ viburnum, Pearlific™ viburnum

Viburnum rhytidophyllum; *V.* × *rhytidophylloides* 'Alleghany'; *V.* × *pragense*

Vitex Blue Puffball™ chaste tree

Conifers

Abies concolor 'Candicans'

Cephalotaxus 'Fastigiata', 'Duke Gardens', 'Golden Dragon'

Chamaecyparis thyoides 'Heatherbun', 'Shiva', 'Red Star', 'Blue Sport', 'Glauca Pendula'

Cryptomeria japonica 'Yoshino', 'Black Dragon'

Picea orientalis 'Gowdy', 'Nana', 'Skylands', Bergmen's Gem'

Thuja plicata 'Atrovirens', 'Virescens'; *Thuja* 'Green Giant'

Broadleaved Evergreens

Distylium racemosa × *D. myricoides* hybrids: Blue Cascade® Isu tree, Emerald Heights® Isu tree, 'Vintage Jade', Coppertone™ Isu tree

Illicium Banana Appeal® anisetree, 'Florida Sunshine', Pink Frost™ anisetree, Shady Lady™ anisetree, 'Swamp Hobbit, 'Woodland Ruby'

Berberis (syn. *Mahonia*) × *media*, 'Charity', 'Underway', 'Winter Sun'; *B. eurybracteata* 'Soft Caress', 'Narihira'; *B. fortunei*; *B* × 'Beijing Beauty', 'Marvel'

Combined Proceedings IPPS 72:263-274. 2022.

Cornus (Dogwood) Breeding at Rutgers University

Thomas J. Molnar

Department of Plant Biology, The School of Environmental and Biological Sciences, Rutgers University, 59 Dudley Road, New Brunswick, New Jersey 08901, U.S.A.

thomas.molnar@rutgers.edu

Keywords: woody ornamental breeding, clonal propagation, budding, seedlings, novel cultivars

Summary

The Rutgers Woody Ornamental Breeding program began in 1960 and continues to this day. The breeding of big-bracted dogwoods has been a focal point of the program since Dr. Elwin Orton, the original breeder, pioneered the crossing of *Cornus florida* with the Asian *C. kousa*, which led to the successful Stellar® series of hybrid dogwoods. In 2006, the dogwood program transitioned over to new leadership and breeding efforts were expanded, building from a large collection of unique trees developed by Dr. Orton. In a fortunate turn of events, trees with vivid pink floral bracts were recovered in the new *kousa* and hybrid dogwood breeding populations leading to the 2015 release of *C. kousa* 'Rutpink' Scarlet Fire® dogwood. This manuscript describes some of the history of the dogwood program, followed by the lessons learned growing and selecting dogwoods at Rutgers for 15 years and what comes next after Scarlet Fire® dogwood.

INTRODUCTION

Dr. Elwin Orton was hired in 1960 to develop a Woody Ornamental Breeding Program at Rutgers University located in New Brunswick, NJ. A summary of his work is described in Molnar and Capik (2013). In brief, for the first 10 years his primary focus was breeding American holly (*Ilex opaca*) for the holiday cut-branch market. The goal was to combine the glossy leaves and beautiful berry displays of English holly (*I. aquifolium*) with the cold hardiness of native *I. opaca.* While the crossing of the two species proved unfruitful, it did lead to the development and release of several excellent *I. opaca* cultivars such as 'Jersey Princess' and 'Jersey Knight'. The American holly project later evolved into a program to develop a variety of unique landscape plants within the *Ilex* genus. Interspecific hybridization was a core component of the program as Dr. Orton enjoyed crossing and intercrossing a number of different holly species. The success of the program is exemplified by the release of *Ilex* × 'Rutzan' Red Beauty® holly, a cross of *I.* × *meserveae* and *I. pernyi* that is known for its deer browse tolerance, semi-compact form, and excellent leaf color and berry display.

Dr. Orton began his work with big-bracted dogwoods (*Cornus* sp.) in 1970 where his passion for interspecific hybridization continued. He pioneered crossing the eastern U.S. native *C. florida* with the Asian *C. kousa* to develop a series of hybrid cultivars released in the early 1990s and marketed under the Stellar Series® trademark name (Orton, 1993). The most widely grown cultivar from this series is 'Rutgan' Stellar Pink® dogwood, known for its vigorous growth, disease resistance, and abundance of blush-pink blooms. Note that this hybrid combination was given the formal species name *C.* × *rutgersensis* to help clear up nomenclature confusion in the nursery trade (Mattera et al., 2015). Dr. Orton also used the Pacific Northwest native *C. nuttallii* in his hybridization program with *C. kousa.* From this combination of species, 'KN30-8' Venus® dogwood was developed, a plant exhibiting excellent dark green foliage free of disease, with the largest floral bracts of any known dogwood. Venus® dogwood is thought by many to be his crowning breeding achievement (**Fig.1**) (Orton and Molnar, 2005; Eberts, 2011). Note that the hybrid cross between *C. kousa* and *C. nuttallii* was given the formal species name *C.* × *elwinortonii* in honor of Dr. Orton's breeding legacy (Mattera et al., 2015).

Figure 1. Dr. Elwin Orton and *Cornus* × *elwinortonii* 'KN30-8' Venus® dogwood in May 2014.

Continuing Dogwood Breeding Efforts

In 2006, the woody ornamental breeding program began its transition to new leadership although Dr. Orton remained in an active role for several years even after his retirement in 2008. It was decided that hybrid dogwoods would be the primary focus of the program going forward, but with a targeted goal of developing hybrid and kousa cultivars with dark pink or red floral bracts similar to those exhibited by *C. florida* forma *rubra*. Obtaining this color trait was a goal of Dr. Orton's since the program's inception in 1970. However, despite dozens of controlled crosses aimed at enhancing color and 1000s of seedlings grown to maturity, he was not able to recover color at the level he desired. In the end, his best pink-bracted plants only displayed minor improvements over existing pink kousas in the nursery trade such as 'Satomi' (**Fig. 2**). 'KN144-2' Rosy Teacups® dogwood is a great example. It is a beautiful tree with unique, large blooms, but the pink color is not significantly different than what was already available in the nursery trade (**Fig. 3**). Thus, the question remained of how to enhance bract color in the hybrid dogwoods.

Figure 2. Floral bracts of *Cornus* × *elwinortonii* 'KN30-8' Venus® dogwood next to blooms of *C. kousa* 'Satomi'. Photo credit Wolfgang Eberts.

Figure 3. Floral display of *Cornus* × *elwinortonii* 'KN144-1' Rosy Teacups® dogwood.

Knowing that dogwoods are highly self-incompatible, it was decided that we should grow out a large population of open-pollinated trees from Dr. Orton's collection of elite breeding selections. The collection consisted of a group of about 50 trees planted in a private location isolated from other dogwoods. Each tree was special in one or more attributes and collectively represented decades of breeding and selection at Rutgers. Most were fertile advanced generation interspecific hybrids; some had blush pink and light pink blooms, while others were white in color. Over the next four years, open pollinated seeds were collected from nearly all the trees in the block; seeds were germinated and over 3,000 seedlings grown to maturity. As each tree was genetically unique, the hope was we might recover a new color combination and/or other valuable traits in the offspring that would help move breeding efforts forward (Molnar, 2017).

To our great surprise, this open-pollinated breeding approach worked—in 2012, 6 years after the project was started, we finally recovered several trees with exceptionally dark-pink bract color! While only a few out of the 3,000 grown had this improved color, we knew it was something special based on the intensity of the pigment and how the bracts glowed in the landscape at a distance. They were unlike any dogwoods we, Dr. Orton included, had ever seen before! One of the trees stood out clearly from the pack based on its vivid pink, almost fuchsia bract color, as well as its fantastic number of blooms at a young age, overall tree health, and tree vigor. We propagated it through budding in 2012 and after only three more years of evaluation decided to file a US plant patent application and release it to the nursery industry.

The tree was named *C. kousa* 'Rutpink' Scarlet Fire® dogwood in honor of the Rutgers mascot the Scarlet Knight combined with how the blooms glowed like "pink fire' in the landscape on a sunny day (**Fig. 4**; Molnar et al., 2017; PP 28,311).

Figure 4. *Cornus kousa* 'Rutpink' Scarlet Fire® dogwood showing bright pink bloom display from a distance. Picture taken at Rutgers University in May 2020.

Note that in our excitement over the tree we also initiated it in tissue culture (micropropagation), which allowed us to plant a stock block of nearly 50 trees prior to its release. These stock trees then allowed for the distribution of a large amount of bud wood early on to help increase numbers for commercial propagation and sale in a short period of time (for a tree species). We were also fortunate that the propagated trees grew quickly in nursery production, had a well-branched habit with a strong central leader, and continued to bloom heavily at a young age (a trait seemingly absent in the prior Stellar Series® releases). These traits, in addition to its bloom color, supported its selection as a Gold Medal Plant from the Pennsylvania Horticultural Society in 2022, only 10 years after the original seedling first bloomed! These traits also set the bar very high for any subsequent releases from the Rutgers dogwood program.

Next Steps After Scarlet Fire® Dogwood

Two questions quickly arose: is this new dark pink floral bract color heritable? Can we use it in subsequent breeding? To investigate, starting in 2013 we crossed our few dark-pink bracted trees with each other and with additional trees in the program to grow out new breeding populations. We also culled the less desirable trees from our large seedling nurseries as part of the selection process and in doing so created new "crossing blocks" where the remaining trees (select trees that showed interesting breeding value) intercrossed naturally, and from which we grew additional open-pollinated seedlings. We continued this approach for several years, building up a new population of over 3,000 "next generation" trees for evaluation.

The first group of these trees bloomed in 2017 and many more in 2018. Again, we were blown away by what we saw! The dark pink bract color was indeed heritable, and we recovered many trees exhibiting a range of colors from blush pink to vivid bright pink all the way to pinkish purple (**Fig. 5**). We also observed significant variation in floral bract shape, size, number, and bloom phenology (timing), as well as tree growth habit, tree vigor, and leaf color. We now had literally hundreds of new trees with exciting bloom colors to select from and 1000s of younger trees waiting to flower in our fields. The questions now became: how do we most effectively and efficiently narrow our large pool of new trees down to a manageable level of superior individuals? With all this variation beyond just floral bract color, what tree types would be of the greatest value to the nursery and landscape industry? Answering these questions is still a work in progress.

Figure 5. Young hybrid *Cornus* breeding selection at Rutgers University exhibiting floral bracts with unique hues of dark pinkish purple.

WHAT HAVE WE LEARNED AFTER 15 YEARS OF GROWING DOGWOODS?

To date, we have grown more than 20,000 seedling dogwood trees (planted about 1,500 per year) of which over 10,000 have bloomed and been evaluated (and most cut down). A small percentage of these trees have been propagated for replicated trials and these and others have been used for further breeding. In 2022, we observed the first blooms on seedling trees that are two generation removed from Scarlet Fire®; in other words, grandchildren of Scarlet Fire®

and its cohorts! From these breeding and selection efforts we have identified a number of very promising trees, some of which are expected to be released as cultivars in the near future (and whose details we cannot disclose at this time); we are still deliberating on which specific trees to move ahead with. Further, along the way we have learned a lot about the dogwood breeding process and continue to refine our objectives and approach to developing improved trees. Following the motto of the IPPS "To Seek and To Share", we describe some of what we have learned below.

1. For greater impact, new cultivars should be easily differentiated from those already in the nursery trade. This is an obvious statement but one not always followed by plant breeders. To date, Scarlet Fire® dogwood continues to increase in popularity and numbers of trees sold each year which we appreciate as royalties returned from tree sales help support further breeding efforts. We have made it our goal that new releases from the program should be easily differentiated from Scarlet Fire® dogwood to help expand the market for new dogwoods instead of cutting into already existing sales with lookalike trees, e.g., dogwoods with floral blooms that are only marginally different (**Fig. 6**). We envision releasing a series of dogwood cultivars with unique traits that complement one another in the landscape, creating a desire to plant not just one specimen dogwood tree in a single location but a group of different cultivars planted together for added appeal.

Figure 6. *Cornus kousa* 'Rutpink' Scarlet Fire® dogwood on the left and its "lookalike", next generation seedling selection on right. They exhibit very similar bloom traits and precocious blooming habits, although with a slightly different shade of pink.

Following this logic, our breeding and selection goals are targeted on trees with unique attributes including (but not limited to) bract color and shape, growth habit, and time of bloom display. One example includes trees that exhibit novel floral bracts, such as double bracts or those that are highly dissected (**Figs. 7 and 8**) which can be combined with different shades of color spanning white, blush pink to vivid pink or nearly red. They can also differ considerably in leaf color, especially during phases of rapid growth in the spring (**Fig. 9**).

Figure 7. Novel double bract trait that appears in a small percentage of hybrid *Cornus* breeding selections.

Figure 8. Novel dissected floral bract trait that appears in a very small proportion of dogwood breeding selections.

Figure 9. Two-year-old field of hybrid *Cornus* seedlings showing a range of leaf colors from pale green to dark green to reddish orange to purplish red.

2. Precocious trees (those that bloom heavy at a young age) have added landscape appeal. Again, an obvious statement but one that presents some breeding hurdles when selection on trees has generally been done once they enter their mature blooming phase. Most kousa and hybrid dogwood trees look gorgeous once they are old and heavy blooming; their branches tend to extend far from the trunk and lean over to display abundant blooms which sit as rafts of color atop the leaf canopy. This example contrasts with most young trees that tend to grow vigorous, upright shoots that hold the blooms pointed toward the sky; they also tend to have fewer blooms per branch area which reduces their impact in the landscape. A breeding challenge arises under this scenario as mature plant phenotypes take a very long time to observe (slows the breeding cycle). Further, they are not always a reflection on how propagated plants respond as young trees in the nursery and the subsequent early "juvenile" years in the landscape. This early phase of tree growth is important as homeowners and other land managers may lack the patience to wait for their trees to grow for a decade before they have a notable display of blooms. Note that the Stellar Series® hybrids have received criticism along these lines. They are known to be spectacular mature specimens but may take 8-10 years or longer before they bloom heavily in the landscape.

Fortunately, within our breeding populations we have found trees that exhibit precocity; in other words, they bloom heavy at a young age. And in most cases, we have found this precocious blooming trait is carried over to propagated plants (budded trees tend to also bloom heavy at a young age). This has also been observed to be a heritable trait, where precocious blooming trees tend to have a higher proportion of precocious blooming offspring. Thus, we can target precocious blooming in our breeding and selection efforts.

In our current breeding nurseries, 15-20% of the seedlings bloom in their 4th growing season from germination. We start "selection" this year through a process of elimination where we immediately remove any trees whose floral bracts do not meet our minimum criteria for color, shape, display habit, and/or size. The next year (year 5), about 75% of the total population blooms. We continue selection by the process of elimination and cull trees based on color expression as well as bract shape, bloom density, tree growth habit, and leaf quality. Selection continues throughout the bloom period taking into consideration response to high temperature fluctuations where there is significant variation in response per tree (some trees lose color once temperatures reach over 90° F [32° C]). Note that trees that do not bloom in year 5, unless they stand out for some other unique trait, are generally culled from our populations. This helps keep our focus on those with the propensity to bloom as young trees.

In year 6, the remaining trees are evaluated once again. To avoid elimination this year, they must not only have superior bloom and other traits, but they must now bloom heavy (>100 flower heads) and blooms must not be hidden behind leaves or positioned as such where they are showy only from above. Now that the breeding nurseries have fewer trees, sight paths are open. We take a step back and judge them for their landscape appeal at a distance. Only those rare trees that are unique in bloom traits and other attributes while also being visually striking at landscape level survive the selection process. The top selections in each age cohort will then be clonally propagated for replicated evaluation (those chosen for propagation are generally less than 2% of the starting population).

Once propagated, trees are field planted and evaluated alongside Scarlet Fire® and other cultivars of a similar planting age. Evaluation continues for the next 5 years. Only those breeding selections that develop into attractive, heavy blooming trees at an early age are considered for release. Some breeding selections will bloom their first growing season, which adds to their likelihood for release. Those that take more than 3 years to bloom heavy as propagated trees are considered less interesting and may be removed from the list.

3. Pruning breeding populations like propagation nurseries aids selections efforts. Early on, we thought it would be best to grow our seedling trees un-pruned to observe their natural growth habits. This led to many trees with multiple leaders, wide crowns, and unruly branching. While providing insight on the trees' natural habits, this approach was not only harder to manage at our research farm due to messy, diverse trees, but we later realized it was also not reflective of how the trees are grown in commercial nursery production. And subsequently not how they will be grown later in the landscape.

Dogwood cultivars are typically propagated through chip budding low to the ground on seedling understock. The developing scion is carefully managed as a single trunk tree that has its branches removed from the bottom 2 to 3 feet or more from the crown. Under this production system, it is important that a tree has a strong central leader, good apical dominance, and sufficient branching to make an attractive and full tree canopy at a young age. The tree should also have good branch angles (not too wide and not too narrow unless a fastigiate form is desired) and tend not to be "leggy", i.e., have very long internodes which show a lot of stems and not a lot of branches and foliage, especially on young rapidly growing trees.

Earlier in our breeding program, when seedling trees were not pruned, it was easy to overlook those trees that had weak apical dominance (multiple competing leaders) and poor branching, as we focused primarily on bloom traits and not growth structure. Thus, a number of plants we selected from the earlier populations of trees as superior for bloom traits ultimately performed poorly under propagation. Once budded, they were found to develop crooked trunks, many competing leaders with weak crotch angles, and/or unappealing leggy growth with reduced branching. Note that discarding a tree from the potential release list only after we learn of its poor propagation performance (a three-year process) wastes considerable time and resources.

Today, we use a different approach. We now prune all our seedlings (~1,500 trees planted per year) to single trunks and lift the bottom 2-3 feet of their branches in years 2 and 3, more similar to how budded trees are managed. Although labor intensive, this pruning is very helpful for weed management in our nurseries while allowing us to better examine trees for their single trunk growth habit. We now eliminate those with poor apical dominance, weak branching, and leggy growth that could have been masked in the presence of multiple leaders found in unpruned seedling trees.

We are fortunate that dogwood growth habit appears to be under strong genetic control (a seedling with a strong central leader tends to maintain this trait once clonally propagated). We have found that our recent trees selected from our pruned breeding nurseries generally perform very well as budded trees (only those that look good as pruned seedling trees make it to the next stage of evaluation).

4. Bloom phenology is an important selection tool. Most of the trees in our program, despite their interspecific hybrid background, tend to look much like kousa dogwoods and bloom during the time of kousa dogwoods (late May into June in New Jersey). However, there is variation in bloom time (date of peak bloom size each year) as well as how fast blooms expand to the point where they are showy in the landscape. As an example, Scarlet Fire® dogwood starts out later than some other trees and has small bracts that take several weeks to become showy (**Fig. 10**). Trees that expand their bracts quickly can have a longer display period in the landscape.

Figure 10. *Cornus kousa* 'Rutpink' Scarlet Fire® dogwood blooms in early May. They start off small and take several weeks to expand to full size.

Further, the pigment group responsible for pink and red colors in dogwood bracts is primarily the anthocyanins. Expression of anthocyanins can be affected by temperature, generally with expression of darker colors when air temperatures are cooler and lighter when it becomes hot. We find in central New Jersey the spring temperatures can fluctuate considerably. However, the later we get into May the more likely we will see a stretch of days with air temperatures over 95°F (35°C). This is the temperature when we tend to see pink colors begin to fade in many of our "dark-pink" dogwood selections including Scarlet Fire® dogwood. However, we found that if a tree has sized up its bracts before the high temperatures occur, they tend to better maintain the color through the heat without a lot of visible degradation. This contrasts with those that still have small, developing bracts which are more affected by the heat and may never reach their peak color if it gets too hot during the bract expansion period.

Thus, since we now have large, diverse breeding populations, we can select for dark-pink colored trees whose bracts size up earlier than the average (7-10 days prior to Scarlet Fire®). We have found that these trees tend to maintain pink color longer during the bloom season and into hotter weather. They might not be inherently darker pink regarding concentration of pigment and color of blooms in cooler spring seasons, but by sizing up earlier they can have more intense colors (purplish red colors), and then a subsequently a longer display of that color. In this case, early phenology of bloom gives a visible boost in color related to temperatures during which the peak bloom size and coloration occurs.

CONCLUSIONS

Today, our breeding program and its successes stands on the shoulders of Dr. Elwin Orton and his decades of dogwood breeding work at Rutgers University. We have pedigrees in our newest dogwood selections that reach back to 1970 and span seven generations of breeding and selection. Based on Dr. Orton's germplasm in combination with help from insect pollinators and some good luck, we recovered a breakthrough dark-pink bract color in hybrid and kousa-type dogwoods that led to the release of Scarlet Fire® dogwood. Fortunately, this dark pink color is heritable, and we have used it in breeding to grow large populations of plants from which to select new trees several of which should be forthcoming as new cultivar release in the near future. Along the way, we have fine-tuned our breeding approach and learned more about dogwood trees. This platform allows us to share some of that knowledge, which we hope will be of value to others working in ornamental tree breeding and those interested in the path taken to develop new dogwoods at Rutgers University.

LITERATURE CITED

Eberts, W. (2011). How *Cornus* 'KN30-8', Venus® hybrid dogwood made it to Europe. Comb Proc. Intel. Plant Prop. Soc. *61*:366-371.

Mattera, R., Molnar, T., and Struwe, L. (2015). *Cornus* × *elwinortonii* and *Cornus* × *rutgersensis* (Cornaceae), new names for two artificially produced hybrids of big-bracted dogwoods. PhytoKeys *55*:93–111 https://doi.org/10.3897/phytokeys.55.9112

Molnar, T.J. and Capik, J.M. (2013). The Rutgers University woody ornamentals breeding program: past, present, and future. Acta Hort. *990*:271–280.

Molnar, T.J., Muehlbauer, M., Wadl, P. and Capik, J. (2017). *Cornus kousa* 'Rutpink' (Scarlet Fire®) dogwood. HortSci. 52:1438–1442. 9.

Molnar, T.J. (2017). Going nuts: continuing a 40-year-old woody ornamental breeding program. Acta Hort *1174*:305-312 (Comb Proc. Intel. Plant Prop. Soc.)

Orton, E.R. (1993). New large-bracted dogwoods from Rutgers University. Comb. Proc. Int. Plant Prop. Soc. *43*:51-54.

Orton, E.R. and Molnar, T.J. (2005). Exciting developments in the world of *Cornus*. Comb. Proc. Int. Plant Prop. Soc. *55*:322-325.

Combined Proceedings IPPS 72:275-285. 2022.

Hazelnut (*Corylus*) Breeding at Rutgers University

Thomas J. Molnar

Department of Plant Biology, The School of Environmental and Biological Sciences, Rutgers University, 59 Dudley Road, New Brunswick, New Jersey 08901, U.S.A.

thomas.molnar@rutgers.edu

Keywords: eastern filbert blight, disease resistance, filbert, clonal propagation, seedlings, novel cultivars

Summary

The Rutgers University hazelnut (*Corylus* spp.) breeding program was started in 1996 by turfgrass breeder Dr. C. Reed Funk as one component of a project focused on temperate nut trees. In 2006, hazelnuts emerged as the target group of species due to a number of attributes that includes their small tree size, ease of making controlled crosses, relatively short generation time, and increasing demand for their kernels. In collaboration with Oregon State University, wide germplasm collection and evaluation efforts were undertaken to help identify trees with resistance to eastern filbert blight, the primary limiting factor of cultivation in the eastern U.S.A. This manuscript provides an overview of the Rutgers hazelnut breeding program starting from its inception and spanning over twenty years to the release of the first cultivars in 2020. It also describes collaborative efforts to develop "hybrid" hazelnuts adapted to colder regions.

INTRODUCTION

The hazelnut breeding program at Rutgers University was started by Dr. C. Reed Funk in 1996 as part of a larger project on temperate nut trees. Dr. Funk, who already had an esteemed 35-year career breeding cool season turfgrasses (Meyer and Funk, 1989), decided to shift his focus when nearing retirement age. He turned the turfgrass program over to a new plant breeder and used the resources and knowledge he had amassed at Rutgers to develop a new project on nut producing trees, inspired in part by J. Russell Smith's "Tree Crops: A Permanent Agriculture" (Smith, 1950). The author of this manuscript, Tom Molnar, began working with Dr. Funk at that time.

Following the same principles that proved very effective in the turfgrass breeding program (Meyer et al., 2017), an extensive germplasm collection effort was undertaken. Trees of many different species were obtained and planted across several Rutgers research farms. Species included black walnuts, Persian walnuts, heartnuts, hickories, pecans, chestnuts, gingko, almonds, pistachio, sweet-pitted apricots, and hazelnuts. The goal was to grow large, diverse populations of trees under low input conditions to help identify which species held the greatest potential for planting in the eastern U.S.A., a region where commercial tree nut production has historically been absent.

In addition to collecting plants already available in U.S.A. and Canada, the acquisition of nut tree germplasm from Central Asia and other parts of the former Soviet Union such as Russia, Ukraine, Moldova, and the Baltic countries was targeted. Seed collection in this region was made possible through the help of plant scientist Dr. David Zaurov who, before coming to Rutgers, was a professor of agronomy in Tashkent, Uzbekistan, and had ties to many institutions in the broader Soviet region. Annual overseas trips were made for nearly a decade, resulting in an extensive collection of seeds in addition to documentation and descriptions of germplasm holdings at former Soviet institutions previously unknown in the U.S literature. For examples, see Mirzaev et al. (2004), Abdushukur et al. (2009), Molnar et al. (2011), Zaurov et al. (2013 & 2015), and Capik et al. (2013). Planting continued for about 10 years and the Rutgers field trials eventually held more than 25,000 trees being evaluated for overall tree health, adaptation, nut quality, and nut yields. It is also important to note that Rutgers' ornamental tree breeder Dr. Elwin Orton and fruit tree breeder Dr. Joseph Goffreda were advisors on the early project, supporting Dr. Funk's transition from perennial grass to tree breeding. This interaction eventually connected Tom Molnar to the ornamental breeding program.

Abundant practical knowledge on growing and managing nut trees was gained over that first decade, and nearly all species showed significant breeding potential and opportunity for genetic improvement (Molnar et al., 2013). However, the expenses put forth to maintain the trees were very large, the field space required was considerable, and the long maturity times of most of the species reduced the ability to show progress in a reasonable time frame (*e.g.*, most pecans from germinated seed took over 10 years to first bloom). With the retirement of Dr. Funk in 2006, a decision was made to reduce the scope of the program to keep it sustainable in the long term. It was then decided to focus primarily on hazelnuts (**Fig. 1**).

Figure 1. Cluster of 'Somerset' European hazelnut (*Corylus avellana*) ready to fall from tree in September. (Photo by author)

Hazelnuts were selected because, in general, they appeared well adapted to the region and produced nuts abundantly with little inputs. Further, they were smaller trees (shrubs) that required less land resources to grow to maturity, which fit constraints at the university in New Jersey where land is a primary limiting factor. They also bloomed at a relatively young age (4-5 years from germination) and are easy to use in controlled crosses compared to many other tree crops. Lastly, hazelnuts have few pests or disease problems, aside from one major exception — eastern filbert blight (EFB) (**Fig. 2**), a serious stem-canker disease caused by the fungus *Anisogramma anomala* that is native to the eastern U.S.A.; however, this disease appeared to be an obstacle that could be managed through breeding. Based on these factors, it was decided that hazelnuts were the species where the greatest impact could be made in the shortest time and with the least resources. In 2008, hazelnuts became the primary focus of the nut tree breeding program at Rutgers with work discontinued on the other species.

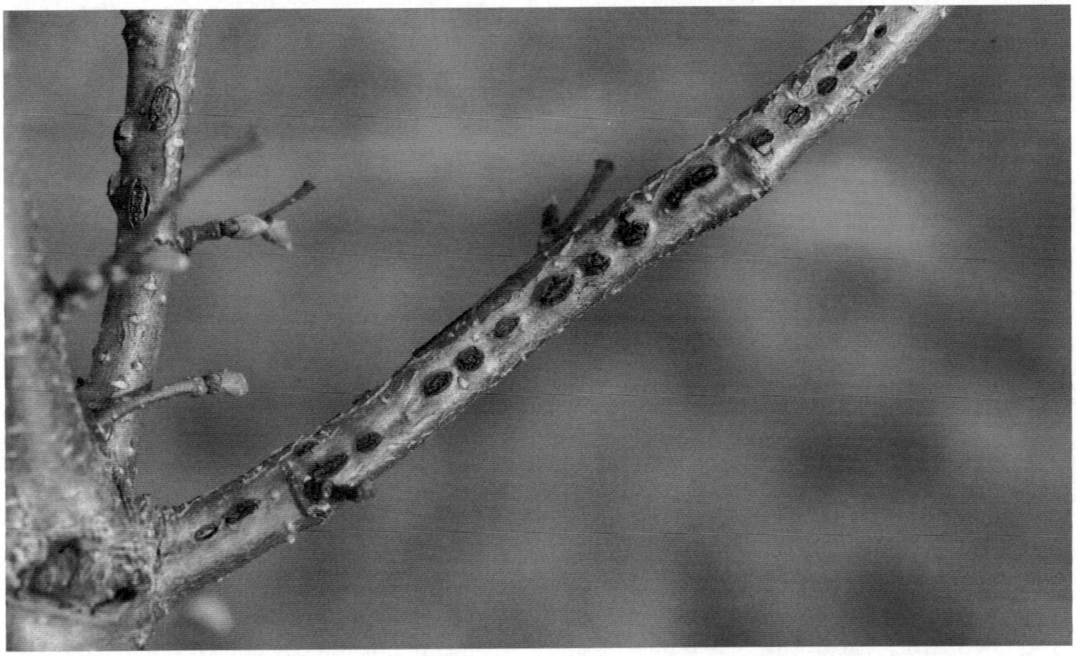

Figure 2. Typical eastern filbert blight canker on European hazelnut exhibiting its "football shaped" stromata. The causal organism is *Anisogramma anomala*. (Photo by author)

Hazelnut Production Worldwide

The *Corylus* genus is recognized to hold at least 13 species native across a wide area of the northern hemisphere; all are monoecious, wind pollinated, and have edible nuts. Of the genus, the European hazelnut (*C. avellana*) is the main species grown commercially for nut production (Botta et al., 2019; Molnar, 2011). While wild *C. avellana* is commonly found throughout much of Europe into the Caucasus region to parts of western Asia, commercial cultivation exists primarily in locations near large bodies of water with mild, Mediterranean-like climates. Major producing countries include Turkey with about 65% of the world's crop followed by Italy (~12-15%), Azerbaijan (~5%), the United States (~5%), and the Republic of Georgia (~3%), with additional production in Chile, China, France, and a few other nations (Food and Agricultural Organization of the United, 2022). In the United States, 99% of hazelnut production occurs in the Willamette Valley of Oregon.

Hazelnut breeding is relatively recent with significant efforts occurring only since the 1960s at Oregon State University (OSU). Most other programs and efforts had been since discontinued except for as described later in this manuscript. Outside of the U.S.A. and Chile, most production orchards are comprised of region-specific, clonally propagated cultivars selected from local plant materials whose origins have been largely lost with antiquity (Mehlenbacher and Molnar, 2021). Studies show that cultivars and wild populations of *C. avellana* remain highly genetically diverse (Gökirmak et al., 2009; Muehlbauer et al. 2014; Oztolan-Erol et al., 2021), which supports opportunities for further genetic improvement (Molnar, 2011).

Eastern Filbert Blight

Attempts to grow European hazelnuts in the eastern United States have historically faltered because of EFB (Fuller, 1908; Molnar et al., 2005). *Anisogramma anomala,* its causal agent, is an ascomycete in the order Diaporthales. It is an obligate biotroph associated strictly with plants of the *Corylus* genus. Its natural host is *C. americana*, the wild American hazelnut, which can be found growing across a wide area of eastern North America east of the Rocky Mountains. Having evolved with the pathogen, *C. americana* is very tolerant of EFB, whereas the European hazelnut is highly susceptible; devastating stem cankers eventually kill most trees lacking genetic resistance (Revord et al., 2020; Capik and Molnar, 2012). The disease is considered the primary limiting factor of hazelnut production in the eastern U.S.A., and since its accidental introduction into Washington in the 1960s and subsequent spread into Oregon, is now the main challenge with growing hazelnuts across all of North America (Johnson and Pinkerton, 2002; Mehlenbacher and Molnar, 2021). Note that *A. anomala* remains confined to North America and strict quarantine rules are in place around the world to help prevent its spread (Jeger et al., 2018)

Breeding for resistance and/or tolerance to *A. anomala* is complicated by its 2-year life cycle that includes a 16–18-month latent period where it generates no outward symptoms (Johnson and Pinkerton, 2002). Further, there exists a considerable amount of genetic diversity among samples of the fungus collected across the U.S.A. and Canada (Muehlbauer et al., 2019). Diversity of the pathogen appears to be limited in the Pacific Northwest, however, where it is not native and spread has been attributed to a single point introduction (Davison and Davidson, 1973; Tobia et al., 2017). In addition, research has shown pathogenic variation is present; some cultivars and breeding selections deemed resistant to EFB in Oregon may succumb to disease in New Jersey and other regions (Molnar et al., 2010; Capik and Molnar, 2012). Adding extra complexity to the system, the pathogen has a giant genome for a fungus (>340 MB), which is composed of >85% repeat regions (Cia et al., 2013).

Collaboration With Oregon State University

Hazelnut breeding has been ongoing at Oregon State University (OSU), Corvallis, OR, since the late 1960s. The breeding program and its associated germplasm collection, when combined with that held at the U.S.D.A. National Clonal Germplasm Repository (also in Corvallis, OR), is considered the largest and most comprehensive in the world (Mehlenbacher and Molnar, 2021). The recent resurgence and expansion of the Oregon hazelnut industry can be credited to the EFB-resistant, high yielding cultivars released by OSU over the past decade.

Collaboration with OSU, specifically with plant breeder Dr. Shawn Mehlenbacher, has been ongoing since the very beginning of the Rutgers project and includes germplasm collection efforts as well as clonal and seedling evaluation. The stressful climate of central New Jersey with relatively cold winters and hot, humid summers positioned within the native range of the EFB pathogen makes for an ideal disease screening location. Dozens of clonal breeding selections and 1000s of seeds from controlled crosses made at OSU have been shared for evaluation. While the primary goal was the selections of improved EFB-resistant plants adapted to local New Jersey conditions, information learned on disease response at Rutgers has been regularly shared with OSU scientists to inform breeding efforts in the current U.SA.. commercial growing region.

Hazelnut Germplasm Collection Efforts

In addition to plants shared for evaluation from the OSU breeding program and germplasm collection, over 5,000 new seedlings from foreign germplasm collections were obtained and evaluated at Rutgers between the years 2002 to 2010. Seeds were collected from Russia, Ukraine, Poland, Moldova, Georgia, Latvia, Lithuania, Estonia, Italy, and Turkey. While most plants eventually died from EFB, about three percent were found to be resistant. Interestingly, these plants spanned nearly all collection locations representing a wide diversity of resistant germplasm (Muehlbauer et al., 2014). Today, when considering hazelnut germplasm at Rutgers and OSU, we have access to over 100 EFB-resistant accessions selected from more than 60 locations equating to a very significant pool of germplasm to support breeding (Molnar et al., 2018). The most promising have been used in controlled crosses with next generation selections now under evaluation. Multiple studies have also shown that most resistance seems to be controlled by only one or a few major genes, although quantitative resistance is also available; *R*-gene mapping projects are underway at OSU and Rutgers with current results summarized in Mehlenbacher and Molnar (2021).

New Cultivars from Rutgers University

In 2020, four cultivars were released from the Rutgers breeding program. They originated from some of the earliest breeding populations grown at Rutgers from controlled crosses made in 2000 and 2004 by Shawn Mehlenbacher at OSU. The new cultivars were selected based on their resistance to EFB as well as their kernel traits and yields in trials at Rutgers. A breeding goal was to release our highest yielding EFB-resistant plants that also produced nuts that would fit the existing world hazelnut market for blanched kernels (confectionary market). This includes kernel with a round shape (not oblong), size of 12-13 mm diameter, thin shells (kernel percent of over 45%), freedom from defects such as molds and split sutures, and very good blanching after roasting as shown in **Fig. 3**. Selection aspects are described in detail in Mehlenbacher and Molnar (2021).

Figure 3. Kernels of 'Monmouth' hazelnut showing round shape and excellent blanching after roasting.

Note that hazelnut cultivars are clonally propagated. In the past this was done by simple layering or stool bed layering, but today this has been largely replaced by micropropagation. However, success with hazelnuts is variable and somewhat genotype dependent with the European hazelnut tending to be easier to work with than the native American hazelnut and its hybrids (Bassil et al., 1992; Pincelli-Souza et al., 2022). The four cultivars described below have been established in axenic culture but vary in their phase of commercial production and availability to date.

'Somerset' (US Plant Patent # 32,494 P2) is the results of a cross of OSU 665.123 × 'Ratoli' (a cultivar with EFB resistance from Spain) made in 2000. It is a high yielding, compact tree with medium size, round kernels that have moderately good blanching. It has self-incompatibility alleles S_3 and S_{10} with S_3 expressed in the pollen. It has notably thin shells and tends to produce good crops even on young trees. Resistance to EFB originates from 'Ratoli' which carries a single *R*-gene that has been shown to provide resistance to *A. anomala* originating from multiple regions (Molnar et al., 2010). 'Somerset' became commercially available from propagation labs in the fall of 2022.

'Raritan' (US Plant Patent # 32,460 P2) is the result of a cross of OSU 539.031 × OSU 616.018 made in 2004. It is a high yielding, vigorous tree that produces medium size, round kernels that blanch well. It has self-incompatibility alleles S_3 and S_{22} with S_3 expressed in the pollen. 'Raritan' exhibits quantitative resistance to EFB, also known as horizontal resistance or tolerance. It is not immune to EFB but only develops very few cankers under high disease pressure, and those it does get are inconsequential and tend to not have fruiting stromata. 'Raritan' appears to be easy to propagate and became commercially available from propagation labs in 2021.

'Monmouth' (US Plant Patent # 32,462 P2) is the result of a cross of 'Sacajawea' × OSU 616.055. It is a high yielding tree that produces medium size, round kernels that blanch very well. It has self-incompatibility alleles S_1 and S_{12} with both expressed in the pollen due to co-dominance. It exhibits quantitative resistance to EFB similar to 'Raritan'. To date, 'Monmouth' is not widely available due to challenges in the multiplication stage in tissue culture.

'Hunterdon' (US Plant Patent # 32 461 P2) is a full sibling to 'Monmouth'. It is moderately high yielding tree that produces medium size, slightly oblong kernels that blanch very well and have a noticeably sweet flavor. It has self-incompatibility alleles S_1 and S_3 with S_3 expressed in the pollen. It exhibits quantitative resistance to EFB but tends to get more cankers than 'Raritan' and 'Monmouth'. To date, 'Hunterdon' is not widely available due to challenges in the multiplication stage in tissue culture.

Hybrid Hazelnuts

European hazelnuts are limited in their adapted range in the U.S. and Canada, mostly to USDA cold hardiness zones 6-8. In contrast, wild *C. americana* can be found growing in much colder regions that include Minnesota, North Dakota, and parts of Manitoba, Canada. The species also expresses resistance and high tolerance to EFB. Fortunately, it is possible to hybridize the two species and select interspecific hybrids that express the best traits of both species. An effort to do so has been ongoing, although intermittently, since the 1920s with *C. americana* 'Rush' hybrids developed in New York and by the USDA. Beyond this, significant progress, especially in adaptation to cold climates and EFB resistance, was made in the 1950s and 1960s by Carl Weschcke in Wisconsin (Weschcke, 1954) and later built upon by breeding at Badgersett research nursery in Canton, Minnesota (Rutter, 1987). Many 1000s of seedlings from Badgersett nursery have been planted across the Upper Midwest from which high-yielding selections have been made (Braun et al., 2019). Additional details on the history of hybrid hazelnut development in North America including other programs are described in Molnar (2011).

Building from the early beginnings of interspecific hybridization, renewed and bolstered efforts have been underway in the past 15 years to develop hybrid hazelnuts as a commercial crop for colder regions. This includes work by the Hybrid Hazelnut Consortium, established in 2008 and today comprised of OSU, Rutgers, the University of Nebraska, Lincoln, the University of Missouri, and the Arbor Day Foundation, as well as the Upper Midwest Hazelnut Development Initiative comprised of the University of Wisconsin, the University of Minnesota, the Savanna Institute, and several other private and public partners. Both groups are working on breeding and selection of improved, EFB-resistant, cold hardy hybrid hazelnuts with a focus on improved nut traits and yields. Regional evaluation trials have been established with advanced plant material now being planted for study across a wide area of the Midwest, Upper Midwest, and northeastern U.S.A.

An exciting early output from the Hybrid Hazelnut Consortium was the release of 'OSU 541.147' "The Beast" (US Plant Patent # 33,561) in 2020. It is the result of a cross of NY 616 (*C. americana* 'Rush' × *C. avellana* 'Barcelona') × *C. avellana* OSU 226.118 made at OSU in 1990 then evaluated at Rutgers since 2000 where it has performed very well. It is a vigorous, high yielding "hybrid" hazelnut tree with small nuts and adequate blanching after roasting. Most kernels are 9–11 mm in diameter and do have a high level of fiber compared to the Rutgers cultivars, but this is removed during roasting. It has S-alleles 8 and 23 with S8 expressed in the pollen, making it a compatible pollinizer for the other Rutgers cultivars. This cultivar is suggested for use primarily as a pollinizer in New Jersey, but growers may find that its high yields of nuts outweigh its small kernel size. Recent tests suggest it can be grown successfully in USDA Zone 5, making it a possible production cultivar in colder regions.

CONCLUSIONS

Although the breeding of multiple temperate nut species was unfortunately discontinued at Rutgers, the hazelnut program was continued and has thrived since its inception. Wide germplasm collection and evaluation in close collaboration with OSU and more recently the Hybrid Hazelnut Consortium and other partners has identified many sources of disease resistance and cold hardiness which support breeding efforts and significant progress. The new EFB-resistant cultivars released from Rutgers in 2020 for the Mid-Atlantic region are becoming available and the first orchards are being planted. Further, one hybrid hazelnut ('OSU 541.147') was also released that is showing promise for the Mid-Atlantic as well as slightly colder regions. The Rutgers, OSU, and Hybrid Hazelnut Consortium breeding pipelines hold many promising breeding selections that combine EFB-resistance and good quality kernels with better cold hardiness. These plants are now under test in multiple regions, and results are eagerly awaited by many growers interested in commercial hazelnut production around North America. The future for greatly expanded hazelnut production in the U.S.A. and Canada remains very bright!

LITERATURE CITED

Abdushukur, A.K., Chernova, G.M., Rakhmonov, A.M., Nikolyi, L.V., Ablaeva, E., Zaurov, D.E., Molnar, T.J. Eisenman, S.W. and Funk, C.R.. (2009). Genetic resources of *Pistacia vera* L. in Central Asia. Genetic Resources and Crop Evolution *56*:429–443. https://doi.org/10.1007/s10722-009-9419-1

Bassil, N., Mok, D.W.S., Mok, M.C., and Rebhuhn, B.J. (1992). Micropropagation of the hazelnut *Corylus avellana*. Acta Hort. *300*:137-140. https://doi.org/10.17660/ActaHortic.1992.300.17

Botta, R., Molnar, T.J., Erdogan, V., Valentini, N., Torello Marinoni, D. and Mehlenbacher, S.A. (2019). Hazelnut (*Corylus* spp.) breeding. In: Advances in Plant Breeding Strategies, Vol. 4. Pg 157-219 Nut and Beverage Crops. Al-Khayri, Jameel M., Jain, S. Mohan, Johnson, and Dennis V. (Eds.) Springer, Cham. https://doi.org/10.1007/978-3-030-23112-5

Braun, L.C., Demchik, M.C., Fischbach, J.A., Turnquist, K. and Kern, A. (2019). Yield, quality and genetic diversity of hybrid hazelnut selections in the Upper Midwest of the USA. Agroforest. Syst. *93*:1081–1091. https://doi.org/10.1007/s10457-018-0209-7

Cai, G., Leadbetter, C.W., Muehlbauer, M.F., Molnar, T.J. and Hillman, B.I. (2013). Genome-wide microsatellite identification in the fungus *Anisogramma anomala* using Illumina sequencing and genome assembly. PLoS One 8(11):e82408. https://doi.org/10.1371/journal.pone.0082408

Capik, J.M. and Molnar, T.J. (2012). Assessment of host (*Corylus* sp.) resistance to eastern filbert blight in New Jersey. J. Amer. Soc. Hort. Sci. *137*:157–172. https://doi.org/10.21273/JASHS.137.3.157

Capik, J.M., Muehlbauer, M., Novy, A., Honig, J.A. and Molnar, T.J. (2013). Eastern filbert blight resistant hazelnuts from Russia, Ukraine, and Poland. HortSci. *48*:466–473. https://doi.org/10.21273/HORTSCI.48.4.466

Davison, A.D. and Davidson, R.M. (1973). Apioporthe and Monochaetia cankers reported in western Washington. Plant Dis. Rptr. 57:522–523.

Food and Agricultural Organization of the United Nations. (2019). FAOSTAT. 15 Oct. 2021. https://www.fao.org/faostat/en/#data/QCL

Fuller, A.S. (1908). The nut culturist. Orange Judd, New York, NY, USA.

Gökirmak, T., Mehlenbacher, S.A. and Bassil, N.V. (2009). Characterization of European hazelnut (*Corylus avellana*) cultivars using SSR markers. Genet. Resour. Crop Evol. *56*:147–172. https://doi.org/10.1007/s10722-008-9352-8

Jeger, M, Bragard, C., Caffier, D., Candresse, T., Chatzivassiliou, E., Dehnen-Schmutz, K., Gilioli, G., Grégoire, J.C., Jaques Miret, J.A., MacLeod, A., Navajas Navarro, M., Niere, B., Parnell, S., Potting, R., Rafoss, T., Rossi, V., Urek, G., Van Bruggen, A., Van der Werf, W., West, J., S. Winter, Boberg, J., Gonthier, P., and Pautasso, M.. (2018). Scientific opinion on the pest categorisation of *Anisogramma anomala*. EFSA Journal *16*:5184. https://doi.org/10.2903/j.efsa.2018.5184

Johnson, K.B. and Pinkerton, J.N. (2002). Eastern filbert blight, p. 44– 46. In: B.L. Teviotdale, T.J. Michailides, and J.W. Pscheidt (eds.). Compendium of nut crop diseases in temperate zone. APS Press, St. Paul, MN.

Mehlenbacher, S.A. and Molnar, T.J. (2021). Hazelnut (*Corylus*) breeding. Plant Breed. Rev. *45*:9-141. https://doi.org/10.1002/9781119828235.ch2

Meyer, W. A. and Funk, C.R. (1989). Progress and benefits to humanity from breeding cool-season grasses for turf. In: D.A. Sleper, et al., editors, contributions from breeding forage and turfgrasses. Spec. Publ. 15. CSAA, Madison, WI. p. 31–48.

Meyer, W.A., Hoffman, L. and Bonos, S.A. (2017). Breeding cool-season turfgrass cultivars for stress tolerance and sustainability in a changing environment. Int. Turfgrass Soc. Res. J. *13*: 3-10. https://doi.org/10.2134/itsrj2016.09.0806

Mirzaev, M.M., Djavacynce, U.M., Zaurov, D.E., Goffreda, J.C., Orton, T.J., Remmers, E.G. and Funk, C.R. (2004). The Schroeder Institute in Uzbekistan: Breeding and germplasm collections. HortSci. *39*:917–921. https://doi.org/10.21273/HORTSCI.39.5.917

Molnar, T. (2011). *Corylus* L. p. 15-48. In: C. Kole (ed.) Wild crop relatives: genomic and breeding resources of forest trees (Volume 10). Springer-Verlag.

Molnar, T.J., Goffreda, J.C. and Funk, C.R. (2005). Developing hazelnuts for the eastern United States. Acta Hort. *686*:609–618, https://doi.org/10.17660/Acta-Hortic.2005.686.81.

Molnar, T.J., Goffreda, J.C. and Funk, C.R. (2010). Survey of *Corylus* resistance to *Anisogramma anomala* from different geographic locations. HortSci. *45*:832–836. https://doi.org/10.21273/HORTSCI.45.5.832

Molnar, T.J., Zaurov, D.E., Capik, J.M., Eisenman, S.W., Ford, T., Nikolyi, L.V. and Funk, C.R. (2011). Persian walnuts (*Juglans regia* L.) in Central Asia. Annual Report of the Northern Nut Growers Association *101*:56–69.

Molnar, T.J., Kahn, P.C., Ford, T.M., Funk, C.J. and Funk, C. R. (2013). Tree crops, a permanent agriculture: concepts from the past for a sustainable future. Resources *2*:457-488. http://www.mdpi.com/2079-9276/2/4/457/htm

Molnar, T.J, Lombardoni, J.J., Muehlbauer, M.F., Honig, J.A., Mehlenbacher, S.A. and Capik, J.M. (2018). Progress breeding for resistance to eastern filbert blight in the eastern United States. ActaHort. *1226*:79-85. https://doi.org/10.17660/Acta-Hortic.2018.1226.11

Muehlbauer, M.F., Honig, J.A., Capik, J.M., Vaiciunas, J.N. and Molnar, T.J. (2014). Characterization of eastern filbert blight-resistant hazelnut germplasm using microsatellite markers. J. Amer. Soc. Hort. Sci. *139*:399-432. https://doi.org/10.21273/JASHS.139.4.399

Muehlbauer, M.F., Tobia, J., Honig, J.A., Zhang, N., Hillman, B.I., Gold, K.M. and Molnar, T.J. (2019). Population differentiation within *Anisogramma anomala* in North America. Phytopathol. *109*:1074–1082. https://doi.org/10.1094/PHYTO-06-18-0209-R.

Oztolan-Erol, N., Helmstetter, A.J., İnan, A., Buggs, R.J.A. and Lucas S.J. (2021). Unraveling Genetic Diversity Amongst European Hazelnut (*Corylus avellana* L.) Varieties in Turkey. Front. Plant Sci. Volume 12. https://www.frontiersin.org/articles/10.3389/fpls.2021.661274

Pincelli-Souza, R.P., Moreira, L.S. and Cohen ,J.D. (2022). Improvements for the micropropagation of hybrid hazelnut (*C. americana* × *C. avellana*). Horticulturae *8*: 849. https://doi.org/10.3390/horticulturae8090849

Revord, R.S., Lovell, S.T., Capik, J.M., Mehlenbacher, S.A. and Molnar, T.J. (2020). Eastern Filbert Blight Resistance in

American and Interspecific Hybrid Hazelnuts. J. Amer. Soc. Hort. Sci.*145*:162-173. https://doi.org/10.21273/JASHS04732-19

Rutter, P.A. (1987). Badgersett Research Farm—plantings, projects and goals. Annual Rep Northern Nut Growers Assn. *78*:173–186.

Smith, J.R. (1950). Tree Crops: A Permanent Agriculture. Devin-Adair: New York, NY, USA.

Tobia, J., Muehlbauer, M., Honig, J., Pscheidt, J. and Molnar, T.J. (2017). Cluster analysis of *Anisogramma anomala* isolates collected from the Pacific Northwest and New Jersey. Phytopathol. *107*:S5.125. (abstr).

Weschcke, C. (1954). Growing nuts in the north. Webb, St. Paul, MN, pp 24–38

Zaurov, D.Z., Molnar, T.J., Eisenman, S.W., Ford, T.M., Mavlyanova, R.F., Capik, J.M., Funk, C.R. and Goffreda, J.C. (2013). Genetic resources of apricots (*Prunus armeniaca* L.) in Central Asia. HortSci. *48*:681–691.
https://doi.org/10.21273/HORTSCI.48.6.681

Zaurov, D.Z., Eisenman, S.W., Ford T., Khokhlov, S., Kenjebaev, S., Shalpykov, K.T. and Funk, C.R. (2015). Genetic Resources of Almond Species in the Former USSR. HortSci. *50*:18–29. 2015 https://doi.org/10.21273/HORTSCI.50.1.18

Combined Proceedings IPPS 72:286-290. 2022.

Production in the Absence of Automation

Alisha Conde

Carolina Native Nursery, 1639 Prices Creek Rd, Burnsville, North Carolina 28714 U.S.A.

alisha@carolinanativenursery.com

Keywords: Production, without automation, native plants, efficiency

Summary

The path of growth at Carolina Native Nursery is discussed. Carolina Native Nursery produces over 200 different native shrubs, perennials, ferns, and grasses. Procedures becoming more efficient through the years are discussed.

INTRODUCTION

Alisha Conde has been the Nursery Manager for Carolina Native Nursery (CNN) in Burnsville, NC since January of 2019. As a Nursery Manager one must be thinking about weather, irrigation, plant diseases, pests, production space and rates, personnel, and equipment maintenance. The list goes on.

All of those responsibilities can be a heavy load, but at the end of the day, working with plants is a fulfilling venture. Additionally, as a professional in this industry, you might be fortunate enough to enjoy the time spent at a few conferences a year. While at these conferences you might see state-of-the-art facilities with machines that were built overseas for a particular nursery's specific needs. Curved glass greenhouses and entire seas of poinsettias are a sight to behold. Carolina Native Nursery has not made it there yet. This nursery has heart, it has tons of beautiful natives, and it has quite a few folks that could talk your ear off about those plants, is one phrase I would not necessarily choose to describe this nursery.

CNN services garden centers, landscapers, parks, and projects throughout the eastern United States. Truckloads of plants are sold from Atlanta to Maine. The nursery also hosts an ever-growing retail area. Our Full Time Staff is 8 people strong, and in the very early spring we bring on an additional 5–6 seasonal employees.

The nursery produces native shrubs, perennials, ferns, and grasses. The specialty of CNN is growing native azaleas from seed. Seeds are started n November/December, and the seedlings start to sprout on heated tables in as little as two weeks.

Our nursery sits on just under 11 acres of what used to be an old tobacco farm. It was purchased in 2002, and propagation started in 2003 with (1) 30 ft × 96 ft wooden propagation house, (2) 20 ft × 96 ft hoop houses, (2) 16 ft × 96 ft hoop houses, and (1) Lath house. There were two workers who potted up about 100 species of shrubs.

In the beginning, employees at propagated from bare root plants and liners on a long table using lugs of soil. The process was to fill a pot from a soil lug on a table, move the potted plant from the table onto a trailer, fill the trailer with plants, drive the trailers off to a hoop house, and drop the freshly potted plants off (**Fig. 1**).

Figure 1. Tractor and trailer setup.

After the plants were loaded into a house, they were top dressed with fertilizer, watered, and tagged. It was a good day if one person could pot up 100 plants in an afternoon (**Fig. 2**).

Figure 2. Potted plants in a hoop house with rice hulls for weed control.

Three years into the business the soil lugs were eliminated, and the switch was made to 18-wheeler soil deliveries. This change removed the time it took to fuss with the soil lugs and made the potting process a bit quicker.

The next step towards efficiency was purchasing a soil hopper. Ten years into the business the nursery acquired a hopper for $1500.00. Some of you might be familiar with these. Essentially it is a giant soil funnel. There are four spaces in which an individual can stand, and it has wheels. It creates a mobile soil pile. The hopper would be hitched to a tractor and pulled around the nursery to fill up hoop houses with a new crop. The hopper is set up in such a way that your hands and arms are moving ever-so-slightly less drastically which adds up to more efficiency over time. Our motto sort of became, "Take less steps, use less time".

There were downsides to this method including spatial limitations, and not having protection from the elements. This method was used until the end of the potting season in 2019. In 2021 a stabilized tent prototype was mocked up. Having a stabilized potting area helped us to eliminate set up time at the start of the workday, and breakdown time at the end of the workday. All of our tools and supplies were within the immediate work range. This newly constructed shelter also helped to keep employees out of the elements.

Since the hopper became stable under the tent, this new plan of attack meant that the plants were the ones that needed to move rather than the hopper full of soil. Standard operating procedures again were changed. Employees no longer were fertilizing, ricing, tagging, and watering in the houses. Instead, they did so on the surface of the trailers that were used to move the plants. This was a back-saving revelation. During this first trail year of having the stabilized potting area, the potting crew only had one tractor. This meant that both sides of the hopper had to share the only means of moving plants. Additionally, the nursery only had two trailers to set the freshly potted plants onto and these trailers would fill up quickly. The goal was to minimize the number of times the trailer left the hopper to drop off plants. Always remember the helpful "Less steps, more efficiency" motto. In 2022, an event tent was acquired to work under (**Fig. 3**). This was 19 years after opening the nursery. Everything is in one place and under cover from the elements. The tent is tall enough to drive a skid steer under so when substrate is getting low in the hopper, and the trailers move out from under the tent to drop plants off to their new homes, someone can take the opportunity to fill the hopper up.

Figure 3. Tent to work under.

An additional improvement to the nursery and the production process was the watering tunnel. (**Fig. 4**). The prior standard operating procedure had been to water the new plants by hand, or water them using overhead irrigation. The new water tunnel method is quick and effective. A second tunnel will be added in 2023.

Figure 4. Our watering tunnel.

The staff can up 252 1-gal pots per hour, and 168 3-gal pots per hour. Heading toward automation is the goal. A potting machine will allow more than 4 people potting and can eliminate steps such as grabbing stacks of pots and filling those pots with soil. The nursery has acquired 12 more acres of land during the tail end of 2022. This means that CNN has more space to fill with the same amount of time in a year.

You may be wondering, why the heck has the nursery not just bit the bullet and acquired a potting machine? The reason was to avoid feeling so overwhelmed by big changes that the final product suffers. When you make the leap to automation it is not just a one-and-done change. You are essentially changing the whole system. Suddenly you are not able to use your event tent. Instead, you will need an actual building to really keep the potting machine out of the elements. You may not be able to use the pots that you have been used to using. Perhaps the pacing of what time of year that your seasonal workers are needed will change. There will be unforeseen changes to your standard operating procedures which you have worked so hard to perfect.

CNN has seen sales increase by: 19%, 18%, 30%, 31% over the past couple of years, adding new growing space along the way. Automation is bound to happen. The number one priority for CNN is to always have beautiful plants. The path to that goal can go any which way, the preference is to focus on creating a workday that is as easy as possible while still reaching production goals.

When our seasonal employees see that we are making changes every year to make the seasonal jobs better they are more likely to come back the following year. When they come back, we don't have to spend as much time onboarding and job training. Those types of employees are so important in helping new members of the company feel comfortable and confident in their first year at the nursery.

Growth can be overwhelming and costly. It is helpful to remind yourself that you can always use what you have got and improve where you can. Whatever you do, never stop analyzing your processes for improvements.

Combined Proceedings IPPS 72:291-294. 2022.

Control of Broad Mite on English Ivy Cuttings with Dip Treatments

Daniel Gilrein[1], Kevin Dichtl[2], and Lucille Siracusano[2]

[1]Extension Entomologist;.[2]Program Assistant, Long Island Horticultural Research and Extension Center, Riverhead, New York, 11901 U.S.A.

dog1@cornell.edu

Keywords: Broad mite (*Polyphagotarsonemus latus*) control, English ivy (*Hedera helix*), dip treatments, horticultural oil treatments

Summary

Research into the control of broad mite [*Polyphagotarsonemus latus*] infesting English ivy (*Hedera helix*) is discussed. In general, mist in propagation as used here appears to be an effective control for broad mite, but in the absence of mist a cutting dip in horticultural oil provides significantly better control of the egg stage and to some extent adults as well compared with other treatments.

INTRODUCTION

Six cutting dip treatments were compared in two greenhouse trials for control of broad mite [*Polyphagotarsonemus latus* (Banks)] infesting English ivy (*Hedera helix*). Treatments included SuffOil-X at two rates (80% mineral oil, BioWorks), M-Pede (49% potassium salts of fatty acids, Gowan), Ultra-Pure Oil (98% mineral oil, BASF), water dip control, and undipped control cuttings. In the first trial, cuttings were maintained in rooting blocks after treatment under mist until rooting. In the second trial, cuttings were kept after treatment under shade cloth in vases with water without mist until rooting.

MATERIALS AND METHODS

This work was conducted at Cornell University's Long Island Horticultural Research and Extension Center, Riverhead, NY from March 5 to March 25, 2014 in one greenhouse for the duration of the trial. English ivy mother plants in a greenhouse showing signs of broad mite infestation (stunted and distorted foliage, bronzing leaves) (**Fig. 1**) were used in this trial. Infestation was confirmed by examining foliage under magnification for both mites and characteristic eggs.

Figure 1. *Hedera helix* showing leaf distortion and stunting due to broad mite.

Terminal cuttings (~12 in.) with symptoms on newest growth were selected from mother plants on March 5 and randomly assigned to treatments noted above. Ten cuttings were used for each treatment in each trial. Cuttings were dipped (5 sec.) to thoroughly wet on March 5 in insecticide preparations, water, or left undipped, then laid out on a bench until dry. One set of cuttings (10 per treatment) was stuck in rockwool blocks (Grow-Cubes, Grodan B.V.) and randomly arranged on a mist bench until rooted. A second set of cuttings (10 per treatment) were stuck in vases of water and placed under shade cloth but without mist on an adjacent bench. Temperatures were maintained at 65- 75°F under ambient light and humidity (ranging 60-80% RH). Plants in both trials were checked for symptoms of phytotoxicity (yellowing leaves; brown, necrotic spots on leaves; leaf drop) and rated on March 10, then examined under a microscope for broad mite eggs and adults on March 14 (both trials), March 17 (cuttings in vases only), and March 25 (both trials). Live eggs and adults found were tallied. ANOVA and pairwise comparisons of transformed or untransformed treatment means were done using Tukey's HSD test. Treatments and results are shown in Tables 1(Trial 1) and 2 (Trial 2).

RESULTS AND DISCUSSION

In the first trial (mist) (**Table 1**), no adult broad mites were found on cuttings on March 14 or 25 (SuffOil 1% and 2%, Ultra-Pure Oil) or at very low levels (undipped, M-Pede, water dip treatments) and treatments were not significantly different. No broad mite eggs were found on cuttings in this trial on either date.

Table 1. Control of broad mite on English ivy cuttings with dip treatments, plants rooted in Grow Cubes under mist, Riverhead, NY, 2014.

		3/10	3/14		3/25	
Treatment	Rate	Phyto[y]	Egg	Adult	Egg	Adult
SuffOil-X	1%	0.0b[z]	0.0ns	0.0ns	0.0ns	0.0ns
SuffOil-X	2%	0.0b	0.0ns	0.0ns	0.0ns	0.0ns
M-Pede	2%	0.9a	0.0ns	0.1ns	0.0ns	0.0ns
Ultra-Pure Oil	2%	1.2a	0.0ns	0.0ns	0.0ns	0.0ns
No dip	0%	0.0b	0.0ns	0.7ns	0.0ns	0.1ns
Water control	100%	0.0b	0.0ns	0.1ns	0.0ns	0.0ns

[z]Means within a column followed by the same letter are not significantly different at p=0.05 (LS means Tukey's HSD). Data were transformed prior to analysis using log(y+1). [y]Phyto (phytotoxicity) rated on a scale of 0 = no damage to 10 = dead plant.

In the second trial (cuttings in water vases under shade cloth, not under mist) (**Table 2**), mites were found but only at low levels on cuttings in all treatments on March 14 except for those dipped in 2% SuffOil; there were no significant differences among treatments.

Table 2. Control of broad mite on English ivy cuttings with dip treatments, plants rooted in water with no mist, Riverhead, NY, 2014.

		3/10	3/14		3/17		3/25	
Treatment	Rate	Phyto [y]	Egg	Adult	Egg	Adult	Egg	Adult
SuffOil-X	1%	0.0b[z]	1.7c	0.6ns	0.0b	0.0b	0.0b	0.0b
SuffOil-X	2%	0.0b	0.2c	0.0ns	0.0b	0.0b	0.1ab	0.0b
M-Pede	2%	0.0b	5.6ab	1.5ns	0.1b	0.1b	0.7ab	0.6b
Ultra-Pure Oil	2%	1.0a	0.0c	0.0ns	0.0b	0.0b	0.0b	0.1ns
No dip	0%	0.0b	13.8a	2.2ns	3.8a	1.3ab	0.4ab	2.4a
Water control	100%	0.0b	1.8b	1.2ns	3.0a	1.3ab	1.3a	0.7a

[z]Means within a column followed by the same letter are not significantly different at p=0.05 (LS means Tukey's HSD). Data were transformed prior to analysis using log(y+1). [y]Phyto (phytotoxicity) rated on a scale of 0 = no damage to 10 = dead plant.

Mite eggs on that date were high on undipped cuttings and on those dipped in M-Pede, with numbers significantly greater in both cases than observed in other treatments. On March 17, a moderate number of eggs and very low numbers of mites were found on undipped cuttings and on those dipped in water, with few or none in other treatments. By March 25 eggs were at low levels in all treatments with slightly but not significantly more on water-dipped cuttings. Mite numbers were low to extremely low in all dip treatments with significantly more found on undipped cuttings. Slight but significant phytotoxicity was noted in both trials on plants dipped in Ultra-Pure Oil and in the mist trial on cuttings dipped in M-Pede. There was no injury observed in any other treatment. In general, mist in propagation as used here appears to be an effective control for broad mite, but in the absence of mist a cutting dip in horticultural oil provides significantly better control of the egg stage and to some extent adults as well compared with other treatments. However, a water or M-Pede dip also appears to reduce mite levels on cuttings compared with not dipping cuttings for propagation in the absence of mist, so even a vigorous wash may provide some control over no insecticide dip treatment at all.

Combined Proceedings IPPS 72:295-297. 2022.

Tissue Culture Panel Write-up for Microplant Nurseries

Jonathan Jasinski

Microplant Nurseries, Inc., PO Box 237, Gervais, Oregon, 97026 U.S.A.

jonathan@microplantnurseries.com

Keywords: Tissue culture, Microplant Nurseries, Inc., production timing, hormone concentrations

Summary

The history of the development of Microplant Nurseries, Inc., is discussed from the beginning with two employees to the present day with 60 employees. Microplant Nurseries, Inc., goals are discussed. Their focus is growing great, healthy plants and learning together how to do things better. The timing of plant material induction to production as well as hormone concentrations and their effect on plant growth are discussed.

INTRODUCTION

Jonathan Jasinski earned a BSc and two MSc degrees in Plant Sciences at the University of Florida. He moved to Oregon in 2015 and is currently serving as the Chief Operating Officer at Microplant Nurseries, Inc. overseeing the production of millions of Stage 2 and Stage 3 tissue culture plants annually.

In the late 1970s, a diverse group of commercial tree growers began funding research on the micropropagation of various tree crops at the Oregon Graduate Center in

Beaverton, Oregon. The group was comprised of Adams Rootstock, Inc. located in Washington State, Stark Brothers in Missouri, Oregon Rootstock, Inc. (later renamed TRECO®) and A. McGill and Son, also in Oregon. As projects developed, the four partners decided to launch a new company, and Microplant Nurseries, Inc. was born.

Microplant Nurseries, Inc. (**Fig. 1**) opened its doors in January 1980 with two goals in mind:

1) To provide large numbers of new and improved fruit tree rootstocks and shade trees to the owners of the company
2) To provide great plants to the general nursery trade for profit

Figure 1. Microplant Nurseries new facility.

Of the founding owners, TRECO and A. McGill and Son remain. Over the past three decades, we have pioneered the technique of commercial micropropagation on a large scale, producing millions of plants each year. While our original focus was on trees, we soon expanded into all kinds of exciting projects, including new ornamental shrubs, blueberries, hazelnuts, raspberries, hops, grapes, perennials, timber products and bulb crops.

We produce only lab product, Stage 2 unrooted microcuttings, and Stage 3 in vitro rooted plantlets (**Fig. 2**). Over the years our company has grown from two employees to a year-round staff of about 60. Most of our employees have been with us for many years. In the year 2000 we built and moved into a new building, as well as digitized recordkeeping for various departments providing increased predictability, efficiency and quality. We continue to invest in new equipment, new methods and our people. A large portion of our work is done under contract, specifically for individual growers.

We have partnered with some of the biggest, best, and most advanced growers in our industry. We appreciate their trust in us. Microplant Nurseries, Inc. has a world-wide reputation for quality and reliability. Our mission is to get you what you want, when you want it, at a price that makes sense for all and also have fun doing it. Our focus is growing great, healthy plants and learning together how to do things better. We walk in the door each day curious about what the plants are going to teach us.

Of particular interest during the IPPS 2022 Conference were questions relating to tissue culture order request timing of plant material, as well as hormone concentrations and their effect on plant growth. In general, most production-scale tissue culture projects require at least one-and-a-half to two years from plant initiation to produce production-run numbers in the thousands. Generally, after a plant has "settled" into culture after multiple subcultures over those 12-16 months, orders placed at least 6-10 months in advance are required for successful multiplication and delivery during the desired timeframe. As a general rule, the less hormone (cytokinin or auxin) you are able to use during the multiplication process, the better. In addition, a question was posed regarding how often plant lines should be re-initiated into culture. This is widely dependent upon the risk tolerance of the person interested in the plant material in terms of mutations. The higher the number of times a plant has been sub cultured, generally the higher likelihood of a somaclonal mutation event may occur. However, this is heavily dependent upon the type of plant that is being cultured in vitro and its genetic stability.

Figure 2. (A left, B center) Lab production facilities and (C right) microcuttings.

Combined Proceedings IPPS 72:298-302. 2022.

New Research Determines Successful and Secure Disposal Method for Greenhouse Waste Infected with ToBRFV

Keith Osborne

Gro-Bark Ltd, 816 Mayfield Road, Caledon, Ontario, L7C 0Y6 Canada

kosborne@walkerind.com

Keywords: Aerated composting, tomato brown rugose fruit virus, ToBRFV, greenhouse waste, stone wool substrates, compost, recycling

Summary

This research evaluated the efficacy of Walker's static aerated composting process in deactivating the tomato brown rugose fruit virus (ToBRFV) in spent stone wool substrates and infected vines in order to create a circular economy for greenhouse waste. It was concluded that Walker's standard 6-week GORE® composting process is 100% effective at deactivating ToBRFV when the cell maintains an internal temperature of over 75°C for 47% of the composting life cycle duration.

INTRODUCTION

Walker Industries (Walker) and the Ontario Greenhouse Vegetable Growers (OGVG) have recently completed a three year-long study as a part of a grant through the Greenhouse Competitive Innovation Initiative (GCII). The study evaluated the efficacy of Walker's static aerated composting process in deactivating the tomato brown rugose fruit virus (ToBRFV) in spent stone wool substrates and infected vines in order to create a circular economy for greenhouse waste.

In Ontario, stone wool slabs are one of the main substrates used in greenhouse vegetable production, however limited processes currently exist to manage the spent slabs in an environmentally friendly manner. Walker, Canada's largest fully-integrated resource recovery company, has developed a method in which to process and recycle spent stone wool slabs by separating the stone wool from the plastic encasement and grinding it to a size that can facilitate its re-use. The primary re-use preference is to add the ground stone wool as a bulking agent in compost, which would then be used to create soil blends, effectively diverting the material from a landfill.

With the discovery and spread of ToBRFV in southern Ontario, the secure disposal of end-of-crop-cycle waste has been critical to reduce further contamination of crops. Walker and OGVG subsequently entered into a coordinated partnership to determine the survival rate of ToBRFV in spend stone wool substrates and infected vines when incorporated with source separated organic (SSO) waste and processed in Walker's existing GORE® composting system (**Fig. 1**).

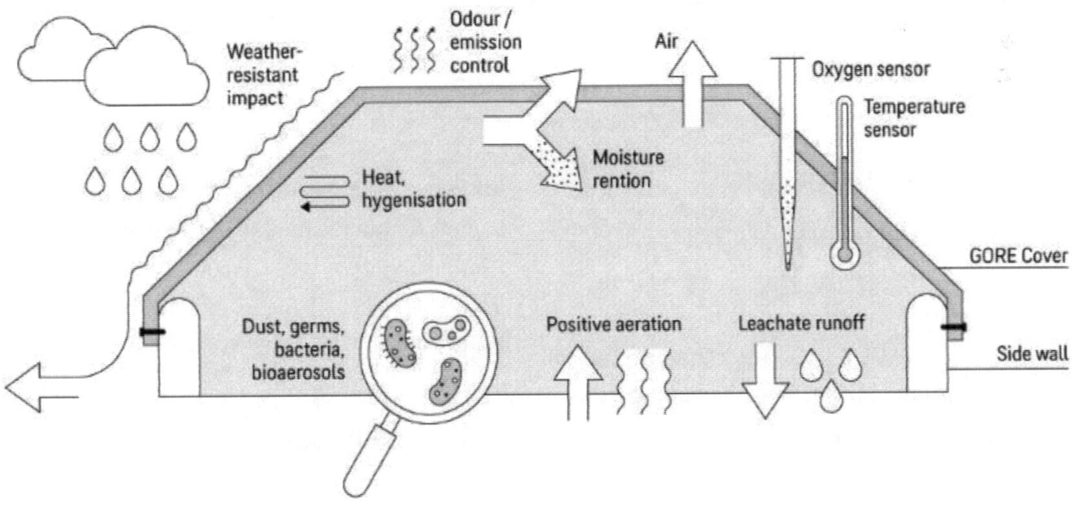

Figure 1. Summary of GORE® system dynamics (www.gore.com).

MATERIALS AND METHODS

Taking place from October 2021 to October 2022, Walker obtained infected Rockwool® substrate and tomato vine waste from Ontario Ministry of Agriculture, Food and Rural Affairs (OMAFRA) to be inserted in mesh bags and incorporated near the center of a GORE® cell at the time of loading. This occurred over seven cell repetitions in which the GORE® cell underwent standard operating procedures for either a 6-week or 8-week composting life cycle. After the composting life cycle was complete, the bags of Rockwool® substrate and tomato vine waste was removed and used to create an inoculum for a plant bioassay study at Walker's Arthur Compost Facility greenhouse.

The plant bioassay procedure consisted of tomato plants infected with several treatment inoculums (**Table 1**) then grown for a 2-week incubation period after which point plant biomass was harvested and sent to a third-party laboratory for ToBRFV detection using one-step conventional reverse transcription polymerase chain reaction (RT-PCR).

Table 1. Composting and bioassay treatment compositions.

Treatment	Components
Composted Rockwool®	- 20% infected Rockwool® - 80% SSO waste
Composted Vine	- 20% infected tomato vine waste - 80% SSO waste
Composted Control	- 100% SSO waste
Positive Rockwool® Control	- 100% infected Rockwool®
Positive Vine Control	- 100% infected tomato vine waste
Negative Control	- 100% phosphate buffer

RESULTS AND DISCUSSION

The initial repetitions of both 6-week and 8-week composting cycles (Rep 1 – 6WK and Rep 2 – 8WK) demonstrated average cell temperatures below 65°C (**Fig. 2**) resulting in only partial deactivation of ToBRFV (**Fig. 3**). However, increased average cell temperatures demonstrated significantly greater ToBRFV deactivation. All subsequent 6- and 8-week composting cycle repetitions of the study resulted in full ToBRFV deactivation due to consistent cell temperatures over 75°C (**Fig. 4**). It was concluded that Walker's standard 6-week GORE® composting process is 100% effective at deactivating ToBRFV when the cell maintains an internal temperature of over 75°C for 47% of the composting life cycle duration (**Fig. 5**).

Due to the success of the GORE® efficacy trial, Walker will further investigate end uses for composted stone wool and explore capital investments to expand Walker's infrastructure for end-of-crop-cycle waste collection and processing in the Windsor-Essex area to offer sustainable alternatives to landfilling and incineration.

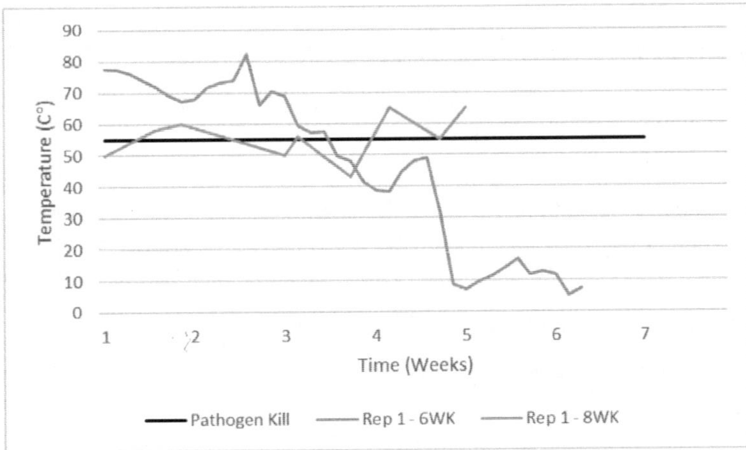

Figure 2. Temperature profiles of material in GORE® cells of the first composting cycle repetitions over time.

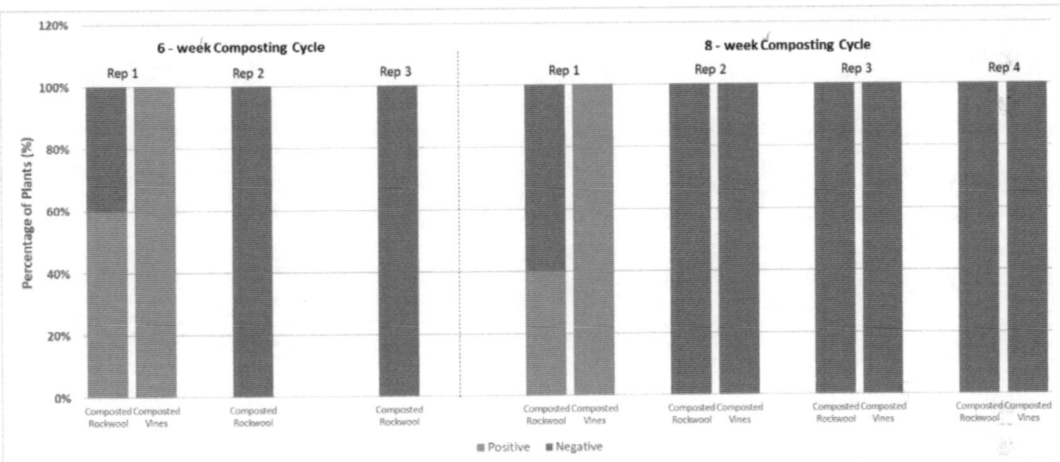

Figure 3. The percentage of plants inoculated with composted Rockwool® and composted vines that tested positive for ToBRFV following a plant bioassay.

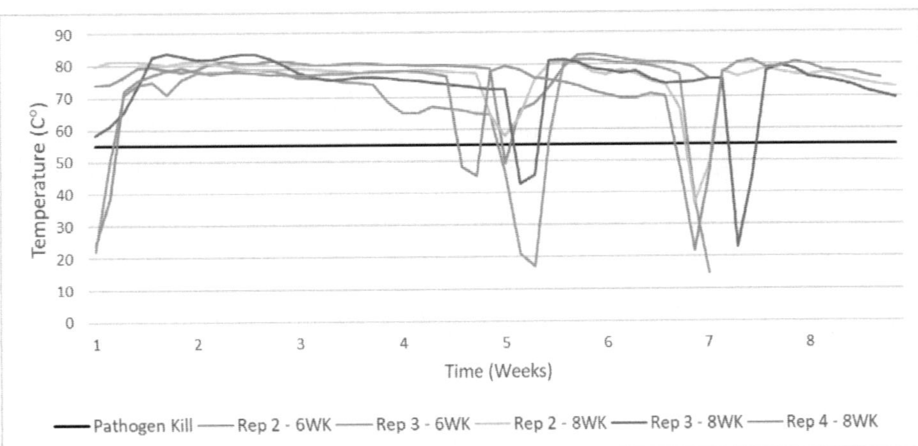

Figure 4. Temperature profiles of material in GORE® cells of composting cycle repetitions over time.

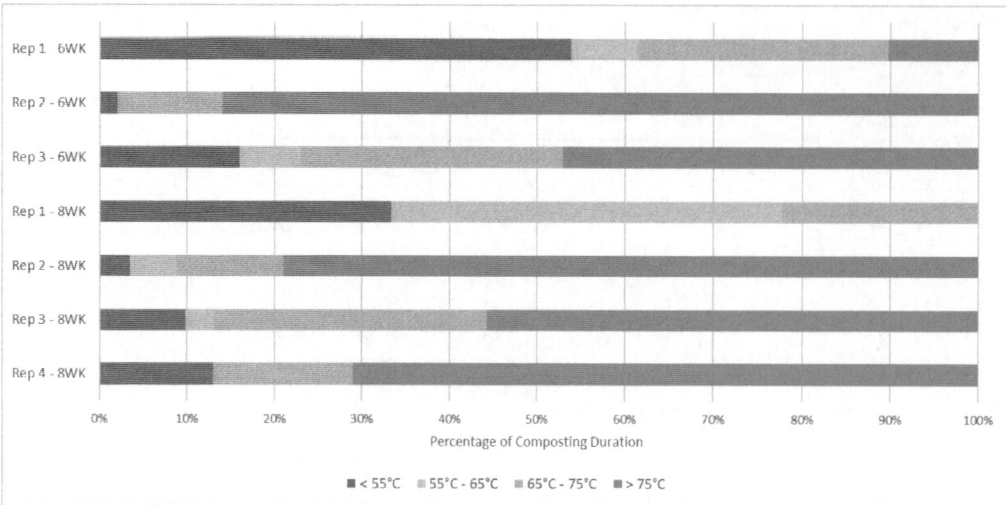

Figure 5. The percentage of treatment time spent within the defined temperature ranges of each repetition.

Combined Proceedings IPPS 72:303-314. 2022.

The IPPS Website: A How To

H. William Barnes

Barnes Horticultural Services LLC, 921 Lakewood Neshannock Falls Rd., New Castle, Pennsylvania 16105 U.S.A.

bill@barnhortservices.com

Keywords: using the website, International Plant Propagation Society, proceedings

Summary

The IPPS website use is explained in detail. Significant changes to the website are explained. The paper explores the IPPS website and members will understand how to use this new technology so that they, the members, can use it to their best advantage.

INTRODUCTION

In recent years the International Office of the IPPS has undertaken a series of significant changes to bring the IPPS on a worldwide basis to the 21st century. One major step forward was to discontinue the publication of the Combined Proceedings of the International Plant Propagation Society, aka "The Black Book". A completely digital format was implemented with all of the proceedings from year one to the present being made available via the Internet on the official IPPS website. This action also required a strong look at the website and decisions were made to completely revamp it as well. Those goals have been accomplished. This missive is to explore the IPPS website and understand how to use this new technology so that you, the members, can use it to your best advantage.

Let's get started. The very first step is to go to: IPPS.org through any search engine. Once there it is necessary to introduce yourself to the system. If you are at all familiar with websites there is the all-encompassing need for the following: User Name and Password. Your user name is the email address the IPPS office has on file for you. For instance: BetsyK@slocumnet.com and then you would have to select in duplicate a password known to you, such as "Catbox12xx" (**Fig. 1**). Once you have accomplished these steps you are into the system and the whole of the IPPS website is available to you (**Fig. 2**).

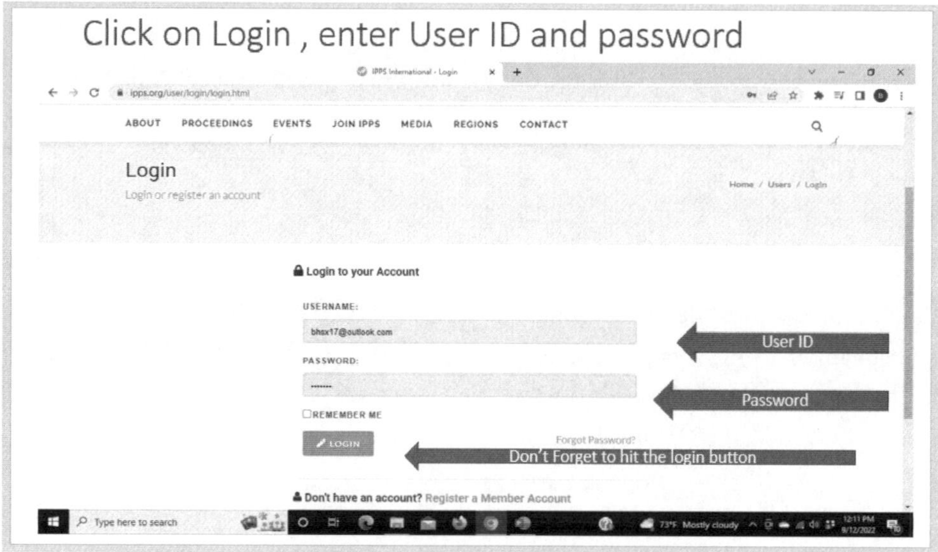

Figure 1. Getting started.

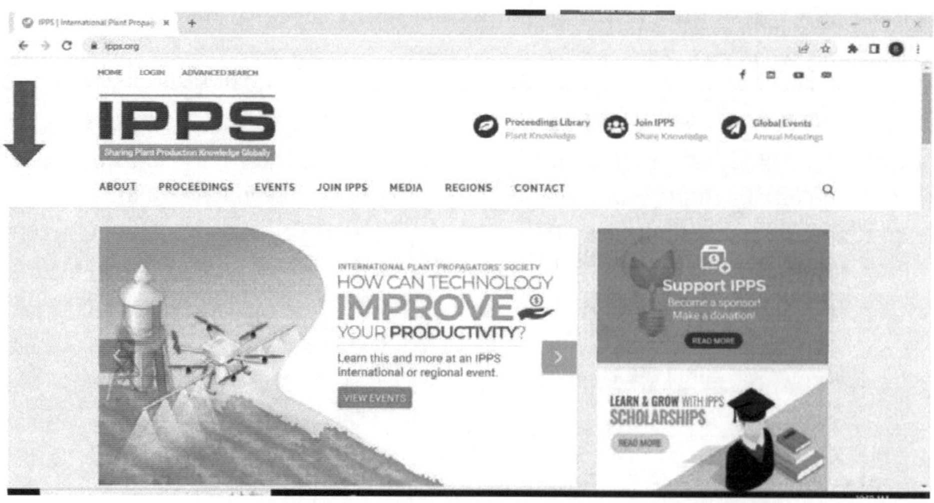

Figure 2. You are now into the system and the whole of the IPPS website is available to you.

The IPPS used to be solely dependent on the "Black Book" for the communications from the various regions of the world. That is no longer the case. The website encompasses a vast reservoir of varied materials from all of the regions in addition to the papers contributed to the proceedings. The quote above, from the 1970s rock band Emerson, Lake and Palmer is more than relevant now. The IPPS website is really a show that never ends. You can read the proceedings from Vol. 1 to the present and then look at current trends via Facebook, YouTube, and various productions, videos and PPTs from specific regions. Something new is added daily. Don't hold back, go have a look.

Figure 3 shows us where to find the proceedings, current, and future events of the various regions, a media section, who and what the regions are and how to reach people in the know. Look at the lower right-hand corner and you will see two videos about online nursery tours in Europe and on the left side bottom is a video about Advances and Innovations in Green and Allied Sciences.

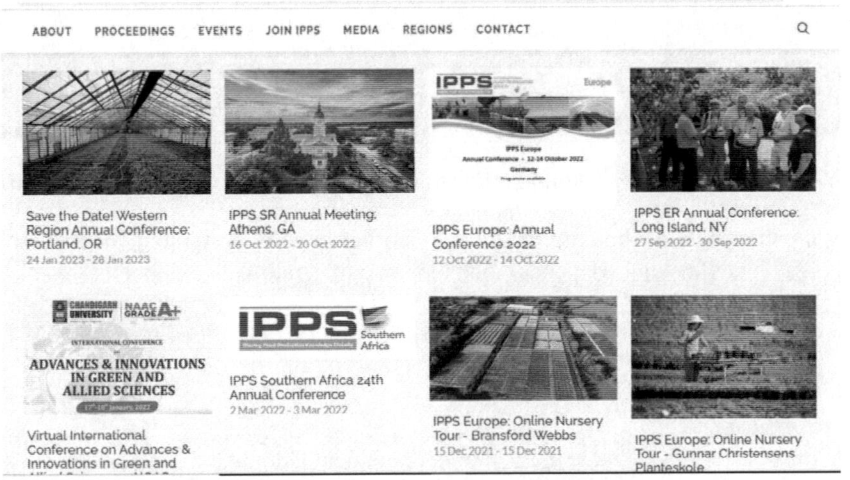

Figure 3. This shows where to find content such as proceedings, regional events, and a media section.

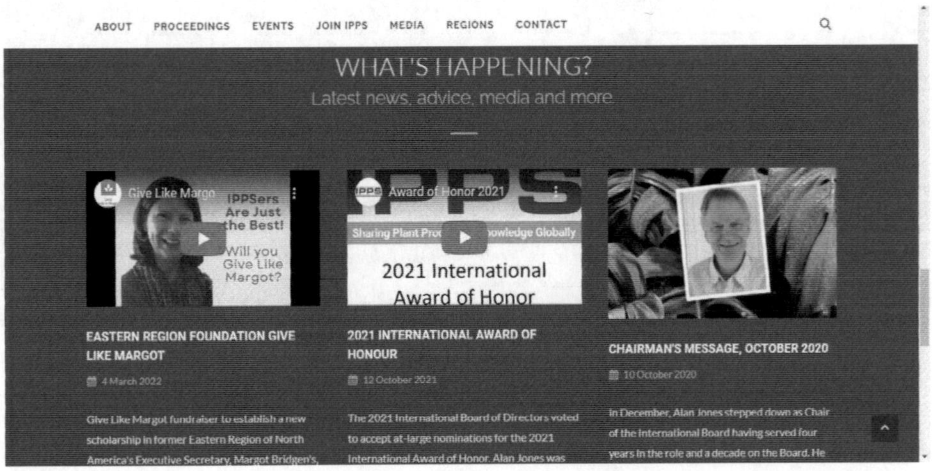

Figure 4. What's Happening Section in the International Office.

Scrolling down the page you will run into the What's Happening Section in **Fig. 4** International Office. You may or may not want these details but they are there and are consistently updated as presented in **Fig. 5**. It changes frequently so a timely suggestion is to review the site periodically, catch what you can that is of interest and come back again for more information.

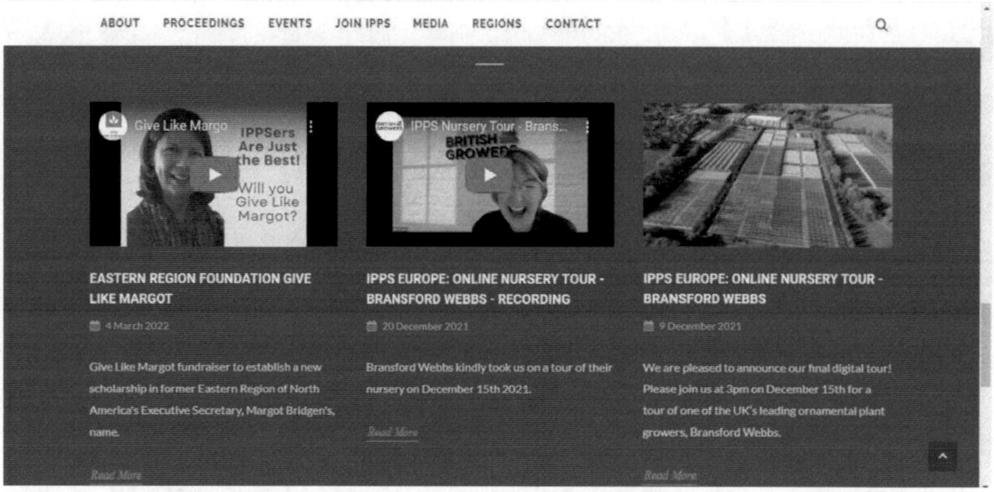

Figure 5. What's Happening Section is consistently updated as shown in this figure.

In your travels on the site, right click your mouse on the About Section, a panel drops down and is fairly self-explanatory (**Fig. 6**).

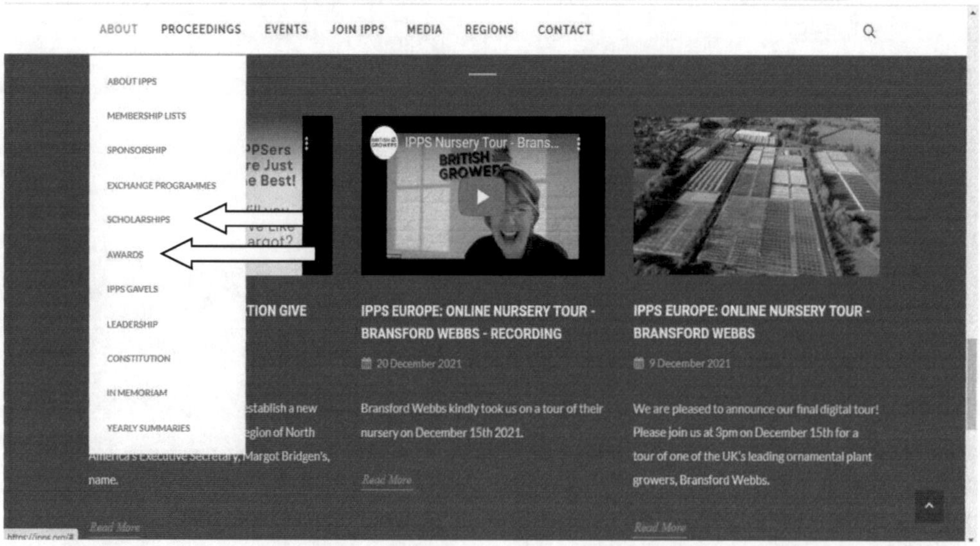

Figure 6. Right clicking your mouse on the About Section a panel drops down.

Clicking on any heading in particular will take you into a more specific explanation about that particular heading. For instance, "Awards" will take you to the various awards from the regions as well as the awards from the International Office. Awards are one of the fundamental ingredients of the IPPS and they are important. It is the organization's method to recognize people for their many contributions to the success of the society. For you students and young people, click on scholarships and see where it takes you. You might want to know that there could be an opportunity waiting for you.

In many ways the Combined Proceedings is some of the best material available to you, our members. It should not be overlooked, there is a wealth of knowledge and information there and it is searchable (**Fig. 7**). Clicking onto the Proceedings heading at the top of just about any page will take you there.

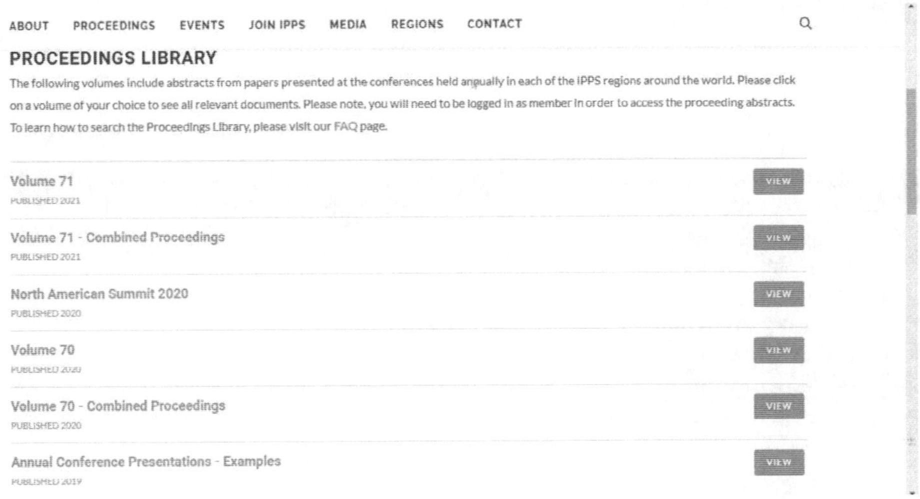

Figure 7. The starting point to explore the Combined Proceedings from the most recent.

Here you see volume available chronologically from the present on down. Notice the bright green buttons on the right. Clicking there will take you to that edition. Take note that there are two headings for the same volume. One is for a quick reference to show what's in there and the other which says Combined Proceedings is a more extensive version of all that is available for that heading. The Quick reference version will suffice if you saw a presentation, you know the title and you can easily search for it and go straight there. The Combined Proceedings is more detailed and extensive (**Figs. 8 and 9**).

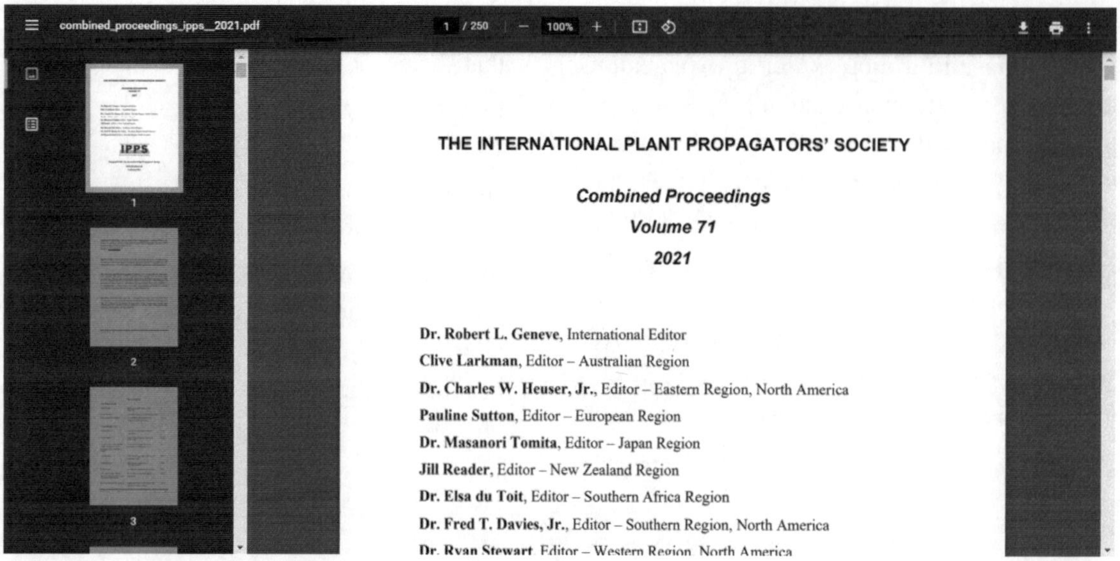

Figure 8. A quick reference to show what's in there.

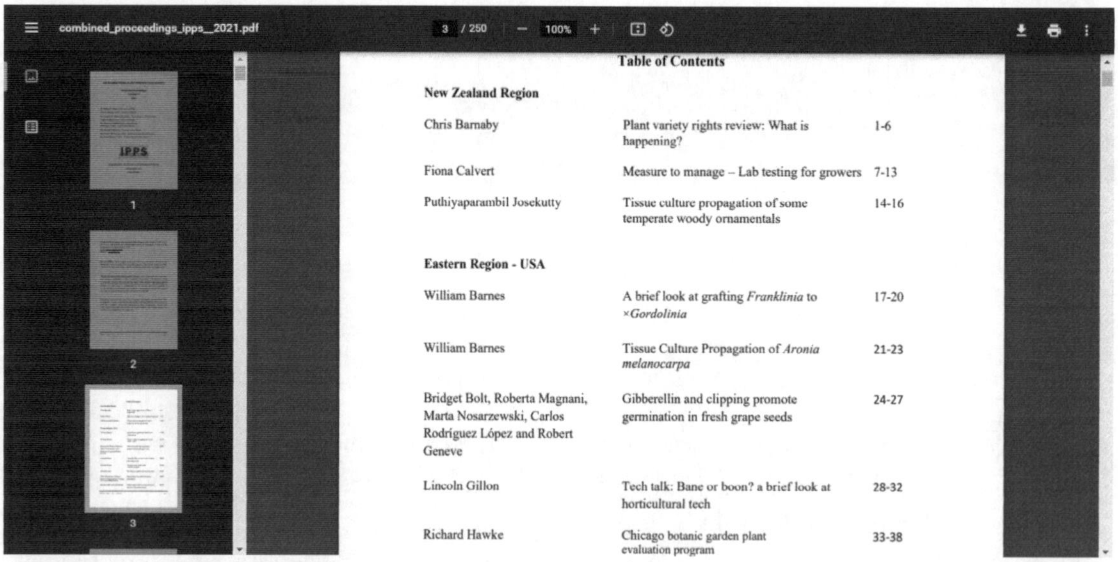

Figure 9. All papers for a given region are alphabetized starting with the 1st author.

Note that all the papers for a given region are alphabetized starting with the 1st author. Also, if you wish you can download a single paper or the whole of the volume. Note too that a printed version is available for a fee but interest in that is not strong and that aspect may not be available in the future. Contact the Regional or International administrative office for details about a printed version. One caveat, it will contain all the information but will not look like the original black book version nor will a printed version have all of the accompanying videos and other media that the website contains.

All of the Regions are represented in a given year's Combined Proceedings

Volume. However, if a Region did not submit any papers, then there will not be any papers available for that region. In the instance that a particular region is not there, the only recommendation would be to contact that region and request a particular paper should you know of it.

One notable achievement beyond the scope of the "Black Book" is that the Proceedings now are in "Living Color". This is a big step forward as now you can see clearly what a particular subject is and you can also interpret graphs with a clarity not found before in the printed version. As an example, you can readily see the graft union of *Franklina* to ×*Gordolinia* is clearly visible (**Fig. 10**).

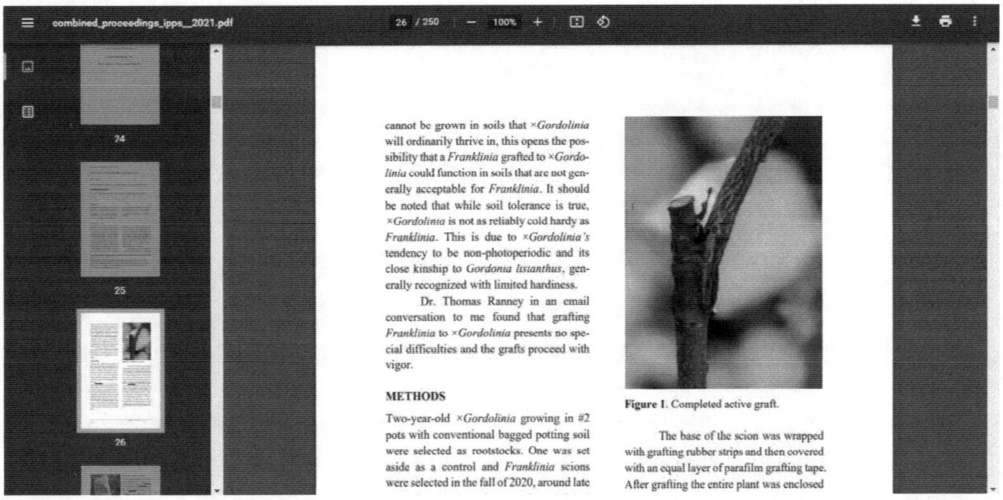

cannot be grown in soils that ×*Gordolinia* will ordinarily thrive in, this opens the possibility that a *Franklinia* grafted to ×*Gordolinia* could function in soils that are not generally acceptable for *Franklinia*. It should be noted that while soil tolerance is true, ×*Gordolinia* is not as reliably cold hardy as *Franklinia*. This is due to ×*Gordolinia's* tendency to be non-photoperiodic and its close kinship to *Gordonia lisianthus*, generally recognized with limited hardiness.

Dr. Thomas Ranney in an email conversation to me found that grafting *Franklinia* to ×*Gordolinia* presents no special difficulties and the grafts proceed with vigor.

METHODS

Two-year-old ×*Gordolinia* growing in #2 pots with conventional bagged potting soil were selected as rootstocks. One was set aside as a control and *Franklinia* scions were selected in the fall of 2020, around late

Figure 1. Completed active graft.

The base of the scion was wrapped with grafting rubber strips and then covered with an equal layer of parafilm grafting tape. After grafting the entire plant was enclosed

Figure 10. The online version is now in color as shown.

Figure 2. Germination in fresh grape seeds after being cut and treated with 2,000 ppm gibberellic acid.

Figure 3. Seedling from seed that was cut and treated with 2,000 ppm gibberellic acid.

26 | IPPS Vol. 71. 2021

Figure 11. The various colors of the figure about the *Vitis* seedling stand out and make interpretation easy to accomplish.

Videos are scattered throughout the website and some show up in the Proceedings (**Fig. 12**). The North American Summit is one such example. It was a new venture to compensate for the total shutdown of IPPS meetings due to Covid 19. An informative and thoughtful presentation of state-of-the-art plant propagation and production activities from a variety of locations, it is a masterpiece that shows just what can be accomplished via our website.

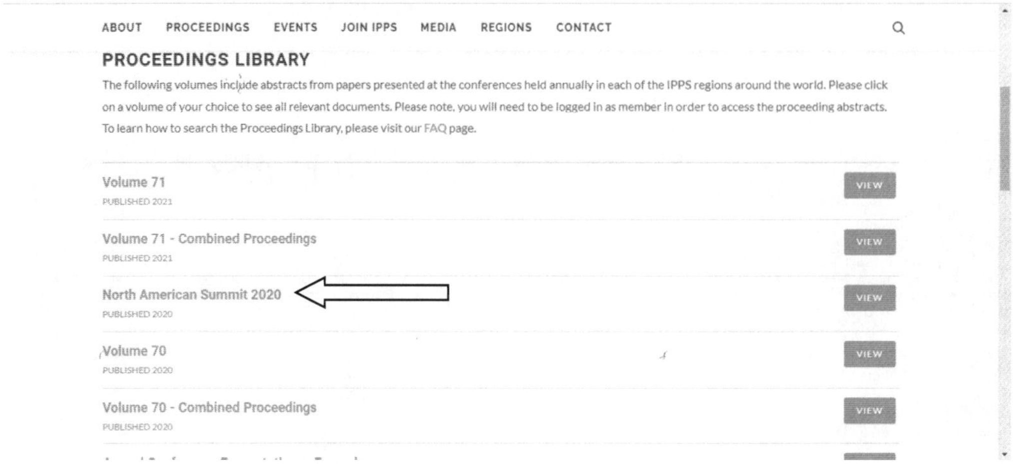

Figure 12. Videos are also found in the website and as part of a proceeding.

Power Points are increasingly becoming available to us and the IPPS website is no exception. Power Points are not commonly found in the Combined Proceedings but there are some and elsewhere on the website such as the media section. **Figure 13** shows one such location within the Proceedings.

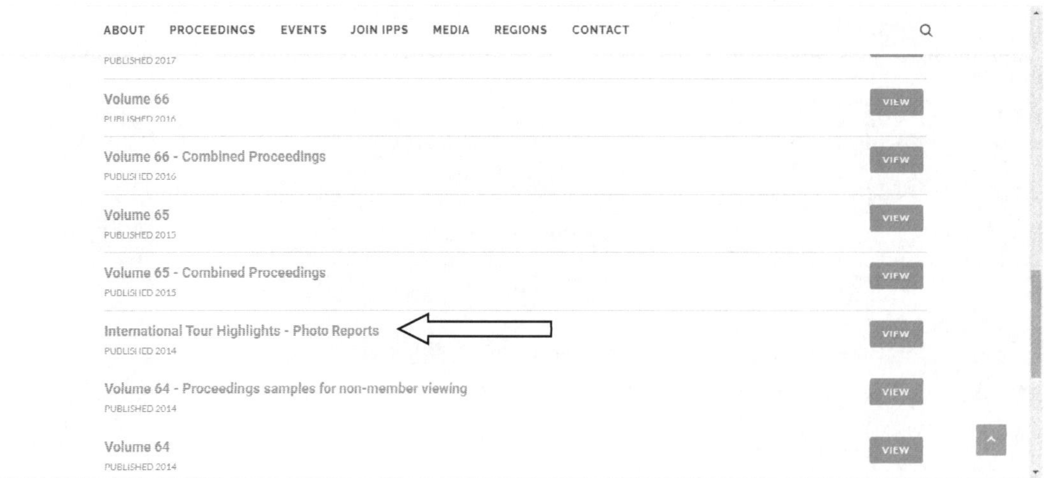

Figure 13. An example of a Power Point in the Combined Proceedings.

Some PowerPoint® presentations on the website are massive and take a considerable amount of time to download. Be patient! In the example below, note there are 44 pages (**Fig. 14**). It's a really big file.

Figure 14. An example of a large PowerPoint® presentation.

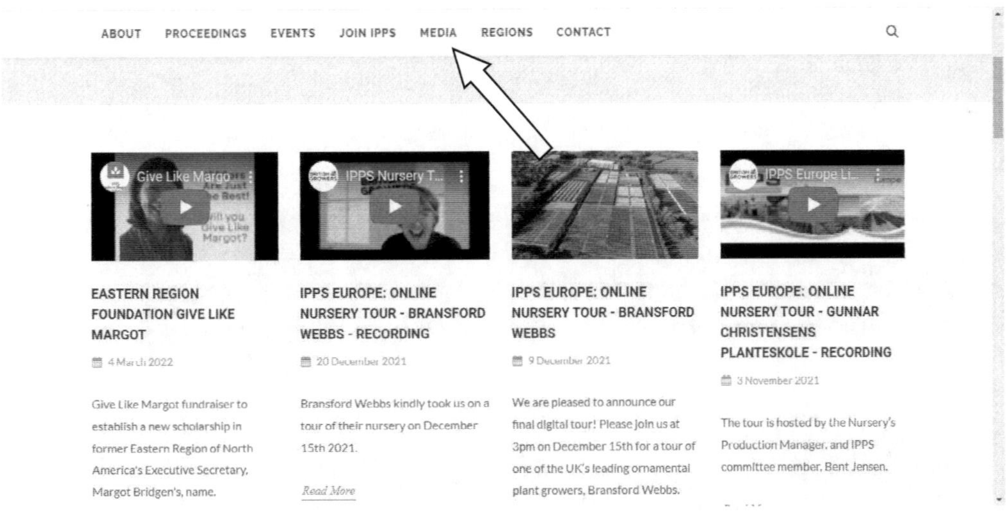

Figure 15. The media section is a good place A place to look for videos, PowerPoint® presentations, and pictures.

The media section has some separate headings such as News and Media and a photo section. **Figures 16 and 17** show how things look.

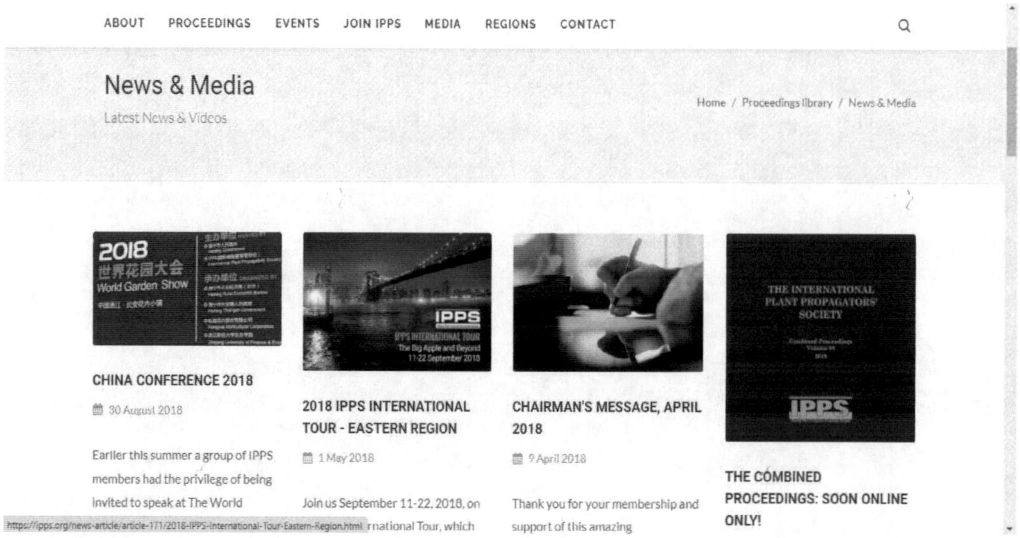

Figure 16. News and Media has a separate section for videos.

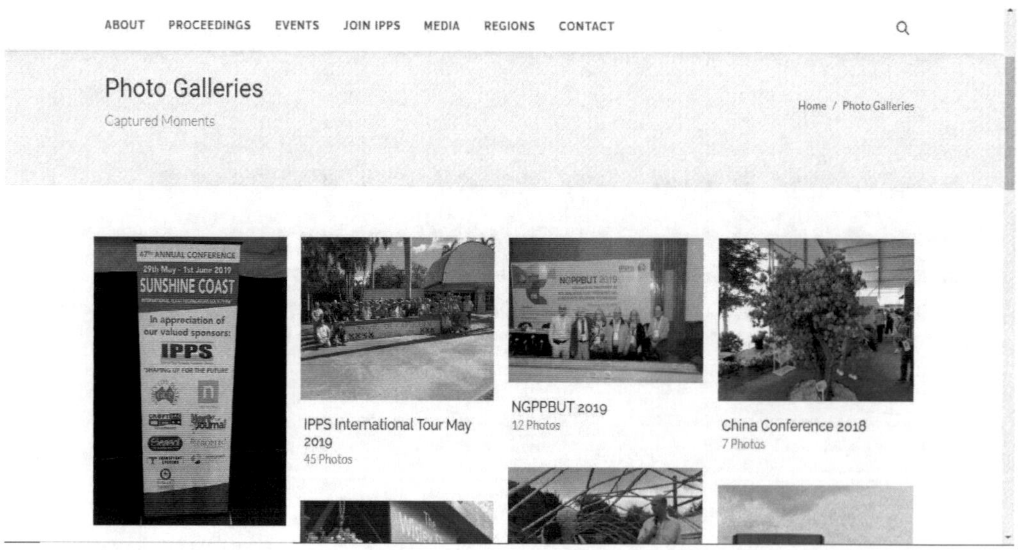

Figure 17. A picture section in the News and Media section.

We also host Facebook and YouTube links on the website (**Fig. 18**). Here is where you can find them. Click on the small (f) or You Tube and this is what you get.

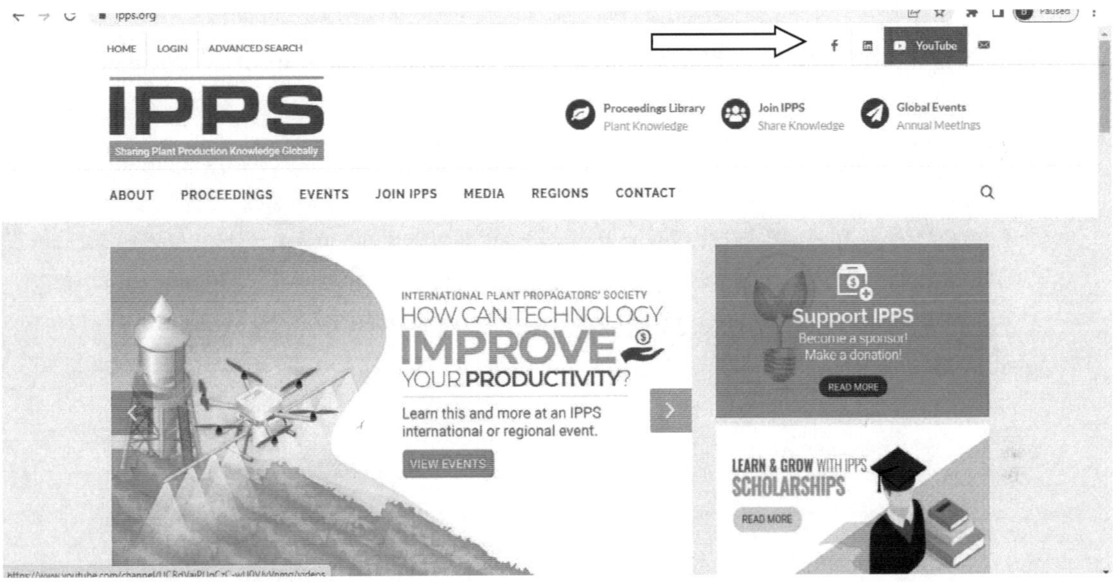

Figure 18. Facebook and You Tube icons.

The IPPS YouTube extravaganza literally explodes in your face, take your pick, get out the headphones and the libation and whatever fascinates you (**Fig. 19**).

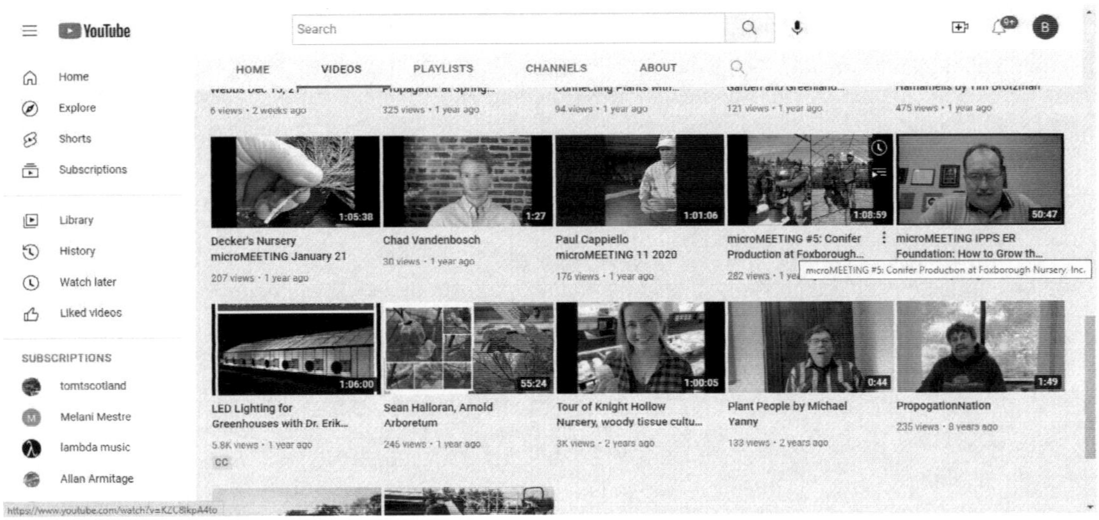

Figure 19. An array of videos many authored by IPPS members.

The IPPS website offers an abundance of details about a variety of subject. It's there, just dig around and you will find it. Can't find it? Call a regional administrator, they can help.

- Tours, both International and access to other regions
- New technologies, new people, new ideas
- A chance to see parts of the world with like-minded people
- Events, Western Meeting Jan 23, Australian Meeting May 23
- Southern Region International Tour, Oct 23, don't miss this
- Get on Board, pack your "kossies" and talk to people from around the world right in your own back yard
- There is more to the IPPS than just the Black Book

IPPS International covers a lot of ground, be a part of it - Eastern NA, Western NA, Southern NA, Australia, New Zealand, South African, Europe, Japan, India (tentatively), China (a ways off).

Combined Proceedings IPPS 72:315-319. 2022.

Breeding for Non-Invasive Nursery Crops: Status of Cultivars and Regulation

Ryan N. Contreras

Department of Horticultural, Oregon State University, Corvallis, Oregon 97331 U.S.A.

ryan.contreras@oregonstate.edu

Keywords: Breeding, noninvasive plants, invasive plants, ecological risks, native plants, cultivars and regulations

Summary

Breeding for noninvasive plants is discussed. The author proposes that the nursery industry should not wait for the attention generated by invasive plants to turn to regulation, but to be proactive and take ownership of the issue. The goal of regulation on this issue should be to prevent the further spread of invasive species but to allow production and sale of plants that have been proven to present little or no ecological risk. Methods for reducing fertility, testing for fertility, and current regulations and the future are discussed.

INTRODUCTION

Most of us would say that good landscape plants are resilient to biotic and abiotic stresses, thrive over a broad range of environments and soils. They are vigorous – certainly producers want plants that finish quickly. We want them to be beautiful with copious flowers and fruit to beautify our landscapes and attract wildlife into the garden. As it turns out, these traits that make great landscape plants are shared by plants that have escaped cultivation and become notorious spreaders and, in some cases, invasive weeds. I shall not debate at what point a plant should be called various names due to degrees of problems – free seeding, nuisance, weedy, naturalized, invasive. Regardless of definitions, there is no dispute that there is much attention on the plants in nursery industry regarding their spread from seed. I propose that we should not wait for the attention to turn to regulation, but to be proactive and take ownership of the issue.

Regulation and the industry's role

I believe the goal of regulation on this issue should be to prevent the further spread of invasive species but to allow production and sale of plants that have been proven to present little or no ecological risk. The proving is critical to this point, and I will address below. Decisions should be based on data that are regionally appropriate, but we need to apply a framework nationally. We are a national and international industry in which plants bred and produced in Oregon are shipped worldwide. The data I generate in Corvallis should not be taken as the final word on how plants may perform in disparate climates and should be tested in regions that differ in important ways such as increased precipitation, longer growing seasons, etc.

It is fortunate that in my experience most state nursery groups have positive relationships with their respective departments of agriculture and we certainly want to maintain those. In addition to the open dialogue with regulators, we should take other steps including to:

1) Not waiting for regulation but addressing the issue within your organization – make it an agenda item to discuss among your nursery and landscape associations
2) Stop growing plants that we all know are highly invasive
3) Support development of and adopt seedless cultivars
4) Educate consumers and the gardening public about the steps our industry has taken on the issue.

We are THE Green Industry, and we should be out in front of folks telling them the actions we are taking to be stewards of the land.

In some circles the discussion surrounding the use of non-native vs. native can get contentious. Individuals on both sides often have strong opinions that are valid and can be backed by science. However, I encourage us all to remember that we all share the same goals when we set out into the garden or urban landscape – that is to install plants that thrive in the environment where they are sited to achieve the ecosystem services that plants render. If you are using the examples of 'Bradford' pear or purple loosestrife as examples of "sterile" plants that became invasive, you are using erroneous and inaccurate examples. The fact is these plants were never sterile, they were self-incompatible and are extremely different from the examples I will provide below that have been developed through modern breeding practices and evaluated to confirm their reduced seed/seedling production.

Methods for reducing fertility

Common methods of reducing fertility in landscape plants includes ploidy manipulation (changing the number of sets of chromosomes), wide hybridization, mutation (gamma radiation, X-rays, chemicals), and biotechnology (transformation, gene editing, targeted mutation). Common examples of triploids include food crops such as seedless watermelon and bananas. Ploidy manipulation has also been applied in landscape plants such as flowering pear (Phillips et al., 2016; *Pyrus* × *triploida* 'NCPX2' PP 30788 Chastity™), maples (Contreras and Hoskins, 2020), spirea (*Spiraea* 'NCSX1' PP 28313 Candy Corn ™), *Hypericum androsaemum* (Trueblood et al., 2010), trumpetvine (Oates et al., 2014), barberry (Brand and Durocher, 2022), althea (Lattier and Contreras, 2022), among others.

Mutation breeding involves exposing plants or plant parts with meristems to physical mutagens such as gamma radiation or chemical mutagens such as ethyl methanesulfonate. There is a long history of using mutation breeding in a wide variety of ornamental crops (reviewed Melsen et al., 2021) and recently we showed efficacy of gamma radiation to reduce seed set of *Galtonia candicans* (Contreras and Shearer, 2020).

Testing for fertility

Regardless of the method of reducing seed production, it is important to properly evaluate plants in an appropriate region(s) and using appropriate methods. We have shown that interspecific hybrids of *Buddleja* are not necessarily less fertile than cultivars of *B. davidii* (data not shown) and Phillips et al. (2016) demonstrated that triploid pears ranged from 0.74% to 13.6% fertility compared to fertile diploids. Thus, the breeders' job is not done when generating a mutant, a hybrid, or a triploid – proper testing is crucial.

To address the issue of testing and introducing seedless or nearly seedless cultivars of weedy/invasive species, we are assembling a working group of individuals from Oregon, North Carolina, Florida, Michigan, and perhaps additional states/regions to prepare a white paper to coordinate and lead on a consistent set of guidelines for testing. The impetus for this is the number of new cultivars that are being introduced that exhibit reduced fertility but there is no coordinated set of rules for evaluating them. This leads to situations in which a plant may be banned in one state but allowed in the neighboring state. While we are not calling for national regulation of species through top-down legislation, we do want a national discussion and framework for breeders and growers to have a common set of standards for evaluation. Interstate commerce is a hallmark of our industry, and we need to have a common understanding of targets, if not common regulation for specific species.

Here is not the place to establish guidelines but there are some general rules that should be followed for testing. First, it is important to document flowering. That may seem obvious but, in some cases, plants can be very slow to flower in a particular region and may give the false impression of sterility. For instance, we have generated many triploid Norway maples over the past 10 years that have yet to flower. These cannot be said to be sterile on the basis that they have not flowered. Plants should be tested in the region where the end user will be growing them. Amur maples from my program have been shown to set no viable seed in Corvallis, OR but considering they will (we hope) be planted in landscapes of the upper Midwest and New England, it is important that we document their reduced fertility in those regions. Plants should be tested under appropriate conditions. I have observed that when irrigated, butterfly bush cultivars may continue flowering and setting seed until frost – often toward the end of October. However, during 2022 we withheld irrigation and plants were nearly done flowering by the first week of September. This change in phenology will have an obvious impact on seed production. Cultivars being tested should be replicated. This means to ascertain reliable information that there needs to be multiple plants of the cultivar being tested, ideally separated into repeated blocks. In cases where plants are insect pollinated, there should be presence of pollinators documented. For all plants, regardless of sex expression (perfect flowers, monoecy, dioecy, etc.), there should be fertile pollinizers of different genotypes present to prevent incompatibility leading to artificially reduced fertility.

Current regulations and the future

There are a number of states that are now working with the industry to adopt amendments that exempt specific cultivars where species previously were banned. In Oregon, OAR 603-52-1200 banned *Buddleja davidii.* This was enacted in 2004, and later in 2009 it was amended to allow for cultivars that exhibit 98% reduction in fertility or were confirmed to be hybrids. The amendment has allowed for the introduction of 14 cultivars to be grown and sold in Oregon. As previously stated, we have found that some hybrids exhibit substantial fertility on par with traditional cultivars. In Ohio there is approval to grow seedless flowering pear with the stipulation that it cannot be grafted onto seedling *P. calleryana.* Based on this common-sense regulation, I expect similar rules may be put in place for budded maples such as Norway maple. There are four barberry (*Berberis thunbergii*) cultivars approved in New York along with two *Miscanthus sinensis* and two *Euonymus fortunei* cultivars. I found an online document for request to exempt a specific cultivar from CMR 01-001 Chapter 273 regulation – while I do not have specifics on their threshold, it provides another example for which state regulators are open to exemptions in whole species bans.

The stakes for such amendments and exemptions are high. It seems to me that if these conditional approvals are done well that we can benefit all parties involved; plants will remain in cultivation where planted and not impact native ecosystems, growers will benefit from economically important taxa, and end users will benefit by resilient and beautiful plants. On the other hand, if it is found that amendments are failing to control new invasions or are found to continue exacerbating the situation, I fear these will be held up as data that cultivar approvals for these species are ineffective and this could endanger future opportunities to introduce cultivars that truly seedless (or nearly so). I believe we need an approach in which the industry is willing to give up cultivars that are weedy, departments of agriculture and other regulators should continue their collaborative approach, and we need to keep generating sound data to help guide regulation and outreach. If we can achieve a set of rules that are stringent enough to prevent future escape but allows cultivars that present no ecological threat then we can have profitable production, resilient landscape plants, while protecting our native ecosystems.

LITERATURE CITED

Brand, M.H. and Durocher, S.N. (2022). Four sterile or near-sterile cultivars of Japanese barberry in three foliage colors. HortSci. *57*:581–587.

Contreras, R.N. and Hoskins, T.C. (2020). Developing triploid maples. Horticulturae *6(4)*:70. https://doi.org/10.3390/horticulturae6040070

Contreras, R.N. and Shearer, K. (2020). Exposing seeds of *Galtonia candicans* to ethyl methanesulfonate (EMS) reduced inflorescence height, lodging, and fertility. HortSci. *55*:621–624. https://doi.org/10.21273/HORTSCI14775-19

Lattier, J.D. and Contreras, R.N. (2022). Flower forms and ploidy levels impact fertility in althea. HortSci. *57*:558–570. https://doi.org/10.21273/HORTSCI16478-21

Melsen, K., van de Wouw, M., and Contreras, R.N. (2021). Mutation breeding in ornamentals. HortSci. *56*:1154–1165. https://doi.org/10.21273/HORTSCI16001-21

Oates, K.M, Ranney, T.G., Touchell, D.H., and Viloria, Z. (2014). *Campsis ×tagliabuana* 'Chastity': A highly infertile triploid trumpet vine. HortSci. *49*:343–345

Phillips, W.D., Ranney, T.G., Touchell, D.H., and Eaker, T.A. (2016). Fertility and reproductive pathways of triploid flowering pears (*Pyrus* sp.). HortSci. *51*:968–971.

Trueblood, C., Ranney, T.G., Lynch, N., Neal, J., Olsen, R. (2010). Evaluating fertility of triploid clones of *Hypericum androsaemum* L. for use as non-invasive landscape plants. HortSci. *45*:1026–1028.

Combined Proceedings IPPS 72:320-325. 2022.

The Cutting Cooler Journey

Rose Daly

North Creek Nurseries, Inc., 388 North Creek Road, Landenberg, PA 19350 U.S.A.

Rose@northcreeknurseries.com

Keywords: Cutting storage cooler, humified cooler, quality liners, cooler construction

Summary

I will be covering our process on how we improved our cutting storage. I will talk about our old cooler, its challenges and what inspired us to change. Then I will segway into our new and improved humidified cooler and the benefits we have gained from the switch. Lastly, we will go into detail about how we built our humidifier and touch on some of the mistakes we learned along the way.

INTRODUCTION

First, let me introduce myself: I am Rose Daly from North Creek Nurseries in Landenberg Pennsylvania. We are a wholesale propagation nursery specializing in perennials, ornamental grasses, ferns, vines, and shrubs with an emphasis on Eastern US native plants. I started as a grower in December of 2018 and got promoted to Production Coordinator in January of 2022 in our Landenberg facility. My primary job is to oversee our production line and aid them in all processes to produce quality liners for our customers.

One of the most important tools for achieving a quality liner is where you store your propagules before they get stuck on the line. You want your propagation types whether it be unrooted/rooted cuttings or tissue culture to be stored for the shortest amount of time possible with consistent cool temperatures and 100% humidity. We were trying to achieve this with a commercial drink cooler. Yes, I said it, a commercial drink cooler!

Even though we were achieving quality flats with what we had, the road to the result was challenging.

Our old cooler- the challenges

We struggled with inconsistent temperatures constantly. The cooler couldn't cool down the cuttings fast enough coming in from our stock houses. With limited space, we would always be at maximum capacity. It was a game of balancing the storage between our inhouse cutting and our over shore buys. Each week we would be stressed about what we should stick first, what couldn't hold over the weekend and what took priority over other cuttings. When we got behind schedule sometimes cuttings started to fail in storage. Our attempts at hydration were subpar, misting inside the bags, tying them shut and poking holes in the sides, trying to create a humidity chamber. The environment we were supplying for our cuttings wasn't ideal, but we made it work with what we had.

Our cuttings relied heavily on our growers to bring them back to full turgidity with over head misters, but it came with a cost. Over misting, melt down, delayed or uneven rooting to name a few.

Stressed cuttings were also more at risk for pest and pathogen attacks due to their weakened immune systems. Something had to change.

Our inspiration

On a rainy day, our growing team got together to watch a presentation on YouTube by Ball Tech on Demand: "Success with Cuttings and Proper Storage = Happy Cuttings!" I highly encourage you to look up this talk series on YouTube. They covered research on the correlation between humidity, temperature, and successful rooting. They found that supplying 100% humidity in an open storage system at ideal consistent cool temperature would greatly increase turgidity and in turn, encouraged uniform rooting. We were amazed at the results from their experiments. After lot of thought and planning we decided to apply the idea to our own nursery practices.

Our new cooler

We decided on a walk-in fridge called a Norlake Kold Locker Indoor Walk-in Cooler with a Russel AC unit (**Fig. 1**). We then humidified the inside with our own home-made customized humidifier that I will go into detail about later. The dimensions of the cooler sit at 6 ft × 10 ft × 7 ft. It has a digital screen that displays the current temperature, a light switch, and a thermostat that consistently holds our temps between 45 and 52°F which is ideal for perennial cuttings.

We then added three wire shelves off Amazon to create storage inside for our cutting bags (**Fig. 2**). The bags are stored completely open, opposite of the old cooler's closed bag system.

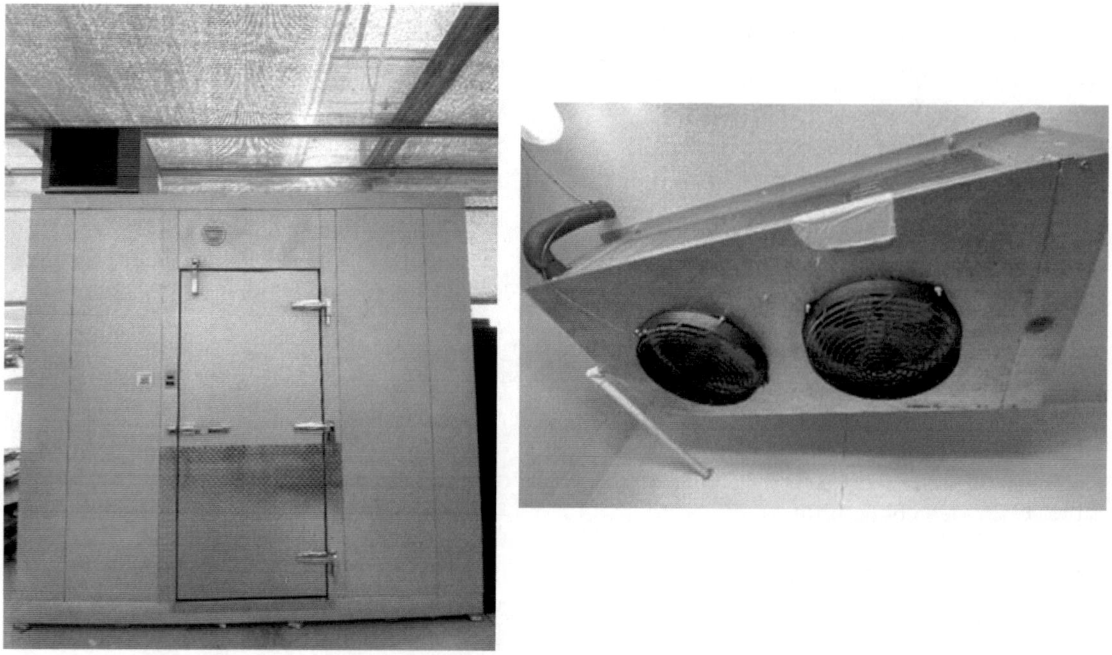

Figure 1. Norlake Indoor walk-in cooler (left) with Russel AC unit (right). www.webstaurantstore.com https://russel.htpg.com/

Figure 2. Wire shelving- Amazon.com.

How we built our humidifier, how it works, and our results

First, we tapped into one of our main water lines in our production line bay with PVC piping and ran it through a drilled hole on the side of our cooler (**Fig. 3**). This pipe is then connected to a 5-gal Lowes bucket. At the end of the PVC pipe inside the bucket, we attached an automatic float valve that stops the bucket from over filling. We wanted to have the bucket automatically refill to avoid employees having to refill it.

Figure 3. PVC tubing – Lowes.

Next, we bought a fan and a misting machine from a company called The House of Hydro (**Fig. 4**).

Figure 4. A 3-disc mist maker with float – The House of Hydro.com.

On the lid of the bucket, we cut a square and installed the fan. Inside the bucket, we placed the floating mister machine on the water surface. Essentially building a fogging machine, last we installed a white plastic PVC tube in the lid adjacent from the fan, so the mist would then leave the bucket into the cooler (**Fig. 5**). We then plugged the fan and the mister machine into a Plusmart outdoor waterproof timer (**Fig. 6**). The timer had a simple pin system that allows us the ability to program how many shots of mist and how long of a duration we want it to run for. The timer is then plugged into an extension cord that runs to a GFCI outlet with a waterproof cover (**Fig. 7**). We can adjust the mist ratio with a dial that came with the fan.

Figure 5. Our humidifier with a white plastic PVC tube in the lid adjacent from the fan, so the mist would then leave the bucket into the cooler.

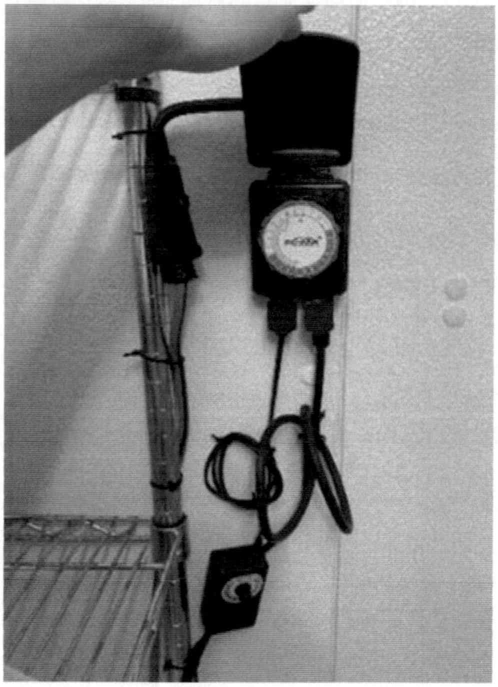

Figure 6. Plusmart Heavy Duty 24 HR outdoor timer – Amazon.com

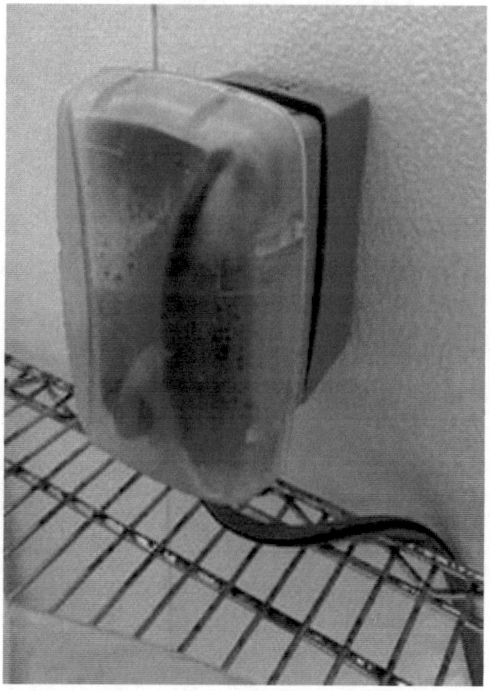

Figure 7. Waterproof outlet cover – Amazon.com.

How the humidifier works: The fan intakes air from the cooler. The air current passes over the mist machine carrying the mist out the tube, humidifying the air. Our device is working so well that when you open the door to the cooler it is completely foggy inside.

Over the next few months, we saw our cuttings health improve significantly. The cuttings have little to no wilt when they come out of the cooler, making it easier to stick on the line which increased production speed. With increased health and turgidity, Cuttings are less stressed, and growers can apply lighter mist which has decreased the risk of meltdown, pathogens, and pest pressures attacks.

Across the board, we are seeing quicker and more uniform rooting. This not only benefits our customers but us as well with faster crop turn over, less unmet ready dates and most importantly quicker picking of orders and less consolidations. The new cooler has also extended our storge limits allowing us to efficiently plan out our week with less worry.

Learning from mistakes

But along with every success comes mistakes to learn from. The biggest lesson we learned came at a chilly cost! Waterproof your circuit boards and thermostat of your cooler extra well. Over a few months, a small drip successfully corroded a circuit board, and the cooler ran all night long, dropping the temps from 50 to 14°F! We arrived in the morning to frozen cuttings. What a nightmare! We quickly learned that some extra proofing was needed, and it is of good practice to have two thermostats in the unit, one as back up, if the first should ever fail again.

Combined Proceedings IPPS 72:326-329. 2022.

Steps to Success with Bareroot Liner Herbaceous Perennials

Laura Robles

Walters Gardens, 1992 96th Ave., Zeeland, Michigan 49464 U.S.A.

Laura.robles.0180@waltersgardens.com

Keywords: field production, breeding, bareroot liners

Summary

Walters Gardens founded in 1946 is a major producer of field-grown herbaceous perennial liners and has an extensive breeding program. The field production program is discussed.

INTRODUCTION

Walters Gardens was founded in 1946 in Zeeland, Michigan by Dennis and Harriet Walters. While it started out with only 5 acres of bareroot field production, it has grown to encompass over 1500 acres of fields and 12+ acres of greenhouse, with construction underway on four additional acres. In 1976, Walters Gardens started one of the first tissue culture labs in the industry, which helped to establish it as a leader in perennial plants, and has also been instrumental in the Walters Gardens breeding program. Walters Gardens partnered with Proven Winners in 2010 and provides the majority of the herbaceous perennial genetics for that program.

Currently, Walters Gardens is a grower of perennial liners and has an extensive breeding program under the direction of John Walters and Hans Hansen.

BAREROOT LINERS

Field Production

Of our 1500+ acres of field space, about 1200 is actively farmed. We rotate crops to rebuild our soils, so in any given year we typically have just under 400 acres planted in perennials (**Fig. 1**).

Figure 1. Field production herbaceous perennials at Walters Gardens

Field crops are started mainly from small 72-count liners planted directly into the fields in spring, with planting beginning in late March and wrapping up by early July. The liners are planted using modified vegetable transplanters, all GPS driven (**Fig. 2**). The harvest begins in late fall and the majority of crops are only in the field for one season before being harvested the same year as they were planted, with a few exceptions that are grown for two or more seasons before harvesting.

Hibiscus is a spring-dig only, as well as fresh dug lavender and other crops to supplement the fall harvest.

Walters Gardens worked with an engineer to design equipment specialized for digging perennials in the sandy soils, and we now have multiple fleets for harvesting our 12-million+ plants that were planted this year in our fields.

Figure 2. Liners are planted using modified vegetable transplanters.

Benefits of Bareroot Liners

Our largest grade bareroot liners come in 25 to a box, which is a nice small quantity that our smaller customers appreciate. Most of the bareroot crops are stored frozen after being harvested and those that aren't frozen are stored in coolers, so they are dormant and acclimated to cold temperatures, which makes them much less susceptible to shipping damage during the winter months than actively growing liners are. Due to this cold storage and dormant state at time of planting, bareroot plants can be grown cooler than many plugs or liners, which leads to energy savings for the grower. They are also a great option for going directly outside in early spring when temperatures are still cold outside but inside heated space is at a premium for many customers. Bareroot liners are also easy to schedule due to the fact that they have been vernalized and will bloom the first year after planting. We have done a lot of research at Walters Gardens to determine finish times at various temperatures. We hold the bareroot liners in freezers/coolers from week 2 to week 26 (beginning of January to end of June), so this allows us to ship bareroot liners to our customers throughout this whole time, enabling customers to cycle plant rather than having to guess at a total number needed from the beginning. A bareroot liner planted later in the spring often catches up to and surpasses earlier planted material due to higher light levels and temperatures, and it also grows to fit the container (Goldfish effect), so a G1 bareroot will fill out a 2-gal container just as well as a 1-gal container with most crops.

Bareroot Liner Inputs and Sizes/Grades

All bareroot liners have been vernalized with the cold treatment beginning in our fields and finishing in our freezers/coolers. Most of the bareroot crops are either a 1-year plant or generously graded divisions of a 1-year plant (**Fig. 3**), with a few exceptions that are held for a longer time in the fields in order to size up properly.

Figure 3. Example of an herbaceous perennial bareroot liner.

While most of the bareroot liners come in the largest G1 size, we do offer smaller G2 and G3 sizes on select crops (**Fig. 4**). These are ideal for using in smaller containers and come in larger quantities per box (100/box G2, 250/box G3).

Handling Bareroot Liners Upon Arrival, Planting

When receiving bareroot liners from Walters Gardens, there are a few steps to take to ensure that the best quality is maintained. Boxes should be opened and inspected to check that roots are firm and relatively dry. Bareroot liners can be stored at cool temperatures (30–45°F) for a few days if planting immediately is not possible, although there are certain crops that should take priority to be planted as soon as possible. Bareroot liners may be trimmed to fit the container, and care should be taken to prevent air pockets around the roots. Various crops have different requirements in terms of what depth the roots should be planted at, and then the best cultural methods should be used to ensure a healthy crop after planting.

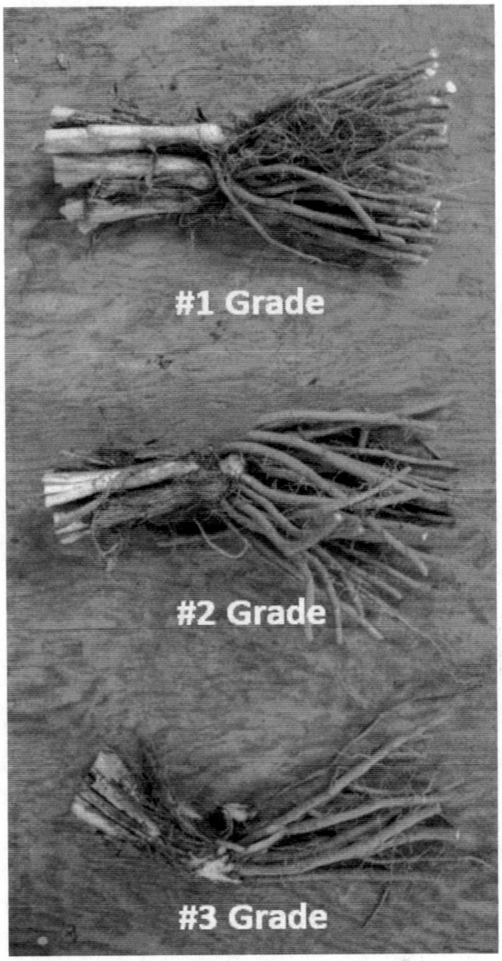

Figure 4. Bareroot sizes and grades.

Combined Proceedings IPPS 72:330-339. 2022.

Are We Propagating Plant Diseases?

Margery Daughtrey

Section of Plant Pathology and Plant-Microbe Biology, School of Integrative Plant Science, Cornell University, Long Island Horticultural Research & Extension Center, Riverhead, New York 11901 U.S.A.

mld9@cornell.edu

Keywords: pathogens, fungi, oomycetes, bacteria, viruses, nematodes, boxwood blight, downy mildews, rose rosette, phytoplasmas

Summary

In this presentation the author discusses examples of a few of the currently notable diseases caused by each of the major groups of pathogens (fungi, oomycetes, bacteria, viruses and nematodes). In the process, the author touches on boxwood blight, downy mildews, rose rosette and other diseases caused by emaraviruses, phytoplasmas, beech leaf disease, and the new vascular streak dieback disease.

INTRODUCTION

What the nursery industry does best is propagation. Over and over again, they start with a bagful of cuttings or a packet of seeds and transform them into a hoop house or a field of saleable plants, making much from a little.

The supervised growth that plants get during their time in the nursery allows a strong start for future landscape plants, attending to their needs for substrate, water and fertilization.

Exclude Pathogens from Propagation

But one of the secrets to success during propagation is keeping the pathogens (disease causing micro-organisms) out of the system. Their impact is greatest when they are there from the start. Pathogens are intimately related to the plants they parasitize and often hitchhike on them without causing obvious symptoms. Bringing many closely related plants into one tightly-spaced growing area sets up the very monocultures that we chide landscape architects for promoting. Optimal conditions for the plants are often also optimal for the pathogens — so, yes, we are often inadvertently propagating the pathogens along with the plants. The best nurseries struggle to perfect their plant and substrate choices, their irrigation methods, their fertility and their sanitation practices in order to grow the crops without favoring the pathogens.

The plant propagation phase is one of the most critical to manage, because plants are close together and irrigation is almost exclusively from overhead. Keeping the pathogens out of the propagation phase of production is critical for maximizing crop quality and minimizing disease management actions later on. The less the grower propagates plant diseases, the more successful the crops will be. Here are some general principles to follow:

- Exclude the pathogen. Be very fussy about the quality of plant material you bring into the nursery and practice careful sanitation. Reusing containers without first disinfesting them will lead to pathogens cycling through the nursery year after year.
- If you have your own stock plants, give them special care and be sure to leave field soil in the field as much as possible: take cuttings high on the plants and avoid the lowest mud-splashed branches. Bringing in cuttings from other nurseries means trusting the other establishments to have high sanitation standards and careful disease management. Cultivate relationships with reliable suppliers, rewarding them with your business.
- In a few specialized cases paying for treated seed may be worthwhile, or even treating it yourself. If you are an herb grower, for example, having your basil seed steam treated will lower the risk of early-season downy mildew losses. If you produce ornamental cabbages and kale, a hot-water treatment of seed done on premises can spare you losses from black rot, which is a bacterial disease caused by the bacterium *Xanthomonas campestris* pv. *campestris*.
- Train workers to inspect all plant material prior to and during sticking, discarding the obviously diseased propagules and perhaps sending samples off to a diagnostic lab. Roguing out the diseased cuttings can reduce future headaches considerably (**Fig. 1**). Keep scouting for disease and insect or mite problems throughout production.

Figure 1. Symptoms of wilting and cankering caused by the fungus *Cylindrocladium scoparium* on rooted cuttings of azalea.

In this presentation we will discuss examples of a few of the currently notable diseases caused by each of the major groups of pathogens (fungi, oomycetes, bacteria, viruses and nematodes).

We'll touch on boxwood blight, downy mildews, rose rosette and other diseases caused by emaraviruses, bacteria (*Xanthomonas* spp.) on asclepias and geranium, phytoplasmas on spirea and echinacea, beech leaf disease, and the new vascular streak dieback disease.

Fungi

Boxwood blight. The boxwood blight disease has been in North American since 2011, but it was known in Europe since 1994 (Daughtrey, 2019). One fungus, *Calonectria pseudonaviculata*, causes the disease in the U.S.A. and Canada, but a second closely-related fungus, *C. henricotiae*, is found in a handful of European countries. It is to our advantage to avoid importing *C. henricotiae*, because it has less sensitivity to some of the fungicides used to manage boxwood blight and also has greater heat tolerance. Shrubs, hedges and topiary are all vulnerable, but the low hedges of an herb garden are especially likely to become infected. A consortium of researchers and industry advisors called BBIG, the Boxwood Blight Insight Group, was formed around an SCRI grant a few years ago; members collaborate with scientists around the world as well as with the boxwood industry and AmericanHort. The Horticultural Research Institute, HRI, manages the BBIG website, www.boxwoodhealth.org, where information and educational presentations, publications and illustrations are stockpiled for everyone's use. Quarterly international seminars are developed and the talks are stored at the website.

Boxwood blight causes tan leaf spots with diffuse black borders, tiny black shoot cankers, and defoliation (**Fig. 2**).

Figure 2. Black cankers and leaf blight caused by *Calonectria pseudonaviculata* on *Buxus sempervirens*.

Pachysandra and sarcococca are also hosts. Periodic fungicide applications and pruning out diseased areas allow maintenance of valuable boxwood plantings under Long Island weather conditions. Most commonly the contact fungicides chlorothalonil or mancozeb and the systemic fungicide propiconazole are used to keep boxwood healthy. Genetic resistance is being used where available in the trade and new plants are under development. New blight-resistant plants from the NewGen series ('Freedom' and 'Independence') are performing well against boxwood blight in university trials and also have boxwood leafminer resistance. A plant breeding effort in Belgium has led to the BetterBoxwood series that should be available in 2023. Symptoms have been determined to develop most easily at 77°F for *Buxus sempervirens* (American boxwood) and *B. sempervirens* 'Suffruticosa' (English boxwood), both highly susceptible. Recently it was reported that for the usually resistant cultivar 'Winter Gem' (Japanese boxwood), temperatures of 59°F markedly increase disease susceptibility (Weiland et al., 2022). Going forward, the research community should investigate how individual cultivars of boxwood interact with *C. pseudonaviculata* and other important fungal pathogen(s) so that boxwood cultivar recommendations may made appropriate for each region of the country.

Oomycetes

Downy mildews. Fortunately, not every crop is plagued by a downy mildew. Some perennials are notorious for them (for example, geum, veronica, phlox, rudbeckia) and a few woodies (notably roses and viburnums) as well. Rudbeckias have had the greatest problems in recent years, showing downy mildew symptoms of purpling or yellowing on the upper side of leaves, and white sporulation on the undersurface of infected leaves (**Fig. 3**).

Figure 3. Purpling on the upper leaf surface and white sporulation of downy mildew on rudbeckia.

Among annuals, the most common hosts of diseases in this group are impatiens, sunflowers, coleus and pansy. The Beacon and Imara XDR series of impatiens have been developed by Ball and Syngenta, respectively, to answer the huge economic threat posed by the downy mildew of impatiens that became widespread beginning in 2011. These plants are highly resistant, but not entirely immune (Daughtrey et al., 2021), so they should be grown away from any susceptible impatiens or else protected with fungicides during production. Impatiens resistant to downy mildew are also being developed by Dr. Mark Bridgen at Cornell's LIHREC. A broad basis of resistance is desirable so that the pathogen (*Plasmopara destructor*) will not be easily able to adapt to the improved plants.

Phytophthora. Different species of *Phytophthora* are problematic on many nursery crops, especially those in the Ericaceae. Pieris, azalea, rhododendron and *Hedera* are some of the common hosts (**Fig. 4**).

Figure 4. Browning foliage and dieback are typical symptoms of *Phytophthora* blight during ivy propagation.

A recent addition to the list is *Phytophthora chrysanthemi*, first described from Ohio only 5 years ago (Lin et al. 2017). This pathogen is favored by high temperature conditions (30°C and above) and infection results in stunting, purpling and death of foliage, often seen on only part of a plant. ImmunoStrip® tests for *Phytophthora* (Agdia, Inc., Elkhart, Indiana) can be used to get an initial indication that this pathogen (rather than *Pythium* or *Fusarium* spp. or a bacterial infection) is responsible for symptoms in production or landscape, and then a diagnostic lab can confirm with culturing at 30°C.

Viruses

Emaraviruses. A new genus of viruses known as Emaraviruses was established in 2012, with the name coming from one of its members, European Mountain Ash Ring-spot-associated Virus (EMARaV) (Mielke-Ehrel and Muhlbach 2012). The disease caused by an emaravirus that is best known to growers of ornamentals is rose rosette caused by Rose rosette virus (RRV) (Windham et al. 2019). There are a few other diseases in this grouping, including fig mosaic and raspberry leaf blotch, but more remain to be discovered. One has been reported from spicebush, *Lindera benzoin*, causing chlorosis (Mollov et al. 2019), and we have recently seen ringspot symptoms on Norway maple that we suspect might be an emaravirus. One trait that a number of the viruses in this genus have in common is that they are spread by tiny eriophyid mites. This is true for Rose rosette virus, spread by the mite *Phyllocoptes fructiphilus*. Symptoms on roses include reddened shoots, hyper-thorniness, witches'-brooms, and decline and death (Windham et al. 2019). Note that rose rosette affects multiflora rose (**Fig. 5**) as well as desirable crops including Knockout ® Rose. Either eliminate these plants from hedgerows around the nursery or else scout them for disease symptoms in the same way that you scout your rose crops. Discarding diseased specimens is the only control, so careful monitoring is important.

Figure 5. A multiflora rose with tiny strap-shaped leaves found adjacent to a nursery with rose rosette in their container rose crop.

Bacteria

Xanthomonas. Various bacterial diseases caused by species in the genus *Xanthomonas* are occasional problem in the nursery industry. Virginia and Ohio have both reported a *Xanthomonas* leaf spot on peony since 2009, and we have also seen this several times in New York (Oliver et al. 2012; Klass et al. 2019). With overhead watering, plants will show reddish purple-rimmed spots on leaves and stems. Early detection is important, because discarding infested plants is the best policy. Treatments such as copper and *Bacillus subtilis* materials are not strongly effective.

Xanthomonas hortorum pv. *hederae* is commonly found on English ivy and other ivies but is frequently confused with the common anthracnose (caused by a fungus, a *Colletotrichum* species). Copper materials will help to reduce both problems while you are seeking a proper diagnosis from a laboratory.

Black rot on ornamental cabbage and kale (*Xanthomonas campestris* pv. *campestris*) is the same disease fought by farmers working with food crops. The most common symptom is Vee-shaped chlorotic wedges showing black veins, which eventually turn necrotic. Seed transmission makes this disease far too common (Daughtrey 2021). Growing plants under a greenhouse roof is ideal to eliminate rainfall and rain splash effects but irrigate early in the day so that the foliage does not sit wet for long periods and scout to eliminate symptomatic plants that could be a source of infection for others. Hot water seed treatments have been developed for cabbage and kale crops but could be more difficult to achieve safely in seeds of ornamental crops: research is needed.

Geranium sanguineum and some other hardy geranium species can be a source of Xanthomonas bacterial blight for greenhouse *Pelargonium* crops, which are highly susceptible to *Xanthomonas hortorum* pv. *pelargonii*. This spring problems with infected cuttings brought in from offshore led to widespread greenhouse geranium losses of zonal and ivy geraniums. This re-emerging of a familiar disease represents a breach in the otherwise effective clean stock procedures which were set up by the international geranium industry to avoid shipment of *Ralstonia solanacearum* Race 3, Biovar 2, a Select Agent, into the United States. Prioritization of sanitation procedures is needed for this important crop. Clean stock programs must be followed to the letter because breaches in plant health security at the top can cause huge losses down the production chain. Agdia ImmunosStrip® tests (Agdia Inc., Elkhart, IN) were helpful this season for quickly identifying the diseased plants. Symptoms begin as small round spots or necrotic wedges at the edge of leaves but will progress to plant death under warm-temperature conditions (Nameth et al. 1999).

Milkweeds are susceptible to a wilt disease caused by a *Xanthomonas* (*Xanthomonas campestris* pv. *asclepiadis*) (**Fig. 6**).

Figure 6. Milkweed with foliar blight after inoculation with the bacterium *Xanthomonas campestris* pv. *asclepiadis*.

Flynn and Vidaver (1995) reported susceptibility of *Asclepias syriaca*, *A. speciosa*, *A. tuberosa*, and *A. erosa*, plus a hybrid of *A syriaca* and *A. speciosa*, but not *A. subulata*; symptoms have been noted in swamp milkweed (*A. incarnata*) in our diagnostic laboratory as well. Procedures for seed treatment have not been worked out for this crop, but symptoms are serious and the disease is very contagious.

Phytoplasmas

The most familiar host of the aster yellows phytoplasma (a phytoplasma is a phloem-dwelling bacterium) is echinacea (**Fig. 7**), although it is possible to see this pathogen affecting other plants in the Asteraceae as well.

Figure 7. *Echinacea* with flower deformation due to infection by the aster yellows phytoplasma.

Typically, the aster yellows phytoplasma, vectored by the aster leaf hopper, will cause stunting, yellowing and peculiar-looking flowers. Both phyllody (conversion of flower parts to leafy tissue) and virescence (greening of flower petals) can occur on aster family members (e.g., marigold, coreopsis and China aster) with aster yellows. Echinacea develops especially dramatic symptoms, and also may show a similar but less dramatic look when infested by a tiny eriophyid mite called the coneflower rosette mite. A diagnostic test for the presence of the phytoplasma may be needed to tell which agent is operating on the plants. Scouting for both of these problems on echinacea and removing them when they are found is the appropriate initial response.

Spiraea spp. are subject to two viruses, Spiraea leaf spot virus and Spiraea yellow leafspot virus. Neither of these was found to be responsible for the plant stunting, shoot proliferation (witches'-brooms) and tiny leaves that were noted in Minnesota a decade ago (Lockhart et al. 2012)—these severe symptoms were instead determined to be due to a phytoplasma (**Fig. 8**). Spirea infection by a phytoplasma had been noted earlier in New York (Griffiths et al. 1994) and this same organism was detected in Minnesota in ornamental *S. japonica* and *S.* × *bumalda* This spirea stunt phytoplasma disease is not uncommon but it generally goes undiagnosed and unreported. Good performance of spirea in plantings will require that this disease no longer be propagated in the trade.

Figure 8. Spirea in the landscape with tiny leaves, witches'-brooming and dieback caused by the spirea stunt phytoplasma.

Nematodes

Beech leaf disease. Beech leaf disease is a new problem caused by an invasive exotic nematode. Symptoms were first noted in Lake County, Ohio, in 2012 but it was not until 2017 that a paper was published showing the link between the symptoms and the presence of a new subspecies of nematode, *Litylenchus crenatae* ssp. *mccannii.* Nursery trade has been one of the inadvertent means of disease spread because symptoms on infested plants are easily overlooked; no vector has yet been identified but new counties and new states in its range are being detected every year. The nematode overwinters in buds and feeds on immature leaves before the buds unfold in the spring. In addition to shallow elongated leaf galls between veins, stunted and distorted leaves and bud death are characteristic of the disease and trees eventually are killed (**Figs. 9A and 9B**). Both the native American beech, *Fagus grandifolia* and the European beech, *F. sylvatica*, are susceptible to the disease. Effective treatments are being sought.

Figure 9. Leaf symptoms of Beech Leaf Disease, caused by the nematode *Litylenchus crenatae* ssp. *mccannii*. A) interveinal bands (leaf galls) seen from above B) view of lower leaf surface of same leaf.

Cause Unknown

Redbud vascular streak dieback. This disease appears to be new, but is it just becoming more noticeable because of stresses associated with global warming or changes in nursery practices (Beckerman et al. 2022)? The disease is causing concern in the southeast/mid-Atlantic area where much of the nursery stock for the Northeast comes from. No plant pathologist has yet demonstrated a definite cause-and-effect for a pathogen (through a process called demonstration of Koch's postulates) but evidence has mounted suggesting that the pathogen is *Rhizoctonia theobromae*. The fungus was previously considered a species of *Ceratobasidium*, *Oncobasidium* or *Thanetephorus*, and is listed as causing dieback in Asia and Australasia on cacao and avocado.

Within the nursery trade there have been reports of yellowing, stunting and severe dieback on redbud (Dismukes, 2022). Epicormic shoots (water sprouts) may form below the dieback, or the dieback may proceed down a branch into the main trunk and lead to death of the tree. The xylem of affected branches may show a dark streaking.

Plants from young to old are affected, in container and field production and in the landscape. Although the streaking suggests Verticillium wilt, no one has isolated a *Verticillium* species from the infected branches. A *Rhizoctonia* is isolated frequently, but this fungus might turn out to be an endophyte (just dwelling within the plant) rather than a pathogen. This pathogen suspect is reportedly difficult to culture and work with.

From the nursery perspective, it is important to avoid symptomatic material when grafting redbuds and to sanitize after each cut when pruning. There are no chemical control recommendations at this time but avoiding stresses such as flooding or growing at too low a pH is important for keeping stress out of redbud culture. Similar symptoms of vascular streaking have also been noted in calycanthus, dogwood, magnolia, red maple, spicebush and wax myrtle.

LITERATURE CITED

Beckerman, J., Creswell, T, Bonkowski, J. and Baysal-Gurel, F. (2022). Vascular Streak Dieback of Redbud: What Plant Pathologists Know So Far. Purdue University Landscape Report 22-11. https://irp.cdn-website.com/5d757b5b/files/uploaded/Redbud%20Vascular%20Streak%20Disease.pdf

Daughtrey, M., Beckerman, J., Davis, W.J., Rane, K., and Crouch, J.A. (2020). Corroboration that highly resistant impatiens cultivars are not immune to downy mildew diseases: a report of crop losses from two California producers. Plant Health Progress *21*:214-216. https://doi.org/10.1094/PHP-05-20-0040-SC

Daughtrey, M.L. (2019). Boxwood blight: Threat to ornamentals. Annu. Rev. Phytopathol. *57*:189-209.

Daughtrey, M. (2021). Examining Xanthomonas. GrowerTalks February. https://www.growertalks.com/Article/?articleid=25076

Dismukes, A. (2022). Vascular Streak Dieback in Redbud. TSU Nursery News to Use. Issue 33, March 8, 2022. https://www.tnstate.edu/agriculture/nrc/documents/NEWS2USE%20March%202022.pdf

Flynn, P., and Vidaver, A. (1995). *Xanthomonas campestris* pv. *asclepiadis*, pv. nov., causative agent of bacterial blight of milkweed (*Asclepias* sp.). Plant Dis. *79*:1176-1180.

Griffiths, H.M., Sinclair, W.A., Davis, R.E., Lee, I.M., Dally, E.L., Guo, Y.-H., Chen, T.A., and Hibben, C.R. (1994). Characterization of mycoplasmalike organisms from *Fraxinus*, *Syringa*, and associated plants from geographically diverse sites. Phytopathology *84*:119-126.

Klass, T.L., Long, J.J., Summers, J.A., Roman-Reyna, V., Koebnik, R., Jacobs, J.M. and Peduto-Hand, F. (2019). First report of bacterial blight of peony caused by *Xanthomonas hortorum* in Ohio. Plant Dis. *103(11)*:2940. doi.org/10.1094/PDIS-05-19-1123-PDN

Lin, S., Martin, D.E., Taylor, N.J., Gabriel, C. K. and Peduto Hand, F. (2017). Occurrence of *Phytophthora chrysanthemi* causing root and stem rot on garden mums in the United States. Plant Dis.*101(6)*. https://doi.org/10.1094/PDIS-12-16-1714-PDN

Lockhart, B., Mollov, D. and Voth-Hulshout, J. (2012). Association of spirea stunt phytoplasma with a disease of *Spiraea* spp. in Minnesota. Plant Health Progress. doi: 10.1094.PHP-2012-1023-01-BR/

Mielke-Ehrel, N., and Muhlbach, H-P. (2012). Emaravirus: A novel genus of multipartite, negative strand RNA plant viruses. Viruses MDPI *4(9)*: 1515-1536.

Mollov, D., Grinstead, S., Fuentes-Bueno, I., Creswell, T., Ruhl, G., Bush, E., Hansen, M.A., Beale, J. and Johnson, D. (2019). New emaravirus identified in spicebush (*Lindera benzoin*).
https://www.ars.usda.gov/research/publications/publication/?seqNo115=360928

Nameth, S.T., Daughtrey, M.L., Moorman, G.W., and Sulzinski, M.A. (1999). Bacterial blight of geraniums: a history of diagnostic challenges. Plant Dis. *83(3)*:204-212.

Oliver, C.L., Cai, R., Vinatzer, B.A., Bush, E.A., and Hansen, M.A. (2012). First report of bacterial spot of peony caused by a Xanthomonas sp. in the United States. Plant Dis. *96(4)*:581.
https://doi.org/10.1094/PDIS-11-11-0919

Weiland, J.E., Ohkura, M., Scagel, C.F., Davis, E.A. and Beck, B.R. (2022). Cool temperatures favor growth of Oregon isolates of *Calonectria pseudonaviculata* and increase severity of boxwood blight on two *Buxus* cultivars. Plant Dis.106:3100-30108. https://doi.org/10.1094/PDIS-04-22-0769-RE.

Windham, A., Windham, M. and Hale, F. (2019). Early detection of rose rosette disease. University of Tennessee Institute of Agriculture Fact Sheet. SP 806 03/19 19-0165 https://extension.tennessee.edu/publications/Documents/SP806.pdf

Combined Proceedings IPPS 72:340-344. 2022.

Impact of Foliar Applied Paclobutrazol in Combination with Auxin on Rooting and Subsequent Shoot Growth in *Angelonia* Cuttings

Anna G. Baloh, W. Garrett Owen, Robert L. Geneve

Department of Horticulture, University of Kentucky, Lexington, Kentucky 40546 U.S.A.

Rgeneve@uky.edu

Keywords: propagation, gibberellin, bedding plant, K-IBA

Summary

Angelonia cuttings were treated with a combination of Bonzi and K-IBA as a quick dip and foliar spray. The foliar auxin spray was not as effective as a basal dip at the concentrations used in this study. A foliar spray using Bonzi increased rooting with or without auxin. Bonzi did not have a significant impact on plant height post-rooting. This study provides initial evidence that a tank-mix of auxin and a gibberellin inhibitor used as a spray application could be an efficient and effective means for application in commercial cuttings.

INTRODUCTION

Angelonia (*Angelonia angustifolia*) is native from Mexico to Columbia (Winhelmann et al., 2018). It is an annual herbaceous plant that produces snapdragon-like flowers used for bedding or mixed containers. Angelonia is typically propagated by auxin-treated cuttings (Dole and Gibson, 2006). Auxin is commonly delivered to cuttings as a basal liquid or talc application (Davies et al., 2018). Foliar auxin spray application has become an alternative application method. that is a cost-effective alternative that can increase worker safety and rooting efficiency (Blythe et al., 2004; Drahn, 2007; Martindell, 2019).

Gibberellin is generally considered inhibitory to adventitious rooting (Davies et al., 2018) and exogenous gibberellin application to cuttings reduces adventitious rooting (Mauriat et al., 2014). Gibberellin biosynthesis inhibitors such as Bonzi (Paclobutrazol) have been shown to increase adventitious root initiation possibly by reducing the negative impact of endogenous gibberellin (Upadhyaya et al., 1986; Wiesman and Riov 1994). Bonzi can also reduce plant height (Davis et al., 1986). Bonzi applied to mungbean cuttings increased rooting and drastically decreased the hypocotyl length in mungbeans compared to the control (Bora at al., 1990).

Bonzi can be applied to cuttings as a foliar spray and therefore the objective of this study was to investigate the impact of Bonzi as a foliar spray alone or in combination with auxin on rooting and subsequent shoot growth in angelonia cuttings.

METHODS AND MATERIALS

Cuttings of angelonia 'Aria Purple' were obtained from Dümmen Orange, Columbus OH. There were nine treatment combinations including:

1. Untreated
2. K-IBA at 1000 mg · L^{-1} basal quick dip
3. K-IBA at 100 mg · L^{-1} foliar spray
4. Bonzi 5 mg · L^{-1} foliar spray
5. Bonzi 20 mg · L^{-1} foliar spray
6. K-IBA at 100 mg · L^{-1} and Bonzi 5 mg · L^{-1} foliar spray
7. K-IBA at 100 mg · L^{-1} and Bonzi 20 mg · L^{-1} foliar spray
8. K-IBA at 1000 mg · L^{-1} quick dip and Bonzi 5 mg · L^{-1} foliar spray
9. K-IBA at 1000 mg · L^{-1} quick dip and Bonzi 20 mg · L^{-1} foliar spray.

Cuttings were stuck in 3 parts Promix / 1 part perlite (v/v) in six-packs. Untreated and basal quick dip cuttings were moved directly to the mist bench. Cuttings receiving a foliar spray treatment were treated prior to moving to the mist bench. Mist applied every ten minutes for 10 seconds. Foliar sprays were applied at 10 mL per six pack from a 50 mL spray bottle which covered the foliage to leaf drip.

Each six-pack was an experimental unit and there were six randomized replicates per treatment (36 cuttings per treatment). After three weeks, roots per cuttings were counted and rooting quality rated on a scale of 1 (poor root formation) to 5 (excellent root formation). The scale considered the length and secondary root formation. Once each cutting was evaluated for rooting, they were placed back into the six-pack cells in the greenhouse. After three weeks, plant height was measured from the container rim to the shoot apex.

RESULTS

Untreated angelonia cuttings had the lowest roots per cutting and the poorest root rating (Table 1). Cuttings treated with 1000 mg · L^{-1} K-IBA quick dip plus 20 mg · L^{-1} Bonzi foliar spray resulted in the highest roots per cutting and root rating quality. All quick dip treatment combinations resulted in higher roots per cutting and root ratings than the foliar spray application combinations (**Table 1**). Bonzi foliar spray alone at 5 or 20 mg · L^{-1} increased the number of roots per cutting compared to untreated cuttings by about two roots per cutting, but only Bonzi at 20 mg · L^{-1} increased rooting percentages (79.7 to 95.8%).

There was no statistical difference in plant height in any cuttings (**Table 2**). Untreated angelonia cuttings had lower but not significantly reduce height compared to treated cuttings (**Table 2**).

Table 1. Root formation in *Angelonia* cuttings treated with K-IBA and Bonzi as a dip or foliar spray.

Application method	K-IBA (mg · L^{-1})	Bonzi (mg · L^{-1})	Rooting percentage	Roots per cutting	Root rating
Untreated	0	0	79.7c	3.9d[x]	1.7b
Quick dip	1000	0	98.1b	13.1b	3.6a
	1000	5	100a	14.2b	3.6a
	1000	20	100a	18.7a	4.1a
Foliar spray	0	5	88.9bc	6.3c	2.9a
	0	20	95.8b	6.1c	3.1a
	100	0	100a	5.9c	3.2a
	100	5	100a	5.6c	2.6ab
	100	20	100a	7.1c	2.8a

[x]means in a column followed by the same letter were not different by Tukey at the 5% level.

Table 2. Plant height in *Angelonia* cuttings treated with K-IBA and Bonzi as a dip or foliar spray.

Application method	K-IBA (mg · L^{-1})	Bonzi (mg · L^{-1})	Height (cm)
Untreated	0	0	6.2
Quick dip	1000	0	7.5
	1000	5	7.3
	1000	20	7.7
Foliar spray	0	5	8.0
	0	20	7.3
	100	0	7.6
	100	5	7.5
	100	20	7.7

DISCUSSION

It is common for gibberellin inhibitors in combination with auxin application to increase rooting in cuttings (Davies et al., 2018). Many of the early studies used bean (*Phaseolus* and *Vigna*) model systems where auxin and a gibberellin inhibitor were applied to cuttings in a basal solution (Bora et al., 1991; Porlingis and Koukourikou-Petridou 1996; Upadhyaya et al., 1986; Wiesman and Riov, 1994). Later studies with woody perennials like *Ligustrum* (Šebánek et al., 1991), *Rhamnus* (Bañón et al., 2003) and *Delonix* (Abdi et al., 2009) reinforced the promotive effect on rooting as a basal treatment combination of auxin and a gibberellin inhibitor. The current study using an herbaceous perennial showed a similar promotive effect of a gibberellin inhibitor used as a foliar spray (Table 1). This provides initial evidence that a tank-mix of auxin and a gibberellin inhibitor used as a spray application could be an efficient and effective means for application in commercial cuttings.

There is also evidence that a gibberellin inhibitor application alone can increase rooting in cuttings (Davies et al., 2018). In the current study, Bonzi at the highest concentration did improve both rooting percentage and root number compared to untreated cuttings. The mode of action could be a reduction of endogenous gibberellin, which could be antagonistic to rooting (Davies et al., 2018). It is becoming more evident that endogenous gibberellin impacts adventitious root formation by altering polar auxin transport reducing the endogenous auxin concentration in the basal rooting portion of the stem (Mauriat et al., 2014). Reducing the endogenous gibberellin titer with a gibberellin biosynthesis inhibitor could be responsible for the increased rooting in cuttings treated with a gibberellin inhibitor like Bonzi (Nagy, et al., 1991; Šebánek, et al., 1991).

Bonzi applied as a foliar spray to cuttings did not play a significant role in controlling the height of the angelonia cuttings after rooting (Table 2). Bonzi typically acts as a growth regulator to reduce plant height in greenhouse crops (Goulston and Shearing, 1985). Additional research is needed to see if Bonzi has similar effects on other herbaceous perennials.

ACKNOWLEDGEMENTS

Cuttings were supplied as a generous gift by Dümmen Orange. The Horticulture Research Institute provided financial support to the foliar auxin research at the University of Kentucky.

LITERATURE CITED

Abdi, G. and Ascari-Raburi N. (2009). Enhancement of IBA, urea-phosphate, paclobutrazol and their combinations on rooting of royal Poinciana (*Delonix regia*) stem cuttings. American-Eurasian J. Agric. Environ. Sci. *6*:132-136.

Bañón, S., Martínez, J.J., Fernández, J.A., Ochoa, J. and González, A. (2003). Effect of indolebutyric acid and paclobutrazol on the rooting of *Rhamnus alaternus* stem cuttings. Acta Hort. *614*:263-267.

Blythe E. K., Sibley, J. L., Ruter, J. M. and Tilt K. M. (2004). Cutting propagation of foliage crops using a foliar application of auxin. Scientia Hort. *103*:31-37.

Bora, K.K., Munot, J. and Mathur, S., (1991). Metabolic changes associated with paclobutrazol induced rooting in hypocotyls cuttings of mung bean. Bot. Bull. Acad. Sin. *32*, 9-14.

Davies, F. T. Jr., Geneve R.L. and Wilson S. B. (2018). Hartmann and Kester's Plant Propagation: Principles and Practices. Boston: Prentice-Hall. Ninth edition.

Davis T. D., N. Sankhla, R. H. Walser, and A. Upadhyaya. (1986). Paclobutrazol- a promising plant growth regulator. Hormonal Reg. Plant Growth Dev. *3*:311-322.

Dole, J.M. and Gibson, J.L. (2006). Cutting propagation: A guide to propagating and producing floriculture crops. Ball Publishing.

Drahn S. R. (2007). Auxin application via foliar sprays. Proc. Intern. Plant Prop. Soc. *57*:44-48.

Goulston G. H. and Shearing S. J. (1985). Review of the effects of paclobutrazol on ornamental pot plants. Acta Hort. *167*:338-349.

Martindell, J.C. (2019). Foliar indole-3-butyric acid rooting hormone application and cost analysis. Proc. Intern. Plant Prop. Soc. *69*:165-168.

Mauriat, M., Petterle, A., Bellini, C. and Moritz, T. (2014). Gibberellins inhibit adventitious rooting in hybrid aspen and Arabidopsis by affecting auxin transport. Plant J. *78*, 372–384.

Nagy, M., Tari, I. and Bubá N, T. (1991). IAA distribution in the hypocotyls and primary leaves of *Phaseolus vulgaris* L. treated with paclobutrazol in relation to rooting capacity. Biochemie Physiol. Pflanzen, *187*, 447-451.

Porlingis, I.C. and Koukourikou-Petridou, M. (1996). Promotion of adventitious root formation in mung bean cuttings by four triazole growth retardants. J. Hort. Sci. *71*:573-579.

Šebánek, J., Klíčová, S., Králík, J., Psota, V., Vítková, H., Kudová, D. and Reinöhl, V. (1991). The effect of paclobutrazol on the level of endogenous IAA in relation to the rooting of cuttings and abscission of petioles. Bioch. Physiol. Pflanzen. *187*, 89-94.

Upadhyaya, A., Davis, T.D., and Sankhla N. (1986). Some biochemical changes associated with paclobutrazol-induced adventitious root formation on bean hypocotyl cuttings. Ann. Bot. *57*:309-315.

Wiesman, Z. and Riov J. (1994). Interaction of paclobutrazol and indole-3-butyric acid in relation to rooting of mung bean (*Vigna radiata*) cuttings. Physiol. Plant. *92*:608-612.

Winhelmann M.C., Grzeça, G.T., Emer, A.A., Tedesco, M., Paris, P., Paolazzi, J., Fior, C.S., and Schafer G. (2018). Rooting of apical cutting of *Angelonia integerrima* Sprengel: Concentrations of indole-3-butyric acid and substrates. Ornamental Hort. *24*:109-115.

Combined Proceedings IPPS 72:345-353. 2022.

Evaluation of Auxin (K-IBA) Concentrations on Rooting Success of maple (*Acer*) Stem Cuttings

Sharika Elahi[1] and Kim Shearer[2]

[1]Plant Breeding Intern, The Morton Arboretum, 4100 Illinois Route 53, Lisle, Illinois 60532 U.S.A.; [2]Curator of Collections, Manager of New Plant Development, The Morton Arboretum.

kshearer@mortonarb.org

Keywords: Auxin, adventitious rooting, *Acer*, vegetative propagation, cutting propagation

Summary

Initial experiments were conducted to evaluate auxin (K-IBA) concentrations on the rooting of selected *Acer* species. the results indicate that water-soluble K-IBA concentrations used in this study do not have a significant effect on rooting success in these seven maple selections after 6 to 7 weeks. Further rooting in all seven maple selections was observed 90 days after initiation, suggesting that greater rooting success could potentially be achieved with extended time in the propagation flats before transplanting.

INTRODUCTION

When selecting trees for the urban landscape, species that exhibit high adaptability and tolerance to tough conditions as well as attractive features are preferable. Many species of maples (*Acer*) have been selected and bred to meet the high demands of dense urban environments (Dirr and Heuser, 2006). A key limiting factor for the commercial availability of some desirable selections is the lack of reliable protocol for vegetative cutting propagation that can be applied on a large scale. Rooting success has been reported by both academic researchers and commercial propagators, however results are too broadly variable for the reported protocols to be considered practical for wide-scale commercial production (Brock, 2014). While other methods of clonal propagation exist, such as grafting or in vitro micropropagation, vegetative propagation from stem cuttings is generally considered the most convenient and cost-effective technique for propagating woody plants (Gabriel et. al, 1961).

Currently, maples are primarily propagated through grafting (Brock, 2014). While this method of propagation is effective on a large-scale, grafting is known to be expensive, labor-intensive, and time-consuming. This method requires technical expertise that is not always readily available in the labor force, and several years are required for grafted plants to reach maturity and produce seed (Tousignant et al., 2003). Another major limitation of grafting is that rootstocks do not always match the hardiness of the scion grafted onto them. Consequently, the broad adaptability of some species relative to cold hardiness and soil conditions may be lost, leading to delayed decline and failure. Additionally, grafted trees are prone to developing suckers that compete with the desired scion (Brock, 2014). If vegetative cutting techniques can be further improved and standardized, this method of clonal propagation has the potential to provide commercially viable selections of broadly adaptable trees for upper Midwest climates while also eliminating other issues associated with grafting (Dirr and Heuser, 2006).

The objective of this experiment was to examine how the following species and cultivars respond to vegetative propagation using varying concentrations of the water-soluble growth hormone Indole-3-butyric acid Potassium Salt (K-IBA): *Acer miyabei* 'Morton' State Street® (miyabe maple), *Acer platanoides* 919-61*1 (Norway maple), *Acer circinatum* × *A. pseudosieboldianum* 'Morton UW' Morning Starburst™ 644-81*1, *Acer pseudosieboldianum* 263-99*1 (Korean maple), *Acer triflorum* 269-2017*1 (three-flowered maple), and two selections of *Acer platanoides* × *A. truncatum* TBN 14.121, TBN 13.128 (purpleblow maple hybrid) (**Table 1**).

MATERIALS AND METHODS

Cuttings were collected from seven *Acer* selections (**Fig. 1**) growing in The Morton Arboretum living collections and tree breeding nursery on June 22 and 23, 2022. Samples were collected between 9:00 a.m. and 12:00 p.m. At the time of collection, daytime temperature ranged from 80°F to 85°F. Collection days followed two days of high temperatures reaching up to 98°F. Semi-hardwood cuttings of 4 to 8 in. were taken at random from the trees and placed into zip locked bags. All samples were stored in a walk-in cooler maintained at 42°F.

Table 1. Descriptions of species and cultivars selected for propagation study.

Species	Cultivar/ Common Name	Accession Number	Description
Acer miyabei 'Morton'	State Street® Miyabe maple	Tree Breeding Nursery (TBN)	- Zone 4-8 - Highly adaptable in urban environments - Upright, oval shaped, limited spread - Potential substitute for Norway maple
Acer platanoides	Norway maple	919-61*1	- Zone 4-7 in East - Zone 4-8 in West - Highly adaptable, shade and sun tolerant, heat and cold hardy - Tolerant of a variety of soils - Great shade tree - Criticized for being overplanted and invasive in the Northeast/Canada
Acer triflorum	Three-flowered maple	269-2017*1	- Zone 4-8 - Best in full sun and moist, acidic, well-drained soil - Consistent fall color - Does well in East, slower growing in West
Acer platanoides × *truncatum*	Purple blow maple hybrid	TBN-C RF 14.121	- Zone 4-8 - Heat and drought tolerant - Can survive heat dry summers of Midwest - Reasonably cold hardy - Tolerant of variety of soils
Acer circinatum × *A. pseudosieboldianum* 'Morton UW'	Morning Starburst™ maple	644-81*1	- Zone 5-8 (tentative, still evaluating) - Vivid crimson fall color
Acer pseudosieboldianum	Korean maple	263-99*1	- Zone 3-8 - Very cold hardy (can survive below -40 F) - Wide spreading
Acer platanoides × *A. truncatum*	purple blow maple hybrid	TBN-C RF 13.128	- Zone 4-8 - Valued as heat and drought tolerant tree - Can survive heat dry summers of Midwest - Reasonably cold hardy - Tolerant of variety of soils

Figure 1. (A) *A. miyabei* 'Morton' State Street® , stock block in tree breeding nursery; (B) *A. circinatum* × *A. pseudosieboldianum* 'Morton UW' Morning Starburst™ , 644-81*1 (C) *A. platanoides* × *A. truncatum* TBN 14.121; (D) *A. pseudosieboldianum* 263-99*1 (E) A. triflorum, 269-2017*1 (F) *A. platanoides* × *A. truncatum* TBN 13.128 (G) *A. platanoides*, 919-61*1; identification numbers preceded by "TBN" are identification numbers associated with individual plants in the Morton Arboretum tree breeding nursery, and other numbers are Morton Arboretum plant identification numbers associated with individuals from the living collections used in this experiment.

Samples were processed June 24, 2022. Cuttings from each individual were treated with varying concentrations of the water-soluble auxin Indole-3-butyric acid potassium salt (K-IBA). Treatment 1 (control) consisted of deionized water. Treatment 2 consisted of 5000 ppm of K-IBA (5 mg/mL). Treatment 3 consisted of 10,000 ppm K-IBA (10 mg/mL). Each treatment was replicated three times in separate 15.75 in. × 15.75 in. × 5 in. Anderson Deep Propagation Flats (Anderson Pots, Portland, Oregon). Each experimental unit was composed of eight cuttings taken from a with the exception of the *A. platanoides* × *A. truncatum* hybrids. For these genotypes, we only had six cuttings per experimental unit available.

Cuttings were stripped of all but 2 or 3 of the topmost leaves to limit evapotranspiration. Wounds were made by scraping the base of each stem with pruners deep enough to expose the cambium layer. Terminal leaves were dipped and swirled in fungicide for ~10 sec. The proximal end of each wounded stem was dipped into a K-IBA treatment for ~10 sec. Cuttings were air dried on newspaper for ~5 min before being inserted at a depth of 2 inches into the propagation substrate (**Fig. 2A**). Propagation substrate was composed of 1 peat: 2 perlite. Cuttings were placed 3 to 4 inches apart in rows of eight according to taxon, and six for both *A. platanoides* × *A. truncatum* hybrids (**Fig. 2B, 2D**). Flats were placed in an overhead fog room at random locations on greenhouse benches (**Fig. 2C**). Continuous fog maintained humidity during the day. Average temperature of the propagation house was ~73.4 °F (Average Max: ~77 °F; Average Min: ~53.6°F). Humidity was close to 100%. Fog was programmed to operate from 60 minutes after sunrise until 90 minutes prior to sunset.

Figure 2. (A) Cuttings being processed, (B) one replication, (C) flats placed in fog room, (D) overhead view of propagation flat.

Rooting results were evaluated 42–50 days after cuttings were initiated (**Fig. 3**). Samples were arranged in randomized block design. Data recorded included percentage of rooting, callusing, and no change/death.

Data on percent rooting, and callusing were subjected to a two-way ANOVA. Fisher's LSD was used for mean separation among taxa. Number of roots and individual root lengths were recorded for each cutting and averaged per treatment.

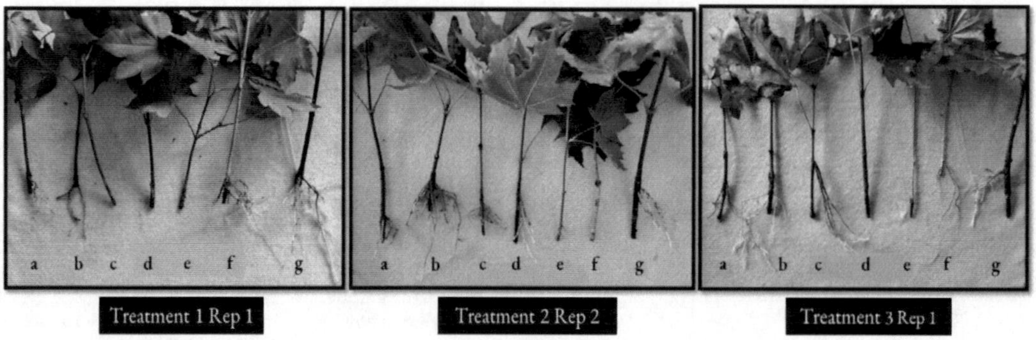

Figure 3. Rooting 60 days after initiation. One sample from each taxon for each treatment. Left to right: a) *A. miyabei* 'Morton', b) *A. platanoides* 919-61*1, c) *A. triflorum* 269-2017*1, d) *A. platanoides* × *A. truncatum* 14.121, e) A. 'Morton UW', f) *A. pseudosieboldianum* 263-99*1, g) *A. platanoides* × *A. truncatum* 13.128.

RESULTS AND DISCUSSION

Blocking had no significant effect on rooting or callusing percentages based on the two-way ANOVA (**Table 2**). It can be assumed that position in the fog room had no significant effect on rooting success. There was no significant difference in percent rooting or percent callusing between treatments. However, there was a significant difference in percent rooting and percent callusing when comparing between taxa ($p < 0.001$, α=0.0.5).

Our results indicate that water-soluble K-IBA concentrations used in this study do not have a significant effect on rooting success in these seven maple selections after 6 to 7 weeks (**Fig. 4**).

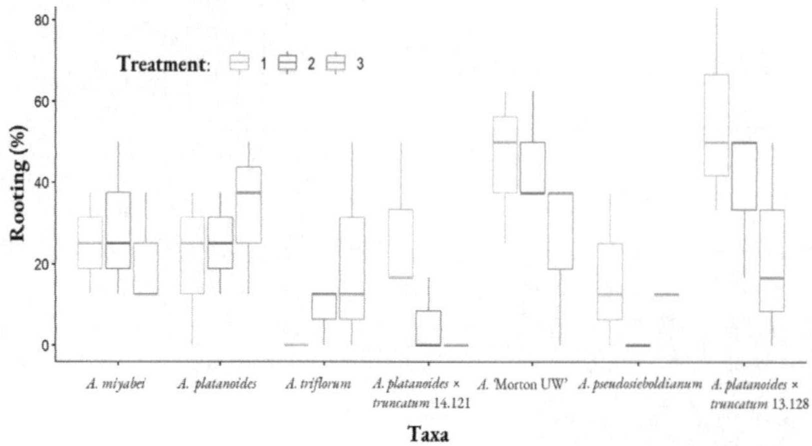

Figure 4. Percent rooting after 42-50 days for maple cuttings treated with 0 ppm (Treatment 1), 5000 ppm (Treatment 2), and 10,000 ppm (Treatment 3) of K-IBA depicted in box plots.

Table 2. Non-significant results after 6-7 weeks for percent rooting, percent callousing, mean root number, and mean root length of maple cuttings treated with varying concentrations of K-IBA.

Taxa	Treatment 1 (Control)				Treatment 2 (5000 ppm)				Treatment 3 (10,000 ppm)			
	Rooting (%)	Callusing (%)	Root No.	Root Length (cm)	Rooting (%)	Callusing (%)	Root No.	Root Length (cm)	Rooting (%)	Callusing (%)	Root No.	Root Length (cm)
A. miyabei **abc***	25.0	33.3	0.29	3.91	29.2	20.8	0.38	3.67	20.8	29.2	1	2.29
A. platanoides **ab***	20.8	79.2	0.67	5.13	25.0	70.8	2.25	4.01	33.3	62.5	1.25	3.76
A. triflorum **c***	0	0	0	N/A	8.33	0	1.08	3.12	20.8	4.2	0.58	7.41
A. platanoides × *truncatum* 14.121 **bc***	2.78	33.3	0.72	5.19	5.56	11.1	0.17	2.27	0	27.8	0	N/A
A. 'Morton UW' **a***	45.8	45.8	1	0.58	45.8	37.5	1	0.82	25.0	50.8	0.46	0.77
A. pseudosieboldianum **c***	16.7	8.33	0.54	4.23	0	8.33	0	N/A	12.5	4.2	0.42	5.32
A. platanoides × *truncatum* 13.128 **a***	55.6	27.8	2.22	7.38	38.9	44.4	0.61	2.12	22.2	72.2	0.33	2.93

* Mean Separation Results from Fisher's LSD Test. Treatments with the same letter are not significantly different.

Results confirmed that different maple genotypes have varying responses to rooted cutting propagation. Selections with the greatest response to rooted cutting propagation after 6 to 7 weeks were *A. platanoides*, *A. platanoides* × *truncatum* 13.128, and *A.* 'Morton UW'.

Further rooting in all seven maple selections was observed 90 days after initiation, suggesting that greater rooting success could potentially be achieved with extended time in the propagation flats before transplanting. Future research could consider data collection at least 90 days or more after initiation (**Fig. 5**).

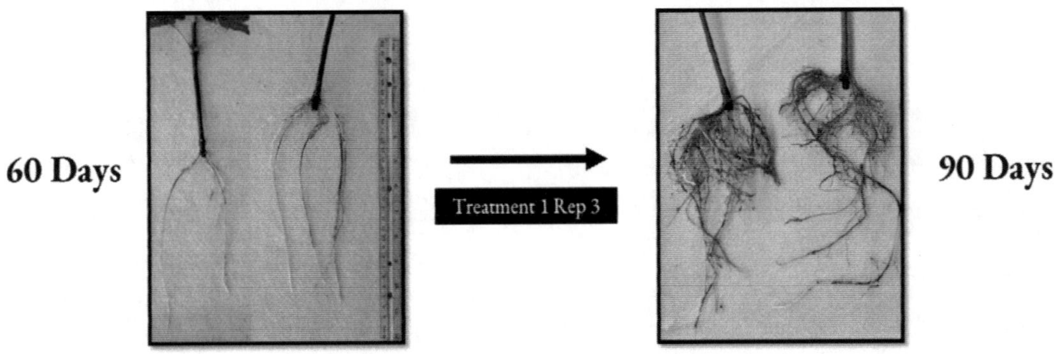

Figure 5. Rooting of *A. platanoides* × *A. truncatum* 13.128 after 60 days compared to 90 days post-transplanting.

Results from previous scientific studies about the impact of auxins on rooting are varied and inconclusive. As reported by Brock (2014), variation in rooting success can often be attributed to genetic differences among trees. The difference in rooting success between the two *A. platanoides* × *A. truncatum* individuals supports this idea. Additional research could explore concentrations of K-IBA both lower and higher than those used in this study or evaluate various auxin application techniques.

According to Coggeshall (1957), Bachtell and Breslauer (1985), Gabriel (1961), and Brock (2014), it is relatively certain that the age of a tree is a huge determinant of rooting success. Cuttings from mature trees tend to be less successful, while those taken from juvenile stock trees are more likely to produce commercially acceptable percentages of root growth. Future research could consider comparing propagation results of cuttings taken from clonally reproduced and coppiced stock plants (ramets) to the parent tree (ortet).

ACKNOWLEDGMENTS

Thanks to the Daniel P. Haerther Charitable Trust Foundation for their generous support of the New Plant Development Program at The Morton Arboretum and to Dr. Todd West and the North Dakota State University Woody Plant Improvement Program for their collaboration. This project was funded by the North Dakota Department of Agriculture through the USDA Specialty Crop Block Grant Program. Thanks to the IPPS-ER Foundation for internship scholarship funds.

LITERATURE CITED

Bachtell, K.R., and Breslauer, L. (1985). Propagating *Acer miyabei* from cuttings. The Plant Propagator. *31*:14–15.

Brock, J.A. (2014). Rooting stem cuttings of shantung maple (*Acer truncatum*), mound layering shantung and caddo sugar maples (*Acer saccharum*), and using eastern redcedar (*Juniperus virginiana*) as a substrate component in stem cutting propagation [Master's thesis, Kansas State University].

Coggeshall, R. (1957). Asiatic maples: their propagation from softwood cuttings. Arnoldia *17*:45–56.

Dirr, M.A. and Heuser, C.W. (2006). The reference manual of woody plant propagation: from seed to tissue culture. Varsity Press, Athens, Georgia.

Dunn, S. and Townsend, R.J. (1954). Propagation of sugar maple by vegetative cuttings. J. Forest. *56*:426-428.

Enright, L.J. (1958). Propagation of several species of *Acer* by cuttings. J. Forest. *56*:426–428.

Gabriel, W.J., Marvin, J.W. and Taylor, F.H. (1961). Rooting greenwood cuttings of sugar maple: Effect of clone and medium. USDA, Northeastern For. Exp. Sta. Res. Note NE-144. Upper Darby, Pennsylvania.

Snow, A.G. (1941). Variables affecting vegetative propagation of red and sugar maples. J. Forest. *39(4)*:395–404.

Tousignant, D., Richer, C., Rioux, J.A., Brassard, N., abd Mottard, J.P. (2003). Vegetative propagation of sugar maple: Relating stem water content and terminal bud developmental stage to adventitious rooting of stem cuttings. *J. Plant Sci. 83*:859–867.

Combined Proceedings IPPS 72:354-356. 2022.

Asexual Propagation of *Forestiera neomexicana* (A. Gray) Using Semi-Hardwood Stem Cuttings

Alyssa M. Headley, Benjamin P. Desrosier, Taylor R. Mathison, and Brandon M. Miller

Department of Horticultural Science, University of Minnesota, Saint Paul, Minnesota 55108 U.S.A.

headl069@umn.edu

Keywords: Desert olive, IBA (Indole-3-butyric acid), adventitious roots

Summary

Forestiera neomexicana (desert olive) a native plant found throughout arid regions in the southwestern United States. This species could be utilized more broadly within ornamental landscapes in urban settings. Little is known regarding asexual propagation techniques for producing this plant and experiments were undertaken to study this. Results show a numerical trend suggesting that increasing IBA concentration leads to bolstered root length and number of roots. Additional studies are needed.

INTRODUCTION

Forestiera neomexicana (desert olive), also referred to as New Mexico Privet or New Mexico Olive, is a native plant found throughout arid regions in the southwestern United States. Purportedly resistant to both drought and periodic flooding, this species could be utilized more broadly within ornamental landscapes in urban settings as well as in green infrastructure, replacing invasive privets (*Ligustrum* spp.). Desert olive has performed well in the living collections of the Minnesota Landscape Arboretum

(USDA zone 4b) showing promise for use in the Upper Midwest. However, little is known regarding asexual propagation techniques for producing this plant.

Our objectives were to:

1) Determine if *F. neomexicana* requires the application of auxin for adventitious rooting, and if so;
2) Which concentration(s) of IBA maximize rooting and root length of semi-hardwood cuttings.

MATERIALS AND METHODS

Semi-hardwood cuttings were collected from an accession at the Minnesota Landscape Arboretum in Chanhassen, MN on July 22, 2022. The propagules collected were 5-node terminal cuttings, ~9 cm in length. Cuttings were treated with 95% ETOH (control only); 1000 ppm IBA; 2000 ppm IBA; 3,000 ppm IBA (each dissolved in 95% ETOH) using a 3-sec quick dip of the basal ⅓ of the cutting. The cuttings were stuck in individual cells of 72-cell trays using a completely randomized design with four experimental units per treatment comprising six single cuttings each (N=336, n=84). Cells of the trays were filled with 100% perlite. The trays were then placed in a mist bay greenhouse located in Saint Paul, Minnesota, where the intermittent mist sprayed for 8 sec intervals, every 4.5 min in 25°C conditions. The average Photosynthetically Active Radiation (PAR) recorded in the mist bay was 129 $\mu mol \cdot m^{-2} \cdot s^{-1}$.

Data were collected 35 days after cuttings were stuck and placed in the mist bay. A root was counted if it measured ≥ 0.25 cm in length. Data were analyzed using a one-way ANOVA and Tukey's HSD for mean separation.

RESULTS AND CONCLUSIONS

In this study, semi-hardwood stem cuttings rooted at 21%, 62%, 74%, and 87% for non-treated controls, 1000 ppm, 2000 ppm, and 3000 ppm IBA, respectively (**Fig. 1**). Treatments of 1000 ppm, 2000 ppm, and 3000 ppm IBA were all significantly different when compared to the non-treated control for number of roots (**Fig 2**). However, the three IBA treatments were not significantly different from each other. This trend can also be seen for root length (**Fig. 3**).

Figure 1: Above (left to right): Control (non-treated), 1000 ppm IBA, 2000 ppm IBA, and 3000 ppm IBA.

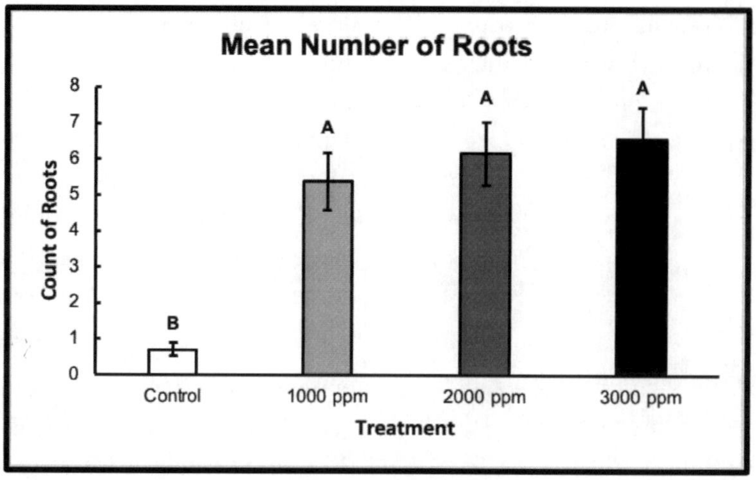

Figure 2: Number of roots on rooted stem cuttings of *F. neomexicana* 35 days after treatment with auxin. Roots ≥ 0.25 cm long were counted. Error bars indicate the standard error of the mean at 5% confidence. Means with the same letter are not significantly different.

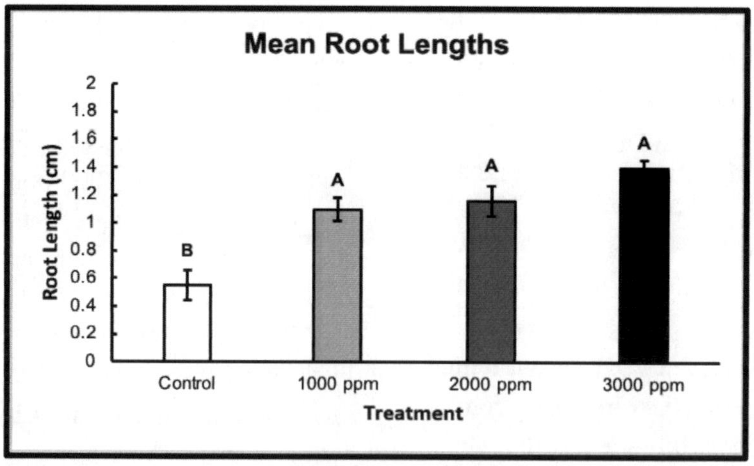

Figure 3: Lengths of roots on rooted cuttings of *F. neomexicana* 35 days after treatment with auxin. Roots ≥ 0.25 cm long were counted. Error bars indicate the standard error of the mean at 5% confidence. Means with the same letter are not significantly different.

Every treatment involving the application of IBA increased the number of roots and root length compared to non-treated controls. Similar rooting was observed across treatments with IBA. Based on these results, we recommend growers apply 3000 ppm IBA on semi-hardwood cuttings of *F. neomexicana* to maximize rooting percentage, the number of roots, and root length. These data show a numerical trend suggesting that increasing IBA concentration leads to bolstered root length and number of roots, therefore, higher concentrations of IBA could be explored to further maximize rooting and root length of cuttings of *F. neomexicana*.

ACKNOWLEDGEMENTS

We thank the Minnesota Landscape Arboretum, the Arboretum Endowed Land Grant Chair fund, and AGREETT for supporting this work.

Combined Proceedings IPPS 72:357-370. 2022.

New Plant Forum 2022 – Eastern Region IPPS

Kim Shearer Moderator

The Morton Arboretum, Lisle, Illinois U.S.A.

kshearer@mortonarb.org

Summary

New plants for 2022 are highlighted and described. This year six IPPS-ER breeders presented herbaceous and woody perennial plants.

PRESENTER

Judson LeCompte
Spring Meadow Nursery, 12601 120th Ave, Grand Haven, Michigan 49417 U.S.A.
Judson@SpringMeadowNursery.com

Hydrangea arborescens **Invincibelle Sublime™ hydrangea PP#34,418; CBRAF.** Invincibelle Sublime hydrangea ushers in a new day for full-sized smooth hydrangeas with cloud-like tourmaline-green mophead flowers floating above the plant on super-sturdy stems. Very dark green foliage sets off the lively green of the blooms. Invincibelle Sublime smooth hydrangea is a versatile native cultivar that can be planted in full to part sun and is hardy down to USDA zone 3. It will mature to 3.5–5 ft tall and wide.

Figure 1. Flowering plant (left) closeup of flowers (right).

Weigela florida **'SMNWFGC', WINE & SPIRITS™ weigela PP#34,358; CBRAF.** This is an update on our classic Wine & Roses weigela that combines even more dramatic dark foliage with crisp white-green flowers. A real showstopper in the garden center and landscape. Sun loving and cold tolerant down to USDA zone 4. It will mature to 3–5 ft tall and wide.

Figure 2. Flowering plant (left) closeup of flowers (right).

Hydrangea macrophylla **'SMNHSME', Let's Dance Sky View™ hydrangea PP#34,327; CBRAF.** Let's Dance Sky View™ hydrangea was selected from the Proven Winners® ColorChoice® extensive Let's Dance® hydrangea breeding program for its ability to conserve its old wood buds in the face of weather challenges and its ability to continue creating new flowers. It's also very easy to turn the flowers blue. The flowers emerge a beautiful light blue with a honeydew-green eye before maturing to a full sky blue. Its nice compact growing habit makes it both a good container and garden plant. Hardy in USDA zones 4–9, will reach heights of 2–3 ft and widths of 2–4 ft.

Figure 3. Flowering plant (left) and easy to change to blue (right).

PRESENTER

Connor Hagemeyer
North Dakota State University, Dept 7670,
PO Box 6050, Fargo, North Dakota 58108 U.S.A.
connor.Hagemeyer@ndsu.edu

***Pinus uncinata* 'GuarDak', Hyland Guard™ mountain pine.** Hyland Guard mountain pine is a unique upright narrow pyramidal evergreen conifer that will reach a mature height that is taller than the currently available upright Mugo pine (*P. mugo*) cultivars. A mature size of 26 to 28 ft and spread of 6 to 8 ft is expected. Needles are evergreen, forest green colored and persisting for five or more years. Needles are 1 to 2 in. long and in pairs. Soil preference is a well-drained soil, pH adaptable and tolerant of higher pH soils. Tolerates clay and high calcareous soils. It originated from a population of *P. uncinata* collected from the Hrubý Jeseník mountain range of Eastern Sudetes in the northern Moravia region near the village of Rejvíz of the Czech Republic. It is in USDA zone 3 to 7.

Propagation is by side grafting onto upright *P. mugo* seedlings or other compatible pine species.

Figure 4. 'GuardDak' Hyland Guard™ mountain pine.

PRESENTER

Arden Pontasch
North Creek Nurseries, Inc., 388 North Creek Road,
Landenberg, Pennsylvania 19350 U.S.A.
arden@northcreeknurseries.com

***Phlox carolina* ssp. *carolina* 'Kim'.** 'Kim' is a fantastic selection that was discovered by Jan Midgley in Alabama. It performs much better than any other cultivar of Carolina phlox in our trial. What sets 'Kim' apart from other members of the species is its lush and vigorous habit (24 in. tall and 48 in. wide) that remains sturdy and disease-free all season long (excellent powdery mildew resistance). The leaves are also a lighter shade of green, almost lime-colored, which can prolong its horticultural interest in the garden. However, the most impressive feature of 'Kim' is its show-stopping light pink flowers which blanket the plant from late May through early June. Flowers carry a mild fragrance. Hardiness zones: 5–9.

Figure 5. *Phlox carolina* ssp. *carolina* 'Kim' flowering plant (left) and closeup of flowers (right).

***Scutellaria* 'Appalachian Blues' PPAF.** 'Appalachian Blues' is a cross of *S. ovata* and *S. serrata*, both plants indigenous to the mountains of West Virginia where breeder Peter Heus resides. Bred by Peter Heus, and brought to market by Plants Nouveau. What's the scuttlebutt about this new skull cap? Bumblebees will love it because it is covered in deep purple flowers. You will love it because it is compact and full, and drought tolerant. The dark green leaves have eggplant-colored edges and are supported by deep maroon stems, making a show-stopping combination. Skullcaps are a great native replacement for salvias and lavenders and this new selection combines perfectly with bright yellow coreopsis, little bluestem, and other drought-tolerant natives. Grow in full sun to partial shade in average garden soil. It has no know pest or diseases. Cold hard in zones 4–9. Grows 12–15 in. tall by 12–20 in. wide.

Figure 6. *Scutellaria* 'Appalachian Blues' flowering plant (left) and flowers (right).

Golden Sunset™ yellow prairie grass was first selected in 2005, and since 2010 propagated and trialed in several locations in Minnesota and the Midwest, was selected for its upright stature, clean olive-green foliage, and numerous early yellow and golden-bronze flowers.

Golden Sunset flowers first emerge in mid-August and remain attractive through the winter. Unlike most yellow prairie grass, Golden Sunset remains upright and does not lodge or fall over. This new patented grass from the University of Minnesota will be a good addition to landscapes throughout the U.S.A. but especially in northern climates.

Plants grow to 4-6 ft tall at maturity but are closer to 4 or 5 ft the first year or two. Large mature plants can be 36 in. wide, with hundreds of flowering stems. Most yellow prairie grass available today has blue-green foliage, Golden Sunset differs in having olive-green foliage that is just over ½ in. wide. Flowers average 9 in. in length and are a showy yellow and golden bronze in color. Golden Sunset has no known pests or diseases.

Selected for upright stature, clean foliage, and plentiful early to bloom golden-bronze flower plumes this gem grows on a wide variety of sites and soils with winter hardiness to zone 3. When in need of a native warm season grass with the verticality of *Calamagrostis* 'Karl Foerster', your search ends with *Sorghastrum* Golden Sunset.

Figure 7. *Sorghastrum* Golden Sunset yellow prairie grass summer stature (left) flower plumes (right).

PRESENTER

Angela Palmer
Plants Nouveau, Box 40125, Mobile, Alabama 36640 U.S.A.
angela@plantsnouveau.com

***Hydrangea paniculata* 'Bokralims' Magical® Lime Sparkle PP 30,098.**
Lime Sparkle is different. It's not too short to be seen or too tall to hide stuff. The flowers begin in early June as lacy lime green panicles and open up and age throughout the summer to apple green and then celery green with a candy apple red blush once the nights get cooler. This panicle hydrangea, in our opinion, is a perfect size. Whether you are looking for backbones for a sunny border, or the ultimate, people-height hedge to gently divide a space or keep the neighbors out, Lime Sparkle delivers what many new introductions cannot. Height and spread: 5–6 ft tall by 5 ft. wide. Hardiness: USDA Hardiness Zones: 3-9.

Availability: JRT Nurseries - liners and finished.

Figure 8. Lime Sparkle hydrangea showing flowering plants with lacy green flowers in June (left) and candy apple red blush as the nights cool (right).

***Hydrangea quercifolia* 'Snowcicle' PP 33072.**

Selected by Richard Davis from a batch of *H. quercifolia* 'Snowflake' in southeastern Virginia. This new, double-flowered form of oakleaf hydrangea was selected for its superior vigor, larger flower panicles, and improved stem strength. It has a spread and height of 4–6 ft. Hardy in USDA Hardiness Zones 5–9.

It is propagated by tissue culture and available from: Knight Hollow Nursery – TC, Richey Nursery – liners, JRT Nursery – liners and finished, Manor View Farms – liners, Heritage Seedlings – liners.

Figure 9. *Hydrangea quercifolia* 'Snowcicle' in flower (left) and fall color (right).

***Picea glauca* 'Kolchomagi' Spruce It Up™ white spruce PPAF.**

Spruce It Up is a faster-growing mite and scorch-resistant sport of *P. glauca* 'Conica' growing 5–7 ft tall by 2–3 ft wide after 10 years. Spruce It Up white spruce was selected because it finishes in a 5-gal pot 1 year faster than the cultivar 'Conica'. This makes the plant much more profitable for growers and especially more profitable for growers who grow many of these for the Christmas holiday. It is drought tolerant and takes full sun like its parent. Hard in USDA Hardiness Zones 3–7b. Available from JRT Nurseries as liners and finished plants.

Figure 10. Spruce It Up white spruce in a 5-gal container.

PRESENTER

Justin Lombardoni
Chicago Botanic Garden, 1000 Lake Cook Rd.,
Glencoe Illinois 60022 U.S.A.
jlombardoni@chicagobotanic.org

***Aster* 'Billowing Pink', *Aster* 'Billowing Violet', and *Aster oblongifolius* 'Cotton Candy'.** The Chicago Botanic Garden is proud to present three new aster cultivars, each with an abundance of flowers sure to provide late season appeal to any garden. 'Billowing Pink' and 'Billowing Violet' are sister hybrid cultivars of *Symphyotrichum novae-angliae* and *S. oblongifolium* that each form a dense mound (30 in. tall and 48 in. wide) and provide a bright display of pink or violet flowers, as suggested by their names. 'Cotton Candy' is an *S. oblongifolium* cultivar and a parent of 'Billowing Pink' and 'Billowing Violet'. It forms a slightly larger mound (36 in. tall and 60 in. wide) than its progeny cultivars and has light pink-violet flowers. All three cultivars are resistant to rust, powdery mildew, and lace bug predation, and the aromatics released from their crushed leaves deter deer/rabbit browsing. These cultivars all bloom in early fall, providing interest when other perennials have retired for the year. As such, these asters should have a place in any sunny, well-drained garden to provide one last display of color before winter arrives. USDA hardiness zones 4-8.

Figure 11. *Aster* 'Billowing Pink' plant (left) and close up of flowers (right).

Figure 12. *Aster* 'Billowing Violet' plant (left) and close up of flowers (right).

Figure 13. *Aster oblongifolius* 'Cotton Candy' plant (left) and close up of flowers (right).

PRESENTER

Kim Shearer
The Morton Arboretum, Lisle, Illinois 60532 U.S.A.
kshearer@mortonarb.org

***Tilia* 'Zamoyskiana' Centennial™ linden.** This new linden from The Morton Arboretum is a chance seedling selection from our Groundcover Garden (1024-40*1). While this tree was accessioned into our collections in 1940, it was originally selected and named by the Kornik Gardens and Arboretum (Poland), and subsequently shared with Morton propagator John van Gemert during a tour of the European continent in the late 1930s. Mr. van Gemert was visiting the Polish arboretum in search of new plants to propagate and include in Morton Arboretum collections. This particular specimen must have caught his eye or been recommended due to qualities it was selected for—gracefully arching branches, vigor, and retention of foliage well into the fall. The original tree was the product of seed collected from a *T. americana* specimen located at Kornik Gardens and Arboretum and thought to be pollinated by a *T. tomentosa* 'Petiolaris' growing nearby. The name 'Zamoyskiana' was given to this selection by former Kornik Director Andrew Wróblewski in honor of Count Wladyslaw Zamoyski, a Polish philanthropist who had donated property to the Polish nation that would be the Kornik Gardens and Arboretum. We have selected the trademark Centennial™ in honor of the centennial anniversary of Morton Arboretum.

But what else? Vigor and graceful architecture are certainly bonuses, however there are other traits that we have considered. The footprint of this selection is relatively narrow compared to other linden selections due to its somewhat weeping branches. More importantly, this selection has demonstrated resistance to Japanese beetle predation. Something in the leaves does not appeal to them. During years of heavy infestation in the Chicagoland region, all other lindens appear to be brown in the landscape due to predation by Japanese beetles while the foliage of this selection remains intact.

While this tree was selected in the early 1900s, it remains virtually unknown. It has not been described in any major literature relative to cultivated trees. Yet, it demonstrates adaptability to the climate of the upper Midwest and great landscape potential due to its resistance to Japanese beetle and architecture. Additional traits include incredibly fragrant flowers which are a source of mid-late June nectar for bees and yellow fall color. Recommended for use as a shade tree or street tree.

In production with Kankakee Nursery (Illinois) and J. Frank Schmidt & Son Co. (Oregon) and seeking additional propagation licensees in eastern North America. Approximate height 70 ft and width 30 ft after 83 years. USDA hardiness zones 4 – 7? Additional cold hardiness evaluation needed in zone 4.

Figure 14. *Tilia* 'Zamoyskiana' Centennial™ linden mature form (left) and leaves (right).

Quercus bicolor **'KB Crystal'**

This oak selection from The Morton Arboretum is a chance seedling selection made by Vice President of Collections and Facilities Kris Bachtell. Originally selected from a group of oaks planted at Orland Park golf course, this tree has noticeably glossy foliage that transitions to yellow-orange fall color and exhibits vigorous growth relative to typical *Q. bicolor*.

In Illinois this selection exhibits resistance to powdery mildew; however, it is heavily infested with powdery mildew when in Pacific Northwest production. For this reason, we are seeking additional propagation licensees in eastern North America for ongoing evaluation. Currently, this is in production at Kankakee Nursery, Illinois.

We recommend this selection for use as a shade tree, street tree, and rain garden specimen. It has a uniform, oval habit reaching 30–40 ft tall and 10–20 ft wide in 20 years. USDA cold hardiness zones 3–8.

Figure 15. Quercus bicolor 'KB Crystal' form and fall color (left) and closeup of glossy leaf (right).

***Platanus ×acerifolia* 'Morton Naper' Monumental™ London planetree.** This is a sibling to the popular and familiar *P. ×acerifolia* 'Morton Circle' (Exclamation!™), this new selection of London planetree from The Morton Arboretum was developed by Dr. George Ware through controlled crosses. The original tree was planted as a sapling in a local cemetery along with others from the same breeding program.

Selected for sycamore anthracnose tolerance, this introduction is estimated to be about 70 ft tall and 50 ft wide at maturity. Mottled bark is stunning in the winter landscape with gray exfoliating plates revealing cream, amber and gray bark and stark white stems in the winter. And the broadly pyramidal, uniform canopy has dense branching making this selection an excellent park and shade tree.

This is a tree for nursery production with consistent caliper, straight stem growth, and vigor. Currently seeking additional propagation licensees (softwood, semi-hardwood, hardwood cuttings; budding; micropropagation), especially in eastern North America. In production with J. Frank Schmidt & Son Co., Oregon. USDA cold hardiness zones 5–9. Cold hardiness evaluation needed for zone 4.

Figure 16. *Platanus ×acerifolia* 'Morton Naper' Monumental™ London planetree showing straight stem growth (left) and bark (right).

PROCEEDING'S PAPERS

AUSTRALIA REGION

Clive Larkman, Regional Editor

Fiftieth Annual Meeting - 2022

Blue Mountains, Australia

Combined Proceedings IPPS 72:371-374. 2022.

Berry Big business: Commercialisation/bulk production of Berry Species

James Burnett

Burnett Family Trust, Morayfield, Australia

Jab342000@gmail.com.au

Keywords: Strawberry, blueberry, raspberry, blackberry

Summary

We spend a lot of time talking about the art of plant breeding, and if you haven't already picked up Plant breeders are a passionate group of artists at that. Without these plant breeders we are likely to be a much smaller industry today.

So, as I am sure you can observe from my robust physique, I am a strong fan of Edible plants and their fruits. Coincidentally but completely unrelated, this has ended up being a factor in propelling my career choices and so I find myself in the wonderful world of berries.

INTRODUCTION

Australia has a robust berry industry with the majority of the production made up of three separate sub categories; strawberry (*Fragaria*) or the humble blueberry (*Vaccinium corymbosum*) or raspberries and blackberries (*Rubus*), - the crop that keeps me up most at night!

In 2021, Horticulture Innovation Australia produced a Strategic Investment Plan for the period 2022-2026. Sourced from this berry SIP we have the following. There is an annual production of over 110,000 tonnes giving a farmgate production value north of One Billion dollars. The *Rubus* section was $216 million which is an increase of over 450% increase in the volume of production over the last eight years.

New Varieties

With the bulk of our berry cultivars coming from international sources, quarantine plays an important part in the Australian berry industry. Australia is an island with many of the world's agricultural pests not present. Hence it has very strong border controls. Anyone who understands the process particularly around Berries and or high-risk imports will know it is expensive and slow, but all for the important reason of protecting the wider Australian environment and primary industries.

The process of plant importation requires patience and time. The are many required treatments prior to export and after arriving in Australia. The time period can be as long as three years in Post Entry Quarantine (PEQ).

I am pleased to report that the quarantine process timeframe for many berry imports has been reduced due to improved testing techniques using every bodies favourite COVID testing method – PCR.

This new testing technique saves substantial cost, and gives importers faster access to commence the cultivar introduction and commercialisation process.

The source of new varieties can be summarised as follows:

Strawberries – There are several successful national breeding programs and many cultivars coming from foreign breeders.
Blueberries – Cultivars coming from increasing number of foreign breeding programs.

Raspberries and blackberries – New varieties are nearly exclusively imported form international breeding programs.

Fun facts aside, the berry sector raises its fair share of challenges. Providing quality planting material that performs every step of the growing cycle is front and centre. As common as the next comments may be to a room of propagators or plant production specialists, they are always worth review especially in high value industries.

Important: Handle with Care

Our plant breeders introduce a fresh new cultivar, what should be done do to ensure the new cultivar can be introduced in the most efficient and rapid way. The key points here are really a KISS principle KEEP IT SIMPLE SHERLOCK and really should apply for most parts of our businesses.

1. Work with reputable partners.

Good material starts with good material and comes from good growers. This is particularly important when importing new material from overseas. Being true to type is critical.

If an overseas IP management firm has not undertaken the appropriate cultural and management practices negative impacts can flow from delays completing establishment in quarantine all the way through to complete failure. Poor plants shipped will generate more investigation by authorities and can make plant establishment in PEQ much harder and riskier.

One potential risk is o receive genetic off types. This is usually something that may not be discovered until well through the overall commercialisation process at which point extensive dollars will already have been outlaid. The initial plants may well be the basis of a whole industry and incorrect forms can cause problems for decades.

2. Cleanliness is next to godliness.

Once you have the material on your site, you must, must, MUST treat the plants with the respect they deserve. We maintain our foundation material in an environment that is insect screened, concrete floored and with added supplemental lighting. There is genus specific fertigation mixes that are regularly tested and adjusted to give optimal performance.

How does that relate to cleanliness? A plant that is grown in optimal conditions is more able to withstand any health pressure that may and WILL arise. Biosecurity Protocols around staff and visitor movements and clean pre entry practices are key, much like we would expect in laboratory conditions. The aim is to ensure no diseases nor other pests enter nor exit the facility. I should also note the importance of keeping backup material in off-site locations, for protection against bushfire flood or other catastrophe. For plants under PEQ this can be an issue and may even involves locations overseas.

3. Know the Minor details.

When it comes to importing, breeding, selecting or other similar practices there are a range of criteria that determine performance. In working with the new ones it is important to understand what you are looking at/for. For berries these are just a few of the differences within a genus and even species group.

Strawberries - Day length - Neutral vs. short day vs. long day.

Raspberries/Blackberries - Primocane vs. long cane v Double cropper.

Blueberries - Northern Lowbush vs. Southern High bush

Once the plants exit PEQ then it is time to bulk up the plants in readiness for release. The above criteria are some of the factors you need to consider when planning out a propagation program. There are many others.

I know of at least one Plant Breeder who sees their selections as their children. Each needs to be coaxed into growing and showing how they differ. As a propagator we need to be the teachers, ask any good teacher how little Johnny behaves in class they can confidently tell you a raft of things you may not already know about little johnny. When working on a new cultivar you really must know the fine details, in order to push the maximum potential out of your material.

4. Understand the end consumer.

As a berry propagator you have two customers; your grower and the final retail buyer who actually eats or use the berries. The above selection may be targeted at the end consumer, but you have to satisfy the grower as well. A high level of pressure comes from our customers, the commercial fruit grower. It's a bug bear of mine, but when the grower has a problem, even one that has no link to any of our processes, we may get the blame for the performance or should I say, lack thereof. It is critical that we understand all the aspects of our new plants and how they grow. The post sale support can be crucial in ensuring return business, or maintenance of your reputation and this is reliant on a complete understanding of how your new plant performs.

Conclusion

Importing and commercialising new varieties is a complex process that has room for great success but also large pitfalls. Something like being in small business. I hope that if nothing else, this presentation may be a good reminder of things to think about in your business if you work in the commercialisation/multiplication/bulking space.

Combined Proceedings IPPS 72:375-383. 2022.

Looking Back Looking Forward: Learning from the Past in Preparing for the Future

John Bunker

Greenlife Solutions, Redland Bay, Australia.

jb@greenlifesolutions.com.au

Keywords: history, plant production, Australia, IPPS.

Summary

This presentation was an overview of the history of the International Plant Propagator's Society in Australia as well as an historical perspective on the Australian nursery industry and how those activities influence our horticultural professions today.

INTRODUCTION

The first farm in Australia was established on the current site of the Royal Botanic Gardens Sydney with seeds and plants brought by the first fleet in 1788. This is the birthplace of Horticulture and Agriculture in Australia. This location in the garden, is still growing vegetables today and is the oldest producing garden in the country.

Horticulture and Agriculture were important to the new settlement, in feeding the population, trying to understand this new land climate and soils, with many failures along the way, which were learning experiments.

It is recorded that William Macarthur commenced the setting out and planting of his Camden Park Garden in 1817 with other garden areas and orchards following. Due to commercial necessity, a commercial nursery was established in the 1840's and soon after, construction of heated greenhouse and propagation houses. These greatly improved the efficiency of the production nursery and plants which were able to be grown, but also improved the efficiency of the plant breeding operations. This nursery became the preeminent nursery in the mid-19th century, responsible for many ornamental and fruit tree introductions. The Camden Park Orchard was one of the largest commercial orchards in Australia through to the late 1970's.

The Nursery Industry associations became formalised in Victoria, South Australia, NSW and Queensland in the early 1900's after federation. They were established as an opportunity to enjoy social interaction on a business-like level, to discuss issues affecting their operations of the day, including supply of stock, stock losses, employment, wages, and other industrial matters.

Soon after, the industry found it necessary to come together on a national basis to discuss common problems and to collaborate on common goals. They became increasingly conscious of the importance of consulting with government for their perceived prosperity and knowledge of legislation. Interestingly, Biosecurity with plant movement between States and banning importation from South Africa were among key areas of focus (Greenlife Industry 2022).

IPPS History

International Plant Propagators Society, IPPS, commenced in 1951, with the first meeting in Cleveland Ohio with 70-80 people in attendance. Rules and guidelines were established which are still in place today. One key area is: "It is not where you work or where you come from, It is you, the propagator, who is important, with the knowledge you have to share."

This was the basis of IPPS motto: **To Seek and Share**.

The foundation was spear headed by Mr Jim Wells, from New Jersey in the United States, who was the first IPPS president. Through his enthusiasm and support from interested people, further regions were established in Australia, New Zealand and Great Britain (Dawkins 1996).

Jim Wells travelled to Australia in 1973 for a meeting with interested people in the aim of establishing an Australian region and attended the first meeting held in October 1973 at Leura, which saw the establishment of IPPS in Australia, with Edward Bunker being elected the inaugural president. May 2022 and we are celebrating the 50th IPPS Conference meeting being held in Australia.

We have had passionate plantsmen and women involved in Horticulture over the years, who have had a quest of continuous improvement, working together, and sharing knowledge that has enabled us to learn from the past in preparation for the future.

Theodore Roosevelt is quoted as saying "I believe that the more you know about the past, the better you are prepared for the future".

Looking forward to the Future: Environmentally sustainable production systems

Education

Education is a key component in developing and preparing future plant propagators and horticulturists. Project based teaching, focussing on problem solving and understanding the principles, should be the basis of any tertiary education, on farm training and developing career pathways.

A key part of my early training was from my Dad, Grandfather, and lecturer Murray Richards at Massey University, in being able to understand the importance of observation and understanding. When solving a problem, it is very easy to focus on the symptom as the cause, but quite often it is another variable which is influencing the symptom, which then through understanding the growth cycle, production cycle or process you can then progressively work through, whether the problem you are dealing with is disease or insect, environmental, mechanical, cultural, nutritional or water management. The plant is not able to communicate verbally but is communicating to us through the signs it is displaying. These principles are important to compliment the use of technology in the future because if you don't understand the principles, you are not going to be able to interpret the answers from the data you receive, or the systems being designed and implemented.

The other key lesson from my Grandfather was that you can gather a group of people together at a starting point and show them a destination, a lot of people will focus on the destination and completely miss the detail in the journey as they travel from A to B, again the importance of observation. The only time you will experience the sunset is if you stop, look up and stand in awe, looking down you will easily miss the beauty and the detail of the close of a day. Observation is a key skill in growing crops and a key life skill.

Biosecurity

It was interesting reading one of the points for the reason for the National Nursery Association, back in the early 1900's, was recognising the importance of what we know call Biosecurity.

The National Biosecurity and Sustainable Plant production Project NY20001 is the single largest research and development pot levy investment which our industry is making. It is a key investment in the future to better manage pest and disease issues on farm and simplify interstate and intrastate plant movement. It has facilitated legislative change to allow on farm certification of crop being shipped interstate, which has been history making, and given the producers involved greater flexibility and cost savings.

Pohlman's Nursery have stated that the program was saving their organisation in excess of $300,000/annum. Pohlman's Nursery was recognised as the Farm Biosecurity Producer of the year for 2021 for their contribution to the development and implementation of Biosecure HACCP on farm, in conjunction with Greenlife Industry Australia (GIA) and Horticulture Innovation Australia (HIA). This has strengthened the recognition of the innovation and importance of this program within Horticulture and is a key investment in future proofing our industry and preparing it for export and Biosecurity incursions (Australian Biosecurity 2021).

Biosecurity will continue to be one of our greatest challenges as an Island nation in preventing insect and disease issues which are not here at present. Horticulture, inclusive of the Nursery and Garden Industry, will continue to focus on monitoring and research efforts to exclude these problems and continue to highlight the most critical pests and diseases that would have major economic damage to our nations horticultural production, along with impacting our native flora and fauna.

Mother Plant production

Viewing ancient vineyards in Germany 3 years ago has given me a clear focus in relooking at our Mother Plant production. Ultimately, healthy roots equal healthy shoots, relooking at irrigation and nutrition management, improving the soil flora, and canopy management are all lessons that we can observe from other horticultural production systems.

Lessons learnt from the past continue to focus on the importance of stock plant management and cutting production as the foundation principle in producing a quality healthy plant with the desirable characteristics of the mother plant.

Observation of field mother stock of *Buddleja* in the Midwest USA, which can get covered by up to 1 metre of snow, demonstrates that the cold kills off the top growth and the plants reshoot from basal juvenile growth. The snow does the equivalent of what we would do in Australia with cutting hard with a chainsaw.

Nutrition management to ensure the correct balance in the harvested cutting to ensure it gets the maximum advantage when harvested and developing roots.

Managing the canopy to engage better light and air movement, as well as making it more effective for spray treatments, harvesting of cuttings and production of juvenile growth.

Soil flora improvement, through additives of bioflora, composting and mulching are the basis of regenerative agriculture, which is a new buzz word, with continuous improvement of soil to aid nutrition and water management - the ultimate aim of producing a healthy root system.

Tissue culture will continue to be expanded and play an importance in the clonal production of high health high volume food and ornamental crops. Tissue culture is playing an important part with Banana production and other clean stock programs including Potato's, Ginger, Sweet Potato, Berries. With the industrial scale and replicability, we will see more automated systems enter this production system when high volumes of similar crop lines are produced.

Tissue culture will continue to play a key role in being able to import new genetics into this country through high health protocols and testing and being able to move larger volumes in smaller cubic capacity.

Growing media

With the supply chain disruption of the last 2 years in preparing for the future, a priority is looking at longer term alternative substrates that are locally sourced and produced.

Alternative substrates that have the nutritional buffering capacity along with air filled pore spaces for root development and moisture management.

This has been a key risk management area identified by the Nursery Industry and has been further highlighted in the last period with the major fires in timber plantations in the Southern States of Australia which removed significant bark production. Together with rising energy costs and some timber mills converting to green energy and burning the wood waste for a higher return on investment rather than being sold for mulch or future growing medias.

There is further competition for wood waste from broad acre horticulture for soil improvement on vegetable crops and orchards/vineyards as compost or mulch.

Peat – how much longer will this be able to be mined before environmental concerns and legislation make it prohibitive?

Coir is a renewable resource. Is there an opportunity to develop a COIR industry in the pacific nations to supplement production from Sri Lanka and to be a primary source for Australia and New Zealand alleviating potential supply chain disruption.

What other renewable products need to be explored that can support container-based production systems?

Waste management

Waste management is a broad cost centre for all businesses. Understanding what is causing wastage and being able to eliminate/reduce in the propagation/production system has a direct bottom line improvement. Understanding the production cycle in all steps is a critical point in being able to understand the cause and eliminate.

With sustainability principles options of reduce, reuse, recycle of waste streams also prevents product going into landfill. The PP5 closed loop recycling project is leading the way amongst horticulture industries. It was foresight by both Norwood Industries and Garden City Plastics nearly 40 years ago to use the same plastic product in the label and pot manufacture which is making this project possible. This project is reducing product going into landfill but has become a valuable resource and raw material component with supply chain disruption and higher oil prices for pot manufacturing (Garden City Plastics 2022)

Wouldn't it be a significant step if the majority of plastic used was from recycled product. With the developments of plant-based resin technology biodegradable pots will become more mainstream but have the challenge of overcoming in our climatic zone temperature and rainfall extremes.

Greenwaste/prunings which traditionally have been burnt are an important resource for providing mulch or compost to improve soil.

Mechanisation and Technology

From planting and harvesting of cuttings, tasks traditionally completed by hand, mechanisation is developing in conjunction with AI to take some systems to the next levels.

Mechanisation ensures replicability of results but will continue to play a vital role due to shortage of labour and people willing to do the more menial tasks performed by hand. With the forecast global population increase and subsequent production required to feed the population, robotics will play a significant role in making timely harvesting possible.

The latest ISO planting equipment is fully automated but has created a production system which requires the correct sizing of cutting to ensure the robots are able to handle the material. With the ability to run extended hours, particularly when seasonal timing is tight, along with labour supply.

The Rapid Antigen Tests which we are using for COVID testing will be developed further for rapid pathogen testing on crops.

Is it possible that we may have chips or nanobots in the vascular system of major crops which is feeding real time data of nutritional, water management and growth of the crop potentially ensuring that there is no stress in the crop? When we visually see changes in a crop it takes time to reverse any detrimental actions which real time data would be ahead of.

Protective cropping will continue to play a greater role where shade was created by using brush or trees to the modern synthetic shade cloths to reduce risk of crop failure and the flexibility of not being hindered by weather events to ensure sales targets are reached and return on investment. Further development and understanding of light spectrum will play a role in disease and insect management as well as aiding plant growth and fruiting development.

The use of drones has only begun, with successful use now in applying pesticides, use of infrared technology in assessing plant health and using of drones in applying beneficials in horticultural crops such as strawberries and fruit tree production. Drones are being used in Cotton production to do flyovers with high-definition cameras for insect monitoring with the data being analysed and forwarded to the grower with updates on insect populations and locations on the farm linked to the electronic mapping system coordinates.

The latest Tevel Aerobotics drone harvester, which is harvesting apples, has attracted the attention of Kubota who have made a direct investment into this company.

Kubota and Yamaha have also made direct investment into an innovative robotic strawberry harvester which is a clear direction in where they see the next stage of advancement in AI assisted autonomous harvesting.

As technology is becoming more affordable imagination will open up more applications in this space.

Robots are currently being used for pot spacing, planting, field mowing and spraying and crop monitoring.

Use of radio and Wi-Fi networks in monitoring field, growth nutritional and water management. I am excited by a recent application we have been able to install using radio for operating field watering and not having to use cabling, it has opened opportunities which previously would have been cost prohibitive.

Golden Grove Nursery through the Hort Innovation funded project 'Digital remote monitoring to improve horticulture's environmental performance' (ST19024) using the Hort Innovation nursery products research and development levy and the Australian Government's National Landcare Program is a collaborator in this project which is using new technology and monitoring devices with the data being uploaded to the cloud which is improving decision making skills, improving production efficiencies and optimise labour and environmental performance. Hitachi Consulting has supplied the technology as a key direction in investing into future Primary production technology which also includes work with autonomous tractors in the sugar industry (GIA 2022)

Targeted research and development, funded through Horticulture Innovation and Horticulture Industry Levies will continue to be an important benefit for Horticulture producers in preparing for the future.

Genetics

In ornamental plant production, work will continue to focus on disease and insect resistance, growth habits that don't require plant growth regulators and uniqueness in being the first breeder in new genera for market advantage.

Exciting breeding work by Queensland Department of Primary Industries is focussing on fruit crops small tree, high yield which started with Apples originally going from 30 tonne/acre through to 130 tonne/acre. Tree crops such as Mango, Macadamia and Avocado are being looked at as well as the architecture of the canopy and orchard layout so that the fruit is accessible for future robotic harvesting and to maximise the yield/ha.

These researchers are turning to plant genetics to help solve DNA mysteries and create the horticultural tree crops of the future under $11.3 million 5-year joint research project.

Delivered through Hort Innovation, this project will develop a breeder's genomic toolkit for tree breeders and researchers to better understand how genes control traits that are valuable to Australian growers, including tree size, yield, disease resistance, and tree maturity (Queensland alliance 2022)

Supply chain disruption

This is a new challenge which has arisen since COVID in 2020 with shipping delays, spare parts and components not being available and new equipment being indefinitely delayed.

It is already seeing more local manufacturing and will see more Aussie ingenuity in being able to repurpose machinery or adapt parts to ensure machines can remain functioning.

It has highlighted how vulnerable we are in being reliant on overseas manufacturing and shipping for some key production inputs.

Energy

We are experiencing significant increases in energy costs, not only at the diesel bowser, but also with power supply and reliability.

It is another key area to continue to monitor, being a key investment in a horticultural business, as in most cases the main use of electricity is for irrigation, rather than cooling and heating.

My first experience with looking differently at electricity was in 2006, as a participant in an Qld Government EcoBIz project, where we looked at an irrigation pump that needed replacing. The pump being replaced had been purchased at an auction for $500 was 4kw it was replaced with a 2.2kw pump which pumped twice as much water.

When we did the calculations the new pump cost $1,200 and was saving $800/ annum in electricity cost/pumped litres of water.

A variable drive pump set, working on pressure starts, achieved a 30% return on investment with reduced energy costs and the added bonus of not damaging older water mains.

In future proofing our businesses we need to be aware of the potential gaps in electricity supply with the decisions that have been taken to decommission base load coal fired power stations in New South Wales and Victoria earlier than planned, which potentially could create power supply issues in the next 5-10 years.

Weather, Resilience and Water

There is a lot of discussion about climate change. In the almost 40 years of growing, I have observed that we have had seasons where there has been no overwintering of two spotted mite because of temperatures being higher and over the last few years drier springs leading into summer and periods of intense rainfall.

In Australia with the limited long range climate records, there is a pattern of climatic cycles. The east coast of Australia weather events of this summer has similarities with 1974 which is in essence a 50-year event.

Since European settlement there have been major floods in the city of Brisbane in the following years of January 1841, March 1890, February 1893 which was three flood events a week apart in conjunction with three separate cyclones, February 1931, January 1974, January 2011 and February 2022.

There is on record, from explorer John Oxley and Major Edmund Lockyer, seeing debris in trees ,100 feet above the normal river level in the area of the current Mt Crosby pumping station. It is suggested that somewhere between Oxley's visit in September 1824 and Lockyer's visit in September 1825 the Brisbane River experienced a flood as great as the February 1893. These flood events are still the largest experienced on the Brisbane River and well before the major development that we see today, and the words climate change being used (State Library of Queensland 2022)

On average a major flood event has occurred every 37 years, with a corresponding link to LaNina weather events, with the current LaNina cycle commencing in 2020 and now expected to continue into the summer of 2022. On average LaNina cycles can run between 2 and 3 years.

In our nation, after every major flood there has been periods of extended dry weather. On our own property in this last decade, we experienced the last water spilling over our dam wall in February of 2013 followed by a period of extremely dry springs and below average rain fall which was enough to continue operating. The next spill of water over the spill way was February 13th, 2020 with the storage capacity during the last three years not going below 70%. The summer of 2022 has seen two significant rain events a month apart, which dropped 200mm in a 4–6 hour period.

Studies and research into understanding these weather patterns in the past and the events leading up to the flooding will better prepare us for future weather events.

We have to be prepared for growing crops in periods of extended wet weather which takes a different skill set to producing in a dry season. Water supply will continue to be one of our nation's biggest challenges with providing the balance of water required for Agriculture, Urban, Industrial, Mining development and Environmental flows. Our greatest challenge is providing water for the urban and Industrial users as we are seeing major development increases and no future plans to increase water storage with the exception of planned irrigation water storages being constructed in Queensland.

As producers we need to ensure that we manage our water resources effectively with irrigation and capture and reuse of irrigation tail water to minimise environmental flows and to maximise crop yield. A continued focus on sprinkler technology, growing media, growing environment, alternative sources of water and monitoring technology will need to be key criteria for producers to focus on.

Environmental

As propagators we are producing living natural photosynthesizing products which convert Carbon Dioxide into carbon. This is essential for feeding the world with food crops or as ornamentals, providing the many positive social, environmental benefits that plant material adds to an interior or exterior landscape.

As an industry the pot levy funded project, which is now known as Greener Spaces Better Places, has been instrumental in funding research, using advocates to positively promote the importance of landscape and Urban Forestry and to see legislative change being implemented in cities urban forestry plans. Ultimately seeing more plants being used in landscape.

In conclusion it is an exciting period to be active as propagators with the challenges and opportunities to feed a growing world population, as well as the raised awareness of the importance of plants in both interior and exterior landscape.

Ultimately, all starter plant material is touched by a plant propagator, with the opportunity to influence the outcomes just as our early settlers did and help make the world a more beautiful place.

Literature Cited

Australian Biosecurity Awards Booklet. (2021). Department of Agriculture, Water and Environment Pohlman's Nursery. 14-17. https://www.awe.gov.au/sites/default/files/documents/qid87094_2021_australian_biosecurity_awards_-_winner_booklet.pdf

Dawkins, D. (1996). Short History of International Plant Propagators Society. Proceedings of the International Plant propagators Society *46*:705.

Garden City Plastics Sustainability https://www.gardencityplastics.com/sustainability

GIA Your levy@Work – Case Study. (2022). Golden Grove Nursery optimising efficiency with Smart Technology.

History of Camden Park. https://www.camdenparkhouse.com.au/horticulture

History of Nursery Industry in Australia. https://www.greenlifeindustry.com.au/about/association/history

Oxley, J. (2011). An overview of Brisbane River Floods. State Library of Queensland. https://www.slq.qld.gov.au/blog/overview-brisbane-river-floods

Queensland Alliance for Agriculture and Food Innovation. (2019). https://qaafi.uq.edu.au/article/2020/04/multi-million-dollar-collaborative-project-design-tree-future

Combined Proceedings IPPS 72:384-388. 2022.

Accelerated In Vitro Breeding Creates Improved Designer Papaya

Puthiyaparambil Josekutty, Mark MacLaughlin, Pui Lam Jay Ma, Teresita Steele, Bongani Ndawana, Candy MacLaughlin, Paul Fagg, Marion MacLaughlin, Ian Mac Laughlin

Skybury Farms, 136 Ivicevic Rd., Paddy's Green, QLD 4880, Australia.

josekutty964@gmail.com

Keywords: *Carica*, tissue culture, somatic embryogenesis, somaclonal variation.

Summary

Papaya (*Carica papaya* L) has very limited genetic variability that hampers conventional breeding. Therefore, we developed, novel, high yielding, more flavoursome designer Skybury papaya lines with high Brix through accelerated in vitro breeding. Skybury farm changed to clonal papaya cultivation, which allowed us to achieve continuous improvement through somaclonal selection from the 300,000 plus papaya grown annually. We developed a high throughput somatic embryogenesis system for Skybury papaya. Induced genetic variability among the embryogenic cell lines, regenerated variant papaya through non-GM method, conducted large scale field trials and recovered unique variants with better agronomic and fruit qualities. We also generated several mutant lines free of dreaded Papaya Sticky Diseases (PSD) caused by the Papaya Meleira Virus (PMeV). We developed a rapid and reliable molecular diagnostic test to screen for PSD and verified the PSD free status of several of the mutant lines using this kit. We field trialled PSD free lines on a large scale to select PSD free lines with improved characteristics in a brief 5 year period compared to other breeding options. Our results confirm that accelerated in vitro breeding approach is the best way for rapid papaya crop improvement in the non-GM agriculture system of Australia. Gene editing technology like CRISPR technology can be handy for further rapid crop improvement (disease resistant e. g. phytophthora resistant) papaya in future.

INTRODUCTION

Skybury farms located in the Tropical Queensland (17° 00' 21.456" S and 145° 20' 18.24" E) is a major, diversified farm producing Papaya, Coffee, Avocados, Lemon, Lime, Oranges, Banana and Passion fruit. The unique Skybury Red Papaya is a market leader with over 50% of the Australian market share. Skybury strives for excellence in all areas of innovative farming, value addition like the award-winning Papaya Vodka and coffee liquors, and marketing tropical fruits. Skybury engages in applied research using conventional breeding as well as biotechnology tools to achieve accelerated crop improvement.

Skybury Red Papaya is a unique hybrid line with high yield and sweet, red flesh, a pleasant aroma and nice texture. It's a vigorous papaya with fast growth and strong thick stem. A potential drawback is the characteristic fruit skin which is not shiny as in other papaya varieties with low sugar levels and strong papaya aroma. Being a hybrid that segregates drastically into all sorts of papaya – ranging from red to-pink to yellow flesh, only cloned Skybury plants can make a stable plantation. Papaya crop improvement is mainly achieved through a conventional breeding approach in Australia (Nandawan et al. 2016), India (Mitra and Dinesh, 2016), Malaysia (Sekeli et al. 2018), Brazil (Pereira et al. 2019) and Mexico (Beans, 2020). However, low genetic variability among members of the genus *Carica* is hampering rapid development of improved papaya hybrids (Da Silva et al. 2007). This is the cited reason for the lack of new improved papaya varieties delivered by the large Australian papaya crop improvement program that has received millions of $ from the Papaya levy fund. On top of this delay in crop improvement, new disease (Papaya Sticky Disease- PSD) became a significant problem for papaya production in recent years. Therefore, Skybury decided to take a more pragmatic approach than conventional breeding to produce a designer papaya that is free of PSD and with better productivity and fruit characteristics through in vitro breeding.

In vitro breeding refers to the crop improvement achieved using in vitro methods. Plants cells are totipotent, meaning a cell is capable of developing into all kinds of tissues and a whole new plant eventually. There exists some variability among the somatic cells (1/ 1,000,000 or so). Therefore, it is possible to capitalise on this variation within the cell population of a plant species to achieve crop improvement through in vitro breeding, provided, high frequency regeneration in vitro can be achieved from cell cultures of the species. Callus based regeneration through organogenesis or preferably through somatic embryogenesis can aid in vitro breeding. Somatic embryogenesis is preferred over organogenesis because of the single cell origin of somatic embryos which reduces potential formation of chimera as in the case of regeneration through organogenesis from callus.

Callus, Cell culture

Callus culture of papaya was initiated from one year old leaf explants of quality assured, superior papaya trees. Friable, rapidly proliferating callus (**Fig. 1**) developed on semisolid MS medium supplemented with Kinetin (1-2 mg/l), IBA and NAA (0.1- 0.5 mg/l). Rapidly proliferating, white, friable callus made an excellent cell suspension culture when transferred to liquid MS medium with similar hormone combinations.

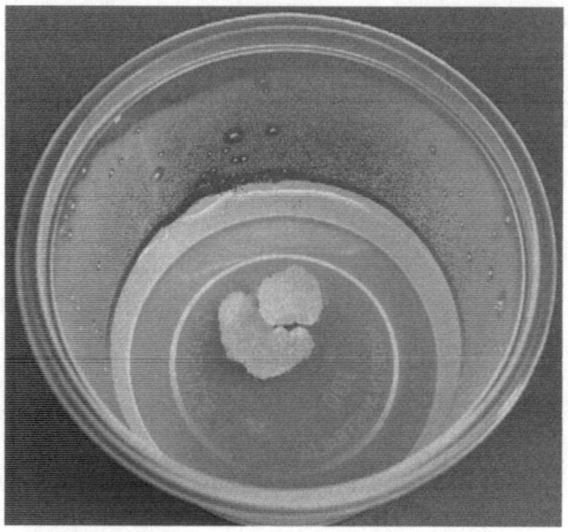

Figure 1 Proliferating friable callus.

Cell cultures were maintained on a shaker (125 rpm) under dark conditions at 25°C. Very fine cell suspension with little clumps and mostly isolated cells or aggregates of 2-3 cells after three to four subcultures at monthly interval. Papaya cell cultures showed a short, 2-day lag phase followed by lag phase of four weeks, then plateaued to the stationary phase (**Fig. 2**).

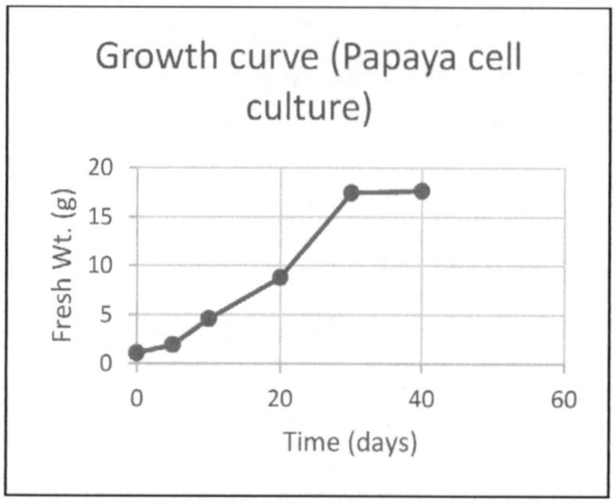

Figure 2. Growth curve (Papaya cell culture).

Cell plating and Somatic embryogenesis

Plated at 10^4 cells per 10 cm Petri dish produced microcolonies in 3-4 weeks when maintained at 25°C in the dark. Microcolonies developed a decent callus mass that became embryogenic upon prolonged incubation (8-12 weeks) in ¼ MS medium + full MS vitamin, fortified with 500 mg/l Glutamine, 60 g/l sucrose and 1-2 mg/L 2,4-D (**Fig. 3**).

Figure 3. Embryogenic callus.

Somatic embryos germinated to produce complete plants under light (2000 lux) on MS medium supplemented with 20 g/l sucrose and no hormones (**Fig. 4**).

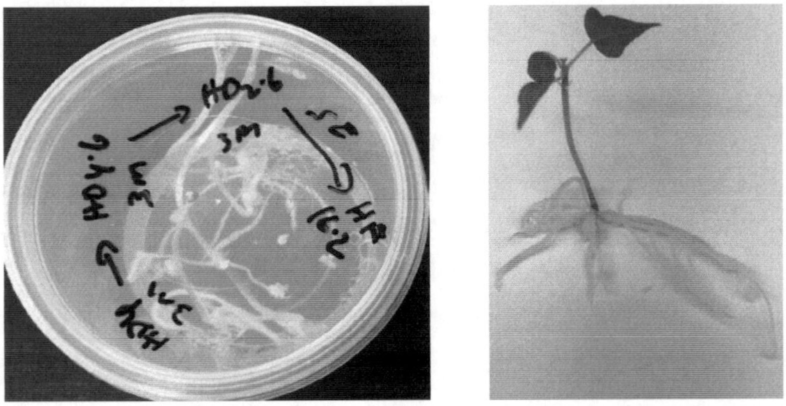

Figure 4. Somatic embryo germination.

Field trials

Key to the success of an invitro breeding program is the large-scale field trial to pick up the useful variants coming out of the in vitro system. Somatic embryo derived plants were micropropagated (30-50 plantlets) and used in the replicated, multiplot field trials. Undesirable variants such as albinos and other visual phenotypes with distorted leaves and spindly stem were rouged out at the in vitro stage or in the early nursery stage. However, many other variants like non-flowering, non-fruiting, stunted types couldn't be identified until the plants are grown to the bearing stage. Similarly, some variants with potential (e. g. hyper vigorous, with excellent fruiting) may turn out to be not ideal if the fruit qualities like shelf life, flavour, flesh colour, texture and brix are not favourable. In our experience, sucking pests like mites and caterpillars loved and appreciated some potential variants with little sap, a good characteristic from picking and packing perspective.

A large population of cloned papayas (300,000 plus) maintained at Skybury farm allowed us to make use of somaclonal variation (culture induced variations) for rapid papaya crop improvement. Although, somaclonal variants are of rare occurrence, we could locate 1-2 superior variants annually from the 300,000+ clonal papaya. Somaclonal selection has aided Skybury to achieve continuous crop improvement over the years.

Induced variations for rapid crop improvement

Gamma rays induced genetic variation is a very useful non-GM tool for rapid crop improvement (Andrew-Peter-Leon et al. 2021). Callus cultures were subjected to gamma radiation to induce variations. Plants were regenerated from gamma treated cell mass through somatic embryogenesis. These variants were tested for Papaya Sticky Disease (PSD) using the molecular diagnostic kit developed specifically to identify PMeV that causes PSD in papaya. Interestingly, gamma radiation at

70 Gy eliminated most of the PMeV and doses Gy 80 and above completely eliminated PMeV (Fig. 5). Some of the superior gamma treated variants selected through the field trial also demonstrated excellent fruit qualities like better shelf life, red flesh colour, texture, flavour and Brix ≥ 17. A great advantage of having an excellent cloning facility at Skybury is that all the improved selections can be rapidly cloned to develop an improved plantation within a year.

Conclusion

- In vitro breeding approach is not hampered by the low genetic variability among *Carica papaya* that hampers conventional breeding.
- In vitro breeding technology can reduce the time needed to generate improved papaya varieties from classic 9-12 years to under 5 years.
- In vitro breeding can be further accelerated by combining with mutagenesis.

Literature Cited

Andrew-Peter-Leon M. T., S. Ramchander, Kumar K. K., Mehanathan Muthamilarasan, M. Arumugam Pillai. (2021).Assessment of efficacy of mutagenesis of gamma-irradiation in plant height and days to maturity through expression analysis in rice. PLoS ONE *16*(1): e0245603. https://doi.org/10.1371/jounal.pone.024503

Beans, C (2020). Crop researchers harness artificial intelligence to breed crops for the changing climate. PNAS, *117*: 27066–27069. https://doi.org/10.1073/pnas.2018732117

Mitra, S.K. and Dinesh, M.R. (2019). Papaya breeding in India - achievements and future thrust. Acta Hortic. *1250*: 23-28. https://doi.org/10.17660/Acta-Hortic.2019.1250.4

Nantawan, U. et al. (2016). Progress in marker-assisted breeding of papaya in Australia. Acta Hortic. 1111. https://doi.org/10.17660/Acta-Hortic.2016.1111.12

Pereira, M. G. et al. (2019). Twenty-two-year papaya breeding program: from breeding strategy establishment to cultivar development. Functional Plant Biology Journal. https://dx.doi.org/10.35418/2526-4117/v1n2a2

Sekeli R., Hamid, M.H., Razak, R.A., Wee, C.Y., Ong-Abdullah. J. (2018). Malaysian *Carica papaya* L. var. Eksotika: Current Research Strategies Fronting Challenges. Front. Plant Sci. *9*:1380. https://doi: 10.3389/fpls.2018.01380

PROCEEDING'S PAPERS

EUROPEAN REGION

Fokko Zur Mühlen, Regional Editor

Fifty-fifth Annual Meeting - 2022

Bad Zwischenahn, Germany

Combined Proceedings IPPS 72:389-390. 2022.

Organic Fertilizers for Container Plants

Heinrich Beltz

LWK Niedersachsen, LVG Bad Zwischenahn, Germany

heinrich.beltz@lwk-niedersachsen.de

Keywords: Controlled release fertilizer, biodegradable, plant nutrients

Summary

Container plants of hardy nursery stock are usually being fertilized with coated mineral fertilizers (controlled release fertilizers) at a rate of 3.0 – 6.0 g per litre substrate. The coating material of all commonly used controlled release fertilizers (Osmocote, Multicote, Nutricote etc.) are made of synthetic polymers. They are usually smaller than 5 mm, so they are microplastic. It is estimated that in the EU 30,000 – 90,000 tonnes of controlled release fertilizer (= 3,000 – 9,000 tonnes coating material) are being consumed (Fraunhofer Institutes 2021).

Following the discussion about possible ecological hazards of microplastic, the European Fertilizing Products and Amending Regulations (EC) of 5 June 2019 will require from biodegradable coatings for controlled release fertilizers by 2026. All controlled release fertilizer producers have been researching biodegradable coatings but it is not sure how such fertilizers perform and what they will cost. And furthermore, considering the current discussions about biodegradable products, it is not sure that the consumers will accept them.

The uncertain future of controlled release fertilizers and the growing organic nursery production, where controlled release fertilizers and other synthetic nitrogen fertilizers are not permitted, led the LVG Bad Zwischenahn to test organic fertilizers since 2017 at full rate applications (without top dressing in summer) for hardy nursery stock in container cultivation.

In the trials, various combinations of organic N- and NPK fertilizers have been tested, at full rate as well as in split applications. The best results were achieved by mixing sheep wool pellets or coarse horn chips (7-12 mm) for the supply of N into substrate containing approximately 10-20% green compost (for the supply of P and K). The necessary rate was quite high (approximately 8.0 g/l = 1,100 mg N compared to 600 – 700 mg N with controlled release fertilizers). Actually, research is being conducted to find out if the binding of microorganisms is the reason for the necessary high rates. So far, the mineralization of N was similar to the N release of controlled release fertilizers as well as N leaching from the pots.

Already several commercial nurseries are using such substrates with horn chips quite successfully. But individuals wishing to change from controlled release fertilizers to organic fertilization should keep in mind that the qualities of different organic fertilizers is very variable. So tests with smaller numbers of plants are recommended. And the supply of microelements in the substrate by fertilizers must not be neglected.

Combined Proceedings IPPS 72:391-392. 2022.

CO_2 Balance of Hardy Nursery Stock

Heinrich Beltz

LWK Niedersachsen, LVG Bad Zwischenahn, Germany

heinrich.beltz@lwk-niedersachsen.de

Keywords: carbon footprint, life cycle assessment

Summary

Discussions about the carbon footprints of the products we consume are popular, so the nurseries are interested in the effects of their plants on climate. It is fundamental to understand that the carbon footprint, the assessment of impacts on the climate, is measured by a life cycle assessment. And such a life cycle assessment is being done following certain rules like ISO 14040.

There are two different types of a life cycle assessment: the longest is the timespan "cradle to grave" which means that all impacts are counted from the start of the production until the end of the life of a product, including its disposal. Measuring the impacts at the consumer's level is often very difficult, so many life cycle assessments are limited to "cradle to gate", from the beginning of the production to the time when it leaves the producer, the "gate of the factory". This difference between "cradle to gate" and "cradle to grave" is often neglected in communication but has an essential influence on the outcome of a carbon footprint. The carbon footprint is measured in CO_2, or, if other greenhouse gases like methane and laughing gas are included, in CO_2 equivalents (CO_2-eq).

There are only few data available about the carbon footprints or CO_2 emissions of the production of hardy nursery stock. In publications from Italy and the USA, the carbon footprints of several examples have been calculated. Following these figures, in several cases the greenhouse gas emissions of container crops were higher than those of plants cultivated in the open ground. In some of publications, the emissions were compared to the CO_2 binding or sequestration (binding and storage) of the plants in their tissue during nursery cultivation ("cradle to gate"). In some, it was higher and in others lower.

An interesting point is the question how to value this CO_2 binding in the plant tissue, especially regarding the far higher amount of CO_2 being bound during the later life of the plants in the garden. Many people and organisations like the German Nursery Association BdB state that this binding is a CO_2 sink and even demand final compensation. But if a life cycle assessment is calculated "cradle to grave", it must be considered that in the state of "disposal", the woody plants are being decomposed or burned and all the carbon in the plant tissue will be emitted as CO_2. So, plants can only bind CO_2 for a limited time, depending on the plant perhaps 10 or 100 years, which is short compared to the sequestration of peat (up to 10,000 years), mineral gas and oil or coal (up to 500 million years). Nevertheless, woody plants are doubtless beneficial for the environment, even if they are not a CO_2 sink. They reduce heat and noise in urban environment, bind fine dust and other air pollutants, deliver food for insects and other organisms and are an important cultural asset.

Combined Proceedings IPPS 72:393-394. 2022.

Biodegradable Tying Materials

Louise Heissel

VuB, Ellerhoop, Germany

Heissel@vub.sh

Keywords: nursery, MAX-pliers, cotton, jute, staking

Summary

In a comparative test, seven different biodegradable tube tying materials and four degradable binding materials for the MAX-pliers were compared.

The hollow cord materials: Jute twine (RHG, Hermann Meyer), black cotton ribbon (RHG, Hermann Meyer), white cotton tree tie (Hermann Meyer), green PLA tube tie (Hermann Meyer), brown PLA tube tie (Hermann Meyer), White PLA bio tube tie 3mm (RHG), green biodegradable PLA tying tube (Agro de Arend).

The MAX-plier materials: Pink PLA tape, brown PLA tape, MAX Paper tape, MAX-biorömerband tape.

Due to its low elasticity, the jute twine can lead to severe constrictions in cultures with strong growth in thickness. However, this material decomposed very quickly in the ground. The materials consisting of cotton (textile tying ribbon and cotton tree tie, 35mm) or paper (MAX-paper tape) showed good durability for a period of 6 months. For longer cultivation times the UV-radiation and overall weather conditions will cause these materials to go porose and they will eventually tear and fall off. The cotton and paper materials decomposed fairly quickly in the ground, after 6 months only small pieces were left to be found.

Most of the PLA tube ties performed very well in handling and durability. Their elasticity allowed the ties to grow with the thickness growth of the trees. The two materials in green and brown from Hermann Meyer started turning porose and falling off the trees after 4 months. These two materials also seem to decompose in the ground after 18 months they were partially degraded. The PLA tying tubes from RHG and Agro de Arend stayed intact longer than 24 months on the trees without causing constrictions and without breaking. After 24 months however, these materials were showing no signs of decomposing in the ground.

All tested MAX-Tapes were overstrained by the thickness growth in the bigger trees. But in the shrubs they performed very well for longer than 12 months.

One of the largest drawbacks by the PLA products is its very slow decomposition. These products are only decomposable under very certain circumstances, and it may take as long as 1000 years for them to compose. Products made of PLA do not fit into plastic recycling systems, since it would contaminate the quality of the oilbased plastics. PLA should, because of the long decomposition time, not enter modern composting-systems, nor should it be left in nature. It stands to mention however, that the material in contrast to oilbased plastics is made of renewable sources such as cornstarch and lactic acids.

Combined Proceedings IPPS 72:395-396. 2022.

Alternatives for Glyphosate in Nursery Production

Britta Zielke

VuB, Ellerhoop, Germany

Zielke@vub.sh

Keywords: herbicide, weeds, seeds

Summary

Glyphosate is widely used in nursery practice. The advantage to Glyphosate is that it is translocated in plants and therefore effective against root weeds, large weeds and overwintering rosettes. In addition, it works in unfavourable weather. In Germany, the most commonly used products (e.g., Durano TF, GlyfosTF Classic) contain 360 g/l glyphosate and TF formulation. There is lots of experience regarding tolerance.

On the contrary, glyphosate is the active ingredient with the world's largest tonnage and it is politically and socially controversial. In Germany it will be banned after January 1st, 2024.

VuB (Versuchs- und Beratungsring Baumschulen e.V.) carried out multiple trials to look for alternative methods:

Autumn seed beds: A change of strategy is necessary to control weed growth, especially the formation of rosettes (shepherd's purse and chamomile). After sowing seeds, weeds have to be prevented in autumn by application of soil herbicides. Depending on tolerance, the following herbicides can be used: Goltix Gold (metamitron), Boxer (prosulfocarb), Spectrum (dimethenamid-P), Stomp Aqua (Pendimethalin), Proman (metobromuron) or Quickdown (pyraflufen) + Toil, all with reduced application rates. Shortly before emergence follow-up treatments with herbicides based on pelargonic acid plus grass herbicides or thermal treatment can be made. In the spring, residual herbicides should only be used if the tolerance is known.

Transplant beds: Glyphosate with reduced rates is sometimes used in autumn to clean weedy conifer beds. This practice is not recommended but can in some cases be used as an "emergency procedure". In trials the herbicide Kerb Flo (a.i. propyzamid) was combined with different partners. VuB looked at efficacy and crop tolerance. Phytotoxicity occured. Herbicides based on pelargonic acid or pyraflufen scorched needles of conifers (*Abies* and *Picea*).

After the use of MaisTer Power (iodosulfuron, foramsulfuron and thiencarbazone) the plants showed yellow needles and a delay in sprouting.

Other combinations like Kerb Flo + Artist (flufenacet and metribuzin) or + Lentagran (pyridat) or + reduced rates of Broadway (florasulam, pyroxsulam) were applied without inflicting damage to the plants. Also, in transplant beds a change of strategy is needed. It has to be a priority to prevent weed growth after planting with residual herbicides. An early treatment with herbicides based on propyzamide + partner is better than a treatment in the late spring.

Directed spraying: A large trial in *Thuja* with 12 treatments and three replicates was presented.

Combinations of herbicides were used against broadleaf weeds and grasses. Some showed quick effects, but with regrowth (e.g. pelargonic acid or pyraflufen in combination with grass herbicide). Good effects on the weeds could be achieved with a combination of Cato (rimsulfuron) and Broadway or with MaisTer Power. The efficacy depended upon weed species. Phytotoxicity did not occur. The larger the trees, the easier it is to find a replacement for glyphosate treatments.

Conclusion

There is no 1:1 replacement for products containing the active ingredient glyphosate. For every intended use the right application has to be found.

Root weeds (couch grass, thistle) should be controlled as long as glyphosate is still permitted. Familiarize yourself with alternatives in good time. Act more preventively in the future, don't let weeds grow large. Right timing is becoming more important. Mechanical and thermal measures should be used where possible.

Disclaimer: Every effort has been made to ensure that the information in this summary is complete and correct at this time. The author does not accept liability for any error or omission in the content, or for any loss, damage or other accident arising from the use of products listed herein. Omission of a product does not necessarily mean that it is not approved and available to use. It is essential to follow the instruction on the approved label before handling, storing or using any crop protection product. Always start on a small scale with new products !

Combined Proceedings IPPS 72:397. 2022.

Mechanical Weed Control – Including Robots

Heinrich Lösing

VuB, Ellerhoop, Germany

Dr.Loesing@vub.sh

Keywords: herbicide, Glyphosate, autonomous weeding

Summary

In times of less availability and restriction on herbicides, especially like Glyphosate, growers are again focusing more on alternatives for example mechanical weed control. Modern machinery is looking totally different than before. The main goal is still the reduction of competition for light, water and nutrients for our plants.

Many developements have been made like finger weeders, power harrows and others. Special machinery for nearly each crop size and row distance is needed. Tool carriers or high clearance tractors are available as 3 or 4-tire systems or even on caterpillar tracks.

Using mechanical weed control includes always a risk of crop damage. Growers want to control as much weeds as possible and getting as close as possible to the crop. Damage can easily happen.

Mechanical sensors and GPS-, camera- or laser-based sytems can already give support to the vehicle driver. They all show advantages and disadvantages in different crops. They are competing with each other and also a combination of different systems in one machine does make sense.

Everybody is waiting for robots doing all the practical handwork in the field. So far, the systems are not able to separate the big number of tree and shrub species from weeds. There is interesting equipment available, but it will take a few more years to go an autonomous weeding robot in the field.